Encyclopedia of Climate Change

Volume I

Encyclopedia of Climate Change
Volume I

Edited by **Mary D'souza**

R CALLISTO REFERENCE

New York

Published by Callisto Reference,
106 Park Avenue, Suite 200,
New York, NY 10016, USA
www.callistoreference.com

Encyclopedia of Climate Change: Volume I
Edited by Mary D'souza

International Standard Book Number: 978-1-63239-221-3 (Hardback)

Printed in the United States of America.

Contents

Preface

The term 'climate change' essentially refers to a change in the distribution of weather patterns, which is determined statistically. This variability can span periods ranging from decades to millions of years. Broadly speaking, the rate at which energy is received from the sun and simultaneously lost ultimately determines the temperature and thus the climate of Earth. Climatic changes have always been happening, but at a minuscule scale and mainly due to natural factors. Now, these changes are happening rapidly, which have become a cause of concern.

Climatic changes are a result of both natural and man-made factors. Natural factors include biotic processes such as variations in solar radiation, volcanic eruptions and plate tectonics among others. Human causes include deforestation and burning of fossil fuels, which trap an excess of carbon dioxide. This trapping of carbon dioxide is referred to as 'global warming'.

Geological evidence such as sediment layers and records of past sea levels are analysed by scientists, to understand climate change. General circulation models and theoretical models are then created to make future projections, and link causes and effects. I'd like to thank all the researchers who've shared their studies with us in this book. They have made consistent efforts to make this book a valuable one by matching the time parameters. I would also like to thank my publisher and my family for their support in me.

Editor

Assessment of Climate Change Impacts in Greece: A General Overview

Maria A. Mimikou, Evangelos A. Baltas

Laboratory of Hydrology and Water Resources Management, Department of Water Resources,
Hydraulic and Maritime Engineering, Faculty of Civil Engineering, National Technical University of Athens, Athens, Greece

ABSTRACT

The climatic and hydrological systems are tightly related and any induced changes cause chained interactions. In an attempt to adequately manage water resources in Greece, a series of experiments were conducted with different GCMs in selected study areas to understand this interplay. This paper is an overview of the studies carried out in the Aliakmon, the Upper Acheloos, the Portaikos, and the Pinios basins, where the regional hydrological cycle was evaluated on river basin spatial scale to assess regional impacts and variability. The impacts of climate change on the water resources are presented in a synthetic quantitative way, in order to draw general conclusions concerning the trends of the hydrological indicators. A good agreement was observed between the different climatic experiments, and the trends on the selected hydrological indicators demonstrate an increase in temperature and PET, reduction in the mean annual precipitation and runoff, and a shifting of the snowmelt period towards the winter, while the snowpack storage was proved to be a controlling factor. It is accentuated that relatively small decreases in the mean annual precipitation cause dramatic increase of reservoir risk levels of annual firm water supply and energy production. As a result, radical increases of reservoir storage volume are required to maintain firm water and energy yields at tolerable risk levels. The adaptive capacity of the country is not that high, and a series of serious actions need to be taken in order to mitigate the effects of climate change and assess its impacts.

Keywords: Basin; Water Resources; Hydrological Indicators; Reservoir; Simulation Model

1. Introduction

The climatic and hydrological systems are interactively related and any changes initiate bidirectional feedbacks. In order to adequately manage water resources, cascading from climate and hydrology, it is essential to understand this interplay and thus, the hydrological cycle needs to be evaluated on a river basin spatial scale in order to assess regional impacts and variability. The climate-driven hydrological changes will be exacerbated when combined with other pressures on water resources (e.g. population growth, urbanization, deforestation etc), changes in life styles increasing water demand and environmental pollution to challenge water management in the 21st century [1]. The importance of the climate change impacts on water resources has been identified from the Intergovernmental Panel on Climate Change (IPCC) which has recognized "water" and "regional integration" as Cross-Cutting Themes (CCT) across the IPCC Working Groups (WGs) of the Fourth Assessment Report (AR4) completed in 2007 [2]. Detecting changes in the hydrological regime that might result from climate changes is complicated by the inherent variability and randomness of the climatic and hydrological variables, and distinguishing between this natural variability and the climate change impacts is challenging [3].

Nowadays, it is widely recognized that Europe is undergoing climate change and the impacts on water resources are obvious; vulnerability of hydrological extremes (floods, droughts), surface water ecosystems at risk, groundwater overexploitation, water quality degradation etc. In the last 100 years the European average temperature increased by 0.95°C and it is projected to further increase 2.0°C - 6.3°C by the year 2100. Precipitation trends have also been altered, demonstrating a high variability as south and southeastern Europe have become drier (up to 20%), while central and northern Europe are receiving more rain than in the past (10% - 40% wetter). This pattern is projected to continue, with the annual precipitation decreasing up to 1% per decade in southern Europe (in summer time 5% decrease is possible), and oppositely increasing in northern Europe by 1% - 2% each decade [4]. Temperature and precipitation extremes are observed as in the past 100 year with the number of cold days decreasing and the number of summer days (T ≥ 25°C) and heatwaves increasing. Wet days have decreased in southern Europe, but increased in the central and northern part, while it is projected that for the terminal year 2080 cold winters will disappear, hot summers will become more frequent, while the frequency of droughts and intense precipitation events is likely to increase. Finally, it is important to note the retreat of glaciers, snow caps and Arctic Sea ice, and the observed sea level at a rate of 0.8 - 3.0 mm/year, which can be intensified by 2.2 to 4.4 times in the future.

2. Impacts of Climate Change on the Hydrological Cycle and the Water Resources Management, and Europe's Adaptation Potential

Climate and land-use changes heavily influence the water balance and water resources of the different systems [5,6]. In an effort to assess the associated impacts different studies have been carried out using and establishing common indicators (such as the mean annual and seasonal runoff, mean monthly evapotranspiration (actual, potential), soil moisture, mean monthly groundwater recharge, extreme events' (flood, droughts) magnitude and frequency etc.) to allow mutual comparison. Following the temperature and precipitation trends of the altered climate, the hydrological variables are effected, thus under most climate change scenarios the mean annual streamflow is expected to increase in northern Europe and decrease in southern Europe [7,8], while a change of 10% is anticipated in mid-latitude Europe although this percentage may be smaller than the one induced by the natural mutli-decadal runoff variability (**Figure 1**). The change in the flow regime is highly affected by the snowfall and snowmelt and thus varies across the continent. In the Mediterranean regions, the range between winter flow and summer flow will amplify, while in the continental and upland areas where much of precipitation is in form of snow, the increased temperature will diminish snow accumulation resulting in higher winter runoff values and spring snowmelt decreases. In more continental and upland areas, where snowfall makes up a large proportion of winter precipitation, a rise in temperature would mean that more precipitation falls as rain and therefore that winter runoff increases and spring snowmelt decreases [9-11]. The above mentioned changes in the streamflow pattern, will lead to an increased river flooding risk, which seasonally will be shifted form spring to winter time. As concern is rising, several studies have been conducted by EU countries on catchment scale, in an effort to assess and mitigate the eminent problem. The effects of the European altered climate on groundwater recharge are not very obvious, but some studies in the UK and Estonia infer that it could increase

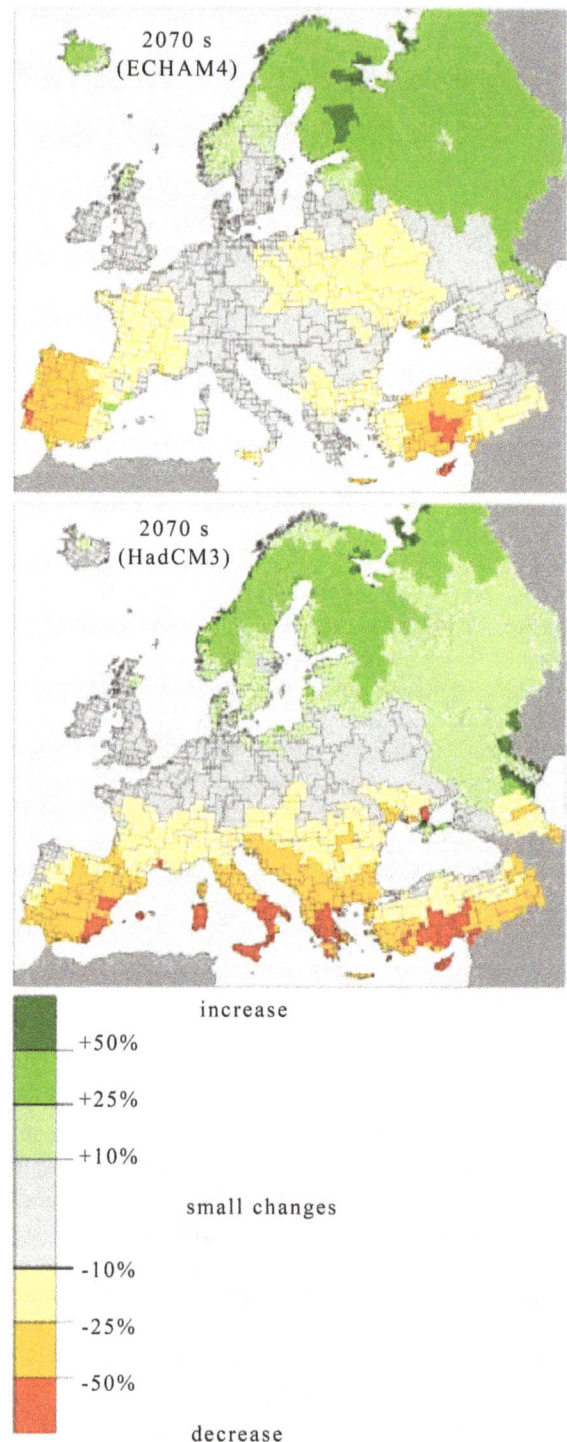

Figure 1. Change in annual average river discharge for European river basins in the 2070s compared with 2000. Note: Two different climate models (ECHAM4 and HadCM3). Source: Lehner *et al.*, 2001.

[12,13]. Finally, concerning river water quality, increases in temperature and reduction in the streamflow will lead to a deteriorated quality as the stream will lose its dilution capacity, algae bloom and biological oxygen demand

(BOD) will increase, and dissolved oxygen (DO) concentrations will diminish. A synopsis of the trends and projections of basic aforementioned indicators is presented in **Table 1**.

Currently, water resources in Europe are under pressure, and proper management is still problematic in many areas, as there is a major lack of communication between policy makers, stakeholders and consumers. This dramatic condition will be exacerbated by climate change [high confidence], and the most stressed systems will exhibit the highest sensitivity and vulnerability.

Climate change impacts on the hydrological cycle can trigger a series of additional impacts on water quantity and quality, such as reservoir risk, power generation inadequacy, hydraulic works failures (dams, culverts etc.), groundwater overexploitation, increased river flooding risk levels, erosion, wetland loss, seawater intrusion in coastal aquifers, water quality deterioration etc. These, in turn, will have multiple effects on the natural, social, and economic European systems [very high confidence, well-established evidence].

The question that clearly arises is: How will European countries be able to adapt to those impacts and mitigate the negative effects of climate change, given the uncertainty and variability that is also associated with climate change? Recently, water managers in many European countries have recognized the importance of acting in this direction, and EU directives are likely to further encourage those initiatives [14]. Generally, the adaptation potential of socioeconomic systems in Europe is relatively high because of the economic conditions (high gross national product and stable growth), the stable population (with the capacity to move within the region), and the well-developed political, institutional, and technological support systems. However, marginal and less wealthy or developed areas will adopt more difficultly, while the adaptation potential for natural systems generally is low (very high confidence) [15]. Vulnerability to climate change differs from region to region, the highestbeing in southern Europe (Mediterranean region) and in the European Arctic [16]. Those areas will face additional impacts due to sea level rise, shore loss, extended droughts, ice cap melting, increased coastal storm surge patterns, erosion etc., with major equity implications and

Table 1. Summary of trends and projections of basic indicators for Europe. Source: EEA Report, No. 2/2004.

	INDICATORS	TRENDS
Meteorological	Annual Precipitation	*for the period* 1900-2000 Northern Europe 10% - 40% wetter. Southern Europe up to 20% drier. *for the projected future* Northern Europe 1% - 2% increase per decade. Southern Europe up to 1% decrease per decade (in summer, decreases of 5% are possible). More frequent droughts.
	Precipitation Extremes	*for the period* 1900-2000 The very wet days' frequency notably decreased in recent decades in many places of Southern Europe, while it increased in Mid and Northern Europe. *for the projected future* The frequency of extreme events (droughts and intense precipitation events) is likely to increase by 2080.
	Temperature Extremes	*for the period* 1900-2000 Decrease of cold and frost days in most parts of Europe. Increase of days with temperatures above 25°C and of heatwaves. *for the projected future* By 2080 cold winters are projected to disappear almost entirely and hot summers are projected to become much more frequent.
Hydrological	Annual River Discharge	*for the period* 1900-2000 Variability attributed to changes in precipitation, with increase in some areas (e.g. Eastern Europe), and decrease in others (e.g. Southern Europe). *for the projected future* Intensification of the changes in annual river discharge, due to the combined effect of the projected precipitation and temperature changes. Northern & Northeastern Europe, increase in annual discharge. Southern & Southeastern Europe, strong decline in annual discharge.
	Flooding	*for the period* 1900-2000 Increase of the annual number of flood events since 1975 (238 events recorded), rise of the number of people affected. Significant decrease of fatal causalities per flood event due to improved warning and rescue measures. *for the projected future* Likelihood of increased extreme flood events' frequency, in particular of flash floods.

economic losses.

3. Description of the Study Area

Greece, located in southern Europe, has a typical Mediterranean climate, with humid winter periods and dry summer periods. The water districts presented in the study, from north to south, are the water district of Western Macedonia, Thessaly and Western Sterea. More analytically:

The water district of Western Macedonia has an extent of 13.624 km^2 and is characterized by intense relief with small flat areas. The climate is continental with harsh winter and snowfalls and the mean annual temperature is 13˚C. This district presents a sufficiency in water. A great part of the water demand is covered from the transnational lakes of Small and Big Prespa, while Aliakmon river is used for the water supply of Thessalonica. The mean annual rainfall depth is 640 mm and the mean annual rainfall volume is estimated at 8.692 hm^3 [17]. The river basin of Aliakmon has the greatest extent (8.847 km^2, 65% of the extent of water district) and the river length is 93 km. A percentage equal to 28% of the total extent corresponds to watersheds with area smaller than 40 km^2. These watersheds drain directly to the sea, they are characterized by ephemeral low flow and do not contribute to the water potential of the district.

The water district of Thessaly has an extent of 13.377 km^2 and includes the greatest flat terrain of the entire country, with the highest crop productivity. The development of the region depends on the promotion of intensive, irrigated agriculture and tourism. There is a dearth of water resources in the region and the demand for water of the existing crops is not covered. Therefore, a special management policy should be applied in order to reinforce the water potential through the rational use of the water resources (extension and modernization of the irrigation network), as well as through the conveyance of water capacities from other water districts. Geomorphologically, the water district is divided into three sections; the eastern coastal and mountainous, with Mediterranean climate, the central flat, with continental climate, the western mountainous, with mountainous climate.

The mean annual temperature varies between 16 and 17˚C. The rainfall depth is greater at the western part, it decreases at the flat area and increases at the eastern mountainous areas. The mean annual rainfall depth is estimated at 858 mm and the corresponding volume at 4.175 hm^3. The surface runoff is estimated at 3.202 hm^3. The Pinios river basin is the greatest of the country, with area equal to 10.628 km^2, which corresponds to 81% of the total area of the water district. The length of the river is 255 km.

The water district of Western Sterea is 10.199 km^2 in area and is located on the high precipitation part of the country. The rivers Acheloos, Evinos and Mornos and the lake Trichonida constitute the major surface water resources of the district, in addition to the underground water resources, which generally stay unutilized. At present, a part (equal to 8.6%) of the district's water potential is used for the water supply of Attica. A small part is also transferred to Thessaly via the Plastiras Lake. Four hydroelectric plants in Acheloos river basin have been constructed for the production of electrical energy. The mean annual rainfall depth is about 800 - 1000 mm at the coastal and flat areas, 1400 mm at the mountainous areas, while it exceeds 1800 mm at high-elevation areas. The mean annual rainfall volume is estimated at 8.680 hm^3, based on the mean annual rainfall depth and the extent of the water district [17]. The surface runoff is about 5.296 hm^3. The river basins with area greater than 1000 km^2 are those of Acheloos (5.635 km^2) and Evinos, which cover 65% of the total extent of the district. A percentage equal to 23% corresponds to the watersheds that are smaller than 40 km^2.

3.1. The Study Basins

Climate change impacts on the water resources and related hydraulic works have been studied for the Aliakmon river basin in northern Greece, as well as for the Upper Acheloos, the Portaikos and the Pinios river basins in central Greece (**Figure 2**). The study area of the Aliakmon river basin is located in northern Greece between 39˚30'S and 40˚30'N, and 20˚30'W and 22˚E.

Three subbasins of the Upper Aliakmon River, the Venetikos, the Siatista and the Ilarion basins were selected for the analysis. The Venetikos subbasin is rather mountainous with steep slopes, while the Siatista basin has

Figure 2. Location of the river basins of the selected case studies.

milder geomorphological characteristics and flood plains. The study area of the Upper Acheloos and Portaikos River basins lies in the central mountainous region of Greece between 39°13'S and 39°42'N. It comprises three drainage basins, the Mesohora and Sykia basins of the Upper Acheloos River, and the Pyli basin of the Portaikos River. The Acheloos basins are purely mountainous, while the Pyli basin lies on the slopes surrounding the plain of Thessaly. Finally, the study area of the Pinios River is situated in the Thessaly plain district. The total drainage area of the river is 9.450 km², with a polymorphic topography varying from narrow gorges to wide flood plains. The Pinios catchments area consists of 15 sub-basins drained by the main channel and its 5 tributaries, and suffers from frequent hazardous storms with consequent flash floods that the natural capacity of the river is inadequate to pass downstream in several locations. Due to lack of sufficient data the study focused solely on the Ali Efenti subasin where reliable and long hydrometeorological time series are available. Some general characteristics of the above mentioned basins and the existing reservoirs are given in **Tables 2** and **3** respectively.

4. Current Situation and Impact Assessment on Water Resources in Greece

The current weather conditions have major effects on the natural, environmental and socioeconomic systems in Greece, exposing their sensitivity and vulnerability to climate, which once altered can exacerbate such effects. This is a well established evidence and calls for proper adaptation strategies in order to adequately protect the water resources of the country.

Water resources development and management is already under pressure given the current conditions, and this situation could become critical if the planners and policy makers do not take into account the projected future climate changes of higher temperature and evaporation rate, reduced streamflow patterns, shifting of the wet period, increase of extreme events (flash floods, droughts). Additionally, as the vast majority of the Greek territory lies in coastal areas, sea level rise and the associated effects of accelerated erosion, land inundation, wetland loss and groundwater regime instability, call for proactive and precautious measures. As the Greek economic conditions (gross national product and growth) are rather unstable, and gaps are observed in the institutional and technological support system, in conjunction with a population increase which leads to an increase in water demand, the adaptation potential is not high, and thus immediate intervention is needed.

Currently, according to studies of the National Observatory of Athens (NOA), an increase of 48% - 52% in the greenhouse gases is foreseen in 2010 relative to the levels of 1990, which is much higher than upper limit of 25% approved in 1997 in Kyoto [18]. To combat the

Table 2. General characteristics of the studied river basins.

River	Subasin	Area (km²)	1	2	3	4	5
Aliakmon	Venetikos	818	1032	10.4	1069.20	17.7	0.0216
	Siatista	2724	1005	10.5	822.3	22.3	0.0082
	Ilarion	5005	917	11	825.1	49	0.0098
Upper Acheloos	Mesohora	633	1390	7.9	1951	24.4	0.0385
	Sykia	1173	1299	8.8	2084	48.4	0.0413
Portaikos	Pyli	134.5	800	11.2	2023	65.1	0.484
Pinios	Ali Efenti	2796.40	539.7	10.1	997	39.05	0.428

Table columns: 1) Mean elevation (m); 2) Mean annual historical temperature (°C); 3) Mean annual historical precipitation (mm); 4) Mean annual historical runoff (m³/sec); 5) Mean annual specific runoff (m³/s/km²).

Table 3. Designed characteristics of the studied reservoirs in the Aliakmon, Upper Acheloos and Portaikos river basins.

River	Reservoir	Storage Capacity (10⁶ m³)	Design Head (m)	Primary energy (GWh/y)	Firm water supply
Aliakmon	Polyfyto	655 (min) - 1160 (max)	146.4	199.4 (min) - 515 (mean)	640×10^6 (m³/y)
Upper Acheloos	Mesohora	228	200	231	-
Upper Acheloos	Sykia	500	137	48.8	5 (m³/s)
Portaikos	Pyli	47	-	-	variable
Portaikos	Mouzaki	530	138.4	370	41 (m³/s)

problem, Greece developed the "Hellenic Action Plan for the abatement of CO_2 and other greenhouse gas emissions" in 1995, with the objective to control the increase of greenhouse gases from all sources to a maximum increase of 12% - 18% of the 1990 level [19]. The adopted measures consider the introduction of natural gas in the industrial, commercial, tertiary and residential sectors, the modernization of the existing energy system and manufacturing units to sustain conservation, the development of cogeneration units in the power plants, the implementation of measures against energy use in transport, and the exploitation of alternative and renewable energy resources. On top of those preliminary actions, additional polices and measures where introduced in the second National Climate Change Program (2000-2010) in order for Greece to meet its Kyoto target. With the goal of reducing emissions to 25% over the time period 2008-2012, natural gas is further introduced in all demand sectors including cogeneration, energy conservation is widely imposed, renewable energy resources for electricity and heat production, as well as energy-efficient appliances are promoted, structural changes in agriculture, chemical industries, transport and waste management are introduced.

Despite the aforementioned measures and mitigation strategies, the effects of climate change on water resources in Greece are already obvious, and demonstrated indicatively by different researchers that studied the phenomenon [20-23]. The hydrological cycle is affected by the climate warming and the increased temperature causes changes in the precipitation and evaporation patterns, and the sea level. Those changes cause a chained reaction affecting the surface runoff, the soil moisture and the groundwater regime. Thus, future water resources management is at risk as the sustainability of the key aspects of water supply, agricultural water use, reservoir failure, energy production, soil drainage, flood protection, and water quality becomes vulnerable and uncertain. To combat these problems, approaches to estimate and assess the regional hydrological impacts of climate change have been attempted for the Greek territory in an effort to draw key conclusions and adequately plan for the future. The Laboratory of Hydrology and Water Resources Management of the National Technical University of Athens (NTUA) has conducted various case studies through EU funded research programs, mainly in central and northern Greece, where the major river systems are located, the results of which will be presented and discussed in the following sections. These studies, aiming in assessing the climate change impacts on selected hydrological indicators, and integrating various climate change experiments in hydrological models' simulations, follow in general the subsequent methodology: 1) Quantitative estimation of climate change using long-term climatic indices of the major climatic variables, and climatic scenarios; 2) Simulation of the hydrological cycle on catchment scale, using the aforementioned scenarios, and estimation of the climatic induced changes on different hydrological variables; 3) Estimation of the impact of the aforementioned hydrological changes on the behavior and reliability of water resources and related water management works (reservoirs, energy production, floodplain design etc.). A summary of the presented case studies' components is presented in **Table 4**. In the following sections the selected case studies are briefly presented, and the results and findings are summarized in combined tables to allow easy comparison.

4.1. The Climate Change Scenarios

To assess the effects of climate change on the study areas, the climate scenarios used where mainly outputs of experiments conducted with different General Circulation Models (GCMs). Climate change scenarios reflecting future global increase in greenhouse gas concentrations have been constructed for Europe (including Greece), at 0.5° latitude/longitude resolution by the Climate Research Unit (CRU) of the University of East Anglia, England. The adopted methodology used the construction by the CRU 1961-1990 baseline climatologies for Europe, the results from different GCMs climate change experiments, and a range of global warming projections calculated by MAGICC (Model for the Assessment of Greenhouse gas Induced Climate Change), a simple upwelling-diffusion energy balance climate model [24,25]. The various climate change models employed in the studied basins, along with the corresponding terminal years are summarized in **Table 5**. Additionally the models ECHAM, CSIRO, CGCM were applied in Ali Efenti basin and the terminal year was 2050.

4.2. The Simulation Models

For all the water balance and water quality simulations of the case studies, three basic models, two of them develed by the Laboratory of Hydrology and Water Resources Management f the NTUA, and one publicly available, were used to assess the water quantity and quality of the studied basins. Additionally, for the assessment of water supply levels and the reliability of energy power producon at the studied reservoirs, a reservoir operation model was applied. The basic information of the models are briefly presented as below:

The WBUDG model is a conceptual, physically based model for the simulation the hydrological cycle using as basic inputs precipitation, temperature, wind velocity, relative humidity and sunshine duration data. The model runs on a monthly time step, and consists of various subutines which function interactively, allowing the estima-

Table 4. Summary of the case studies components.

Components	Indices and Models
Climatic Indices	Temperature (mean annual, seasonal, monthly)
Assessment Indices (hydrological, water quality etc.)	Precipitation (mean annual, seasonal, monthly) Evapotranspiration (mean annual, seasonal, monthly) Total Runoff (mean/min/max annual, seasonal, monthly) Surface, Lateral, Groundwater flow (mean monthly) Flood magnitude changes (for the 10, 20, 100, 1000 yr floods) Nitrate-nitrogen losses (mean monthly) Organic nitrogen losses (mean monthly) Sediment losses (mean monthly) Ammonium NH_4^+ (mean monthly) Biological Oxygen Demand BOD (mean monthly) Dissolved Oxygen DO (mean monthly) Reservoir Primary Energy production (mean/min annual) Risk of Reservoir annual Primary Energy production (mean, min, range) Probability of failure of the reservoir design values for water supply
Climatic Scenarios	equilibrium: UKHI, CCC transient: HadCM2, UKTR, ECHAM, CSIRO, CGCM
Simulation models	WBUDG (water budget), R-Qual (water quality), SWAT (water budget, nitrogen yield), Reservoir (water and energy supply)
Terminal years	2020, 2050, 2100

Table 5. Climate change scenarios applied to the studied basins and the corresponding terminal years.

Subasin	UKHI	CCC	HadCM2	UKTR	Hypothetical T, P changes	
Venetikos	2020, 2050, 2100	2020, 2050, 2100	2050	2020, 2050, 2100	+1˚C, 2˚C, 4˚C	−20% to +20% (P)
Siatista	2020, 2050, 2100	2020, 2050, 2100	2050	2020, 2050, 2100		
Ilarion	2020, 2050, 2100	2020, 2050, 2100	2050	2020, 2050, 2100		
Mesohora					+1˚C, 2˚C, 4˚C	−20% to +20% (P)
Sykia					+1˚C, 2˚C, 4˚C	−20% to +20% (P)
Pyli					+1˚C, 2˚C, 4˚C	−20% to +20% (P)
Ali Efenti	2050		2020 2050 2100			

on of the rain, snow, snowmelt, snow storage, groundter storage, soil moisture, actual vapotranspirations and filly stream runoff.

The R-Qual model [26] is a one-dimension finite difrences instream mathematical model derived by applying the mass conservation principles to mall completely mixed control volumes. Based on the advection disperon equation it is able to simulate temperature, BOD, DO, $N-NO_3$, $N-NO_2$, $N-NH_4^+$, N-Organic, P-Dissolved, P-rganic and Algae, while it runs both in steady state and dynamic mode. The model can be used to assess the imct of waste loads on in-stream water quality, or to idenfy the magnitude and quality characteristics of non-point waste loads.

The Soil and Water Assessment Tool model (SWAT), is a physically based model used to simulate the hydrogical cycle and its influence on the quantity and quality of water and sediments [27]. The model integrates the functionalities of several others, allowing the simulation of climate, hydrology, plant growth, erosion, nutrient transport and transformation, pesticide transport and management practices, erating under the ArcView platrm and running on a daily time step.

The Reservoir model uses the water balance equation on a monthly time step to describe the operation of the reservoirs, taking into account constraints regarding the storage volume, the reservoir outflow nd the energy procction.

5. Results and Discussion

The assessment of the regional hydrological effects of the studied subasins is based on changes observed on selected hydrological indicators, such as the mean annual precipitation, mean annual potential evapotranspiration, mean annual and seasonal runoff etc. The results from the different simulation models, concerning water budget or water quality runs, and their overall performance are summarized in **Table 6**, where the simulation type, caliation period, time step (M denotes monthly, D denotes

Table 6. Summary of simulation models' performance.

River	Subasin	Model	Simulation type	Calibration period	Time step	Nash-Sutcliffe coefficient NTD
Aliakmon	Venetikos	WBUDG	1	A	M	0.860
	Siatista	WBUDG	1	A	M	0.850
	Ilarion	WBUDG	1	A	M	0.850
	Polyfyto reservoir	Reservoir	2	A	M	
Upper Acheloos	Mesohora	WBUDG	1	B	M	0.864
		Reservoir	2	B	M	
	Sykia	WBUDG	1	B	M	0.882
		Reservoir	2	B	M	
Portaikos	Pyli	WBUDG	1	C	M	0.911
	Mouzaki	Reservoir	2	B	M	
Pinios	Ali Efenti	WBUDG	1	D	M	0.6
		R-Qual	3	E	M	
		SWAT	1	F	M,D	0.81 (M)
						0.62 (D)
			4	F	M,D	0.75

Table columns: 1 Water budget; 2 Water and Energy supply; 3 Water quality (BOD, DO, NH_4^+); 4 Nitrogen yield; A: 1961-1990 (30 yrs); B: 1971-1986 (15 yrs); C: 1971-84, 1986 (15 yrs); D: 1960-1996 (36 yrs); E: 1988-1991 (3 yrs); F: 1970-1993 (23 yrs).

daily), and goodness of fit statistics are presented.

As summarized in Table 6, the water budget and quality models that were used in the different cases performed successfully and accurately in reproducing the hydrological conditions, and the calibration results as assessed with the Nash-Sutcliffe coefficient were satisfactory.

The available data used in the calibration process varied in general from 15 to 30 years, which is a good time range to be able to draw conclusions on the reliability on the simulations. The quality and accuracy of the available data was adequate, although more data would give even better confidence on the models' results. An assessment of the hydrological impacts of the different experiments that have been performed for the case studies for the terminal year 2050, and the respective results are illustrated in **Table 7**. In this table the climatically induced changes in a series of indicators, such as: mean annual precipitation, snow, potential evapotranspiration, runoff, summer runoff, winter runoff, maximum annual runoff, minimum annual runoff, flood peaks for the 20, 100, 1000 year return periods, mean annual BOD, DO, NH_4^+, nitrate-nitrogen loss, and sediment loss-are presented in qualitative terms by using the signs "+" to denote an increase, "-" to denote a decrease, and "=" to symbolize no change on the hydrological or water quality indicators.

As inferred from this table, the trends on the indicators' changes are almost consistent for all the examined case studies. Based on these results, and the experience gained from all the case studies, the following conclusions can be derived:

➤ A good agreement was observed between the different climatic experiments. The climatically induced changes on the hydrological variables were consistent, although the transient experiments attributed higher changes of the hydrological indicators in comparison with the equilibrium experiments. It worth noticing that the CCC equilibrium scenario had some discrepancy with the other scenarios, and predicted increases in the maximum annual runoff when the rest produced decreases.

➤ The selected hydrological indicators were adequate, and allowed the intercomparison of the results to lead to some generalized conclusions.

➤ General trends of the climate change hydrological impacts were observed in all basins, such as: temperature increase, potential evapotranspiration increase, reduction in the mean annual and summer precipitation, reduction in the mean annual and summer runoff, reduction in the minimum and maximum annual runoff with extension of the dry summer period, and shifting of the snowmelt period towards the winter.

➤ Snowpack storage is a governing factor. The basins where a significant retention of precipitation occurs in the form of snowpack storage, exhibit a great sensiti-

Table 7. Summary of the results of the case studies.

River	Basins	Climate Scenarios	Precipitation	Snow	Potential Evapotranspiration	Runoff	Summer Runoff	Winter Runoff	Max Annual Runoff	Min Annual Runoff	Yearly Floods (T = 20,100, 1000)	BOD mg/l	DO mg/l	NH$_4$ mg/l	N-N loss	Sediment loss
Aliakmon	Venetikos	UKHI	-	-	+	-	-	-	-	-						
		CCC	-	-	+	-	-	-	+	-						
		HadCM2			+	-										
		UKTR	-	-	+	-	-	-	-	-						
	Siatista	UKHI	-	+	+	-	-	-	-	-						
		CCC	-	-	+	-	-	-	+	-						
		HadCM2			+	-										
		UKTR	-	-	+	-	-	-	-	-						
	Ilarion	UKHI	-	+	+	-	-	-	-	-						
		CCC	-	-	+	-	-	-	+	-						
		HadCM2			+	-										
		UKTR	-	-	+	-	-	-	-	-						
Upper Acheloos	Mesohora	DP = −10%, DT = +2°C	-	-	+	-	-	+	-	-						
	Sykia	DP = −10%, DT = +2°C	-	-	+	-	-	=	-	-						
Portaikos	Pyli	DP = −10%, DT = +2°C	-	-	+	-	-	-	-	-						
Pinios	Ali Efenti	UKHI	-		+	-	-	-	-	-		+	-	+		
		HadCM2	-		+	-	-	-	-		+	+		+	-	-
		ECHAM	-		+	-	-	+	-		+				-	-
		CSIRO	-		+	-	-	-	-		+				-	-
		CGCM	-		+	-	-	-	+	-	+				-	-

vity in temperature increase, which is reflected with serious reduction of mean and minimum annual runoff, even more serious reduction of summer runoff, and a significant increase of the maximum annual and winter runoff. On the other hand, in the absence of significant snowpack storage, temperature increases are restricted to affect runoff only through the evapotranspiration process and those basins are not as sensitive to them.

➤ A magnification factor characterizes the effect of precipitation change on runoff, and can be considered as a measure of basin sensitivity to precipitation change. Regarding annual runoff, the magnification factor seems to be independent of temperature and positively associated with basin aridity, whereas on a seasonal basis it depends on the presence of winter snow cover and consequently temperature increases.

➤ The local physiographic characteristics of the basin are important as they shape basic parameters, such as the runoff coefficient which, when usually high, does not permit other hydrological processes, sensitive to climate change, to be accomplished or even take place at all.

> Climate change has an important, but by no mean determining, role in the growth of floods. The uncertainty while moving in a finer time scale (daily) is still large, abd the use of different GCM outputs and simple downscaling techniques can lead to adverse results. However, there is evidence of increasing flood magnitude.

> Changes in precipitation have a pronounced effect of water quality. Reduced stream flows lead to loss of the stream's dilution capacity, and thus increased BOD and NH_4^+, and decrease DO values are observed, especially during the summer months, as well as reduced losses for both nitrate and organic nitrogen.

Concerning the conducted reservoir risk analyses, a dramatic increase of risk levels of annual firm water supply and energy production (for both the minimum and mean annual) was exhibited under the examined plausible climate change scenarios. The reservoirs were found to be very sensitive to precipitation decreases. Although the probability of failure associated with the design values of W and E is currently well below 1%, when facing reduction in precipitation the probability of failure ranges form 20% up to 60%. Radical increases of reservoir storage volume are required for all the reservoirs in order to maintain firm water and energy yields at tolerable risk levels. Downstream reservoirs are more sensitive than the upstream ones, and this is attributed to the cumulative effect of upstream failures being transferred downstream. The effect o temperature increase is also associated with risk levels, but was marginal when compared with precipitation reductions.

6. Conclusions

Global climate change has major impacts on the water resources and the related hydraulic works. In an effort to assess those impacts, different climatic scenarios are integrated in hydrological models and many studies use this approach to project the future status, draw conclusions and plan adequately. The most common problem is the spatial and time scales mismatch between the global climate models and the regional river basins. Additionally, the selection of suitable hydrological indicators, and the development of robust and realistic hydrological models that produce high confidence simulated results need to be thoroughly examined.

The European continent in undergoing climate change with a projected increase in temperature of 2.0°C - 6.3°C in the year 2100. As a result the south is becoming drier, while the northern and southern areas are receiving more rain, manifesting the exacerbation of the extreme hydrological events (droughts, floods). The impacts on the water resources, which in many regions are already under

pressure, are obvious (reservoir risk, power generation shortage, failure of hydraulic works etc.) and call for adequate policy measures. Although vulnerability to climate change is substantial, the adaptation potential of the European socioeconomic systems is relatively high due to the population stability, the well developed political and institutional support systems, and the economic conditions.

Greece follows the European climate change trends, demonstrating increases in temperature and evapotranspiration, reductions in precipitation and runoff, and a prolongation of dry summer periods. The adaptive capacity of the country though, is not that high, and a series of serious actions need to be taken in order to mitigate the effects of climate change and assess its impacts. It is accentuated that relatively small decreases in the mean annual precipitation cause dramatic dramatic increase of reservoir risk levels of annual firm water supply and energy production. As a result, radical increases of reservoir storage volume are required to maintain firm water and energy yields at tolerable risk levels. Climate change has similar impacts on other hydraulic works, such as groundwater supplies, flood protection measures, erosion prevention structures etc.

Although many studies have so far been conducted with the purpose of assessing the impacts of climate change at European level, mainly through the implementtation of different EU funded projects, they remain scattered and thus their findings are not fully exploited. Coordinated actions need to be initiated from the European Community, focusing, on one hand, on planning a common concrete strategy for studying the effects of climate change on water resources (e.g. selection of common hydrological indicators), and strengthening, on the other hand, the policy initiatives in order to combat the problem and better assess the impacts following the global standards of the IPCC.

REFERENCES

[1] Z. W. Kundzewicz, L. J. Mata, N. W. Arnell, P. Döll, P. Kabat, B. Jiménez, K. A. Miller, T. Oki, Z. Sen, I. A. Shiklomanov, "The Implications of Projected Climate Change for Freshwater Resources and Their Management," *Journal of Hydrological Sciences*, Vol. 53, No. 1, 2008, pp. 3-10.

[2] IPCC (Intergovernmental Panel on Climate Change), "Fourth Assessment Report, "Climate Change 2007: Synthesis Report," Cambridge University Press, Cambridge.

[3] A. J. Askew, "Climate Change and Water Resources," IAHS Publ., Vol. 168, 1987, pp. 421-430.

[4] EEA (European Environmental Agency), "Impacts pf Europe's Changing Climate, an Indicator-Based Assessment," EEA Report No 2/2004, EEA, Copenhagen, 2004.

[5] L. Ryszkowski, A. Kedziora and J. Olejnik, "Potential Effects of Climate and Land Use Changes on the Water

Balance Structure in Poland," In: *Processes of Change*: *Environmental Transformation and Future Patterns*, Kluwer Academic Publishers, Dordrecht, 1990, pp. 253-274.

[6] J. Olejnik and A. Kedziora, "A Model for Heat and Water Balance Estimation and Its Application to Land Use and Climate Variation," *Earth Surface Processes and Landforms*, Vol. 16, No. 7, 1991, pp. 601-617.

[7] N. W. Arnell, "The Effect of Climate Change on Hydrological Regimes in Europe: A Continental Perspective," *Global Environmental Change*, Vol. 9, No. 2, 1999, pp. 5-23.

[8] B. Lehner, T. Henrichs, P. Döll and J. Alcamo, "Euro-Wasser: Model-Based Assessment of European Water Resources and Hydrology in the Face of Global Change," Vol. 5, Centre for Environmental Systems Research, University of Kassel, Kassel, 2001.

[9] J. Hladny, J. Buchtele, M. Doubková, V. Dvorák, L. Kaspárek, O. Novicky and E. Prenosilová, "Impact of a Potential Climate Change on Hydrology and Water Resources: Country Study of Climate Change for the Czech Republic," National Climate Program of the Czech Republic, Czech Republic, Prague, 1997, p. 137.

[10] L. Kasparék, "Regional Study on Impacts of Climate Change on Hydrological Conditions in the Czech Republic," *Prace a Studie*, Vol. 193, 1998, p.70.

[11] K. Hlavcová and J. Cunderlík, "Impact of Climate Change on the Seasonal Distribution of Runoff in Mountainous Basins in Slovakia," In: *Hydrology, Water Resources, and Ecology in Headwaters*, IAHS Publication, No. 248, 1998, pp. 39-46.

[12] N. W. Arnell and N. S. Reynard, "Climate Change and British Hydrology," In: M. C. Acreman, Ed., *The Hydrology of the UK*, Routledge, London, 1999, pp. 3-29.

[13] A. Jarvet, "Estimation of Possible Climate Change Impact on Water Management in Estonia," *Proceedings of the Second International Conference on Climate and Water*, Espoo, 1998, pp. 1449-1458.

[14] EEA (European Environmental Agency), "Climate Change and Water Adaptation Issues," EEA Technical Report No 2/2007, EEA, Copenhagen, 2007.

[15] IPCC (Intergovernmental Panel on Climate Change) Third Assessment Report, "Climate Change 2001: Working Group II: Impacts, Adaptation and Vulnerability," Cambridge University Press, Cambridge.

[16] EEA (European Environmental Agency), "Vulnerability

and Adaptation to Climate Change in Europe," Technical report No 7/2005, EEA, Copenhagen, 2005.

[17] E. Baltas, "Climatic Conditions and Availability of Water Resources in Greece," *International Journal of Water Resources Development*, Vol. 24, No. 4, 2008, pp. 635-649.

[18] Observatory of Athens (NOA). http://www.noa.gr.

[19] Greece 3rd National Communication to the UNFCCC, "Policies and Measures."

[20] G. Lambrakis and G. Kallergis, "Reaction of Subsurface Coastal Aquifers to Climate and Land Use Changes in Greece: Modelling of Groundwater Refreshening Patterns under Natural Recharge Conditions," *Journal of Hydrology*, Vol. 245, No. 1-4, 2001, pp. 19-31.

[21] E. Baltas and M. Mimikou, "Climate Change Impacts on the Water Supply of Tessaloniki," *International Journal of Water Resources Development*, Vol. 21, No. 2, 2005, pp. 341-353.

[22] D. Panagoulia, "Catchment Climatological Data and Climate Change in Continental Greece," *Proceedings of the Intertnational Conference on Development and Application of Computer Techniques to Environmental Studies*, Vol. 2, 1994, p. 29.

[23] E. Varanou, E. Gkouvatsou, E. Baltas and M. Mimikou, "Quantity and Quality Integrated Catchment Modeling under Climate Change with Use of Soil and Water Assessment Tool Model," *Journal of Hydrologic Engineering*, Vol. 7, No. 3, 2002, pp. 228-244.

[24] M. Hulme, D. Conway, O. Brown and E. Barrow, "A 1961-1990 Baseline Climatology and Future Climate Change Scenarios for Great Britain and Europe," Climatic Research Unit, University of East Anglia, Norwich, 1994.

[25] CRU (Climate Research Unit), "Global Average Temperature Change 1856-2003," 2003. http://www.cru.uea.ac.uk/cru/data/temperature/

[26] M. Mimikou, E. Baltas, E. Varanou and K. Pantazis, "Regional Impacts of Climate Change on Water Resources Quantity and Quality Indicators," *Journal of Hydrology*, Vol. 234, No. 1-2, 2000, pp. 95-109.

[27] J. G. Arnold, J. R. Williams, R. Srinivasan and K. W. King, "SWAT: Soil and Water Assessment Tool," US Deptartment of Agriculture, Agricultural Research Service, Temple, 1994.

Regional Climate Index for Floods and Droughts Using Canadian Climate Model (CGCM3.1)

Nassir El-Jabi[1], Noyan Turkkan[1], Daniel Caissie[2]
[1]Department of Civil Engineering, Université de Moncton, Moncton, Canada
[2]Department of Fisheries and Oceans Canada, Moncton, Canada

ABSTRACT

The impacts of climate change on the discharge regimes in New Brunswick (Canada) were analyzed, using artificial neural network models. Future climate data were extracted from the Canadian Coupled General Climate Model (CGCM3.1) under the greenhouse gas emission scenarios B1 and A2 defined by the Intergovernmental Panel on Climate Change (IPCC). The climate change fields (temperatures and precipitation) were downscaled using the delta change approach. Using the artificial neural network, future river discharge was predicted for selected hydrometric stations. Then, a frequency analysis was carried out using the Generalized Extreme Value (GEV) distribution function, where the parameters of the distribution were estimated using L-moments method. Depending on the scenario and the time slice used, the increase in low return floods was about 30% and about 15% for higher return floods. Low flows showed increases of about 10% for low return droughts and about 20% for higher return droughts. An important part of the design process using frequency analysis is the estimation of future change in floods or droughts under climate scenarios at a given site and for specific return periods. This was carried out through the development of Regional Climate Index (RCI), linking future floods and droughts to their frequencies under climate scenarios B1 and A2.

Keywords: Canadian Climate Model; Artificial Neural Networks; Floods; Droughts; Regional Climate Index

1. Introduction

There is currently a broad scientific consensus that the global climate is changing in ways that are likely to have a profound impact on human society and the natural environment over the coming decades. Climate change and its impacts on a global scale are the focus of intense, broad-based international research efforts in natural and social sciences. However, understanding the nature and potential consequences of climate change at regional scales remains a challenge. Moreover, it has been recognized that changes in the frequency and magnitude of extreme weather events are likely to have more substantial and widespread impacts on the environment and human activities than changes in the average climate.

A number of extreme events (with significant impacts on the environment and socio-economic activities) have been observed during the last decade, including severe floods and droughts as well as extreme heat around the world. These extreme events have caused serious risks to the health of populations as well as to ecosystems and have had severe economic consequences on sectors such as agriculture and water resources. Anticipating specific climatic impacts is thus, as much of a challenge in assessing risks and uncertainties than predicting future changes. As such, it is important: 1) to improve our ability to manage extreme climatic risks, 2) to assess the consequences of extreme events over the next decades, and 3) to develop new tools and design criteria to more accurately assess the impact of extreme events on water resources and river discharge (e.g., floods and droughts).

In Atlantic Canada over the next 100 years, the mean surface air temperature is expected to increase by 2 to 6°C [1], contributing to potentially large reductions in streamflow [2,3] and significant impacts on aquatic resources [4]. The demand for water withdrawal from rivers (e.g. irrigation, drinking water, etc.) and maintaining adequate instream flow for the protection of aquatic resources are tenuously balanced, and represent an ongoing challenge in water resources management. Many rivers in eastern Canada have recently experienced record low flow conditions, coupled with record high water temperatures [5-7], and an increase in water withdrawal.

The study was carried out in New Brunswick, Canada (NB), which lies on Canada's Atlantic coast. Rainfall, snowmelt, and groundwater all contribute to river flow, producing seasonal and annual variations. High flows are generally the result of spring snowmelt, although heavy rainfall can also cause high flows and floods, especially in small streams. Low flows generally occur in late summer, when precipitation is low and evaporation is high, and in late winter, when precipitation is stored until spring in the form of ice and snow.

Flood studies in New Brunswick, Canada (NB) were carried out over 20 years ago [8] and were recently updated [9], using two distributions functions, namely the 3-parameter lognormal (LN3) and the generalized extreme value (GEV) distributions.

In Aucoin et al. [9], single station analysis was carried out for the same hydrometric stations analyzed earlier by Environment Canada and New Brunswick Department of Municipal Affairs and Environment [8]. In addition, a regional flood frequency analysis was carried out using both regression equations and the index flood approach. In general, the results were consistent with those from early studies [8,9]. However, it was observed that updating the flood information resulted, for many stations, in an improvement of flood estimates.

Low flows were also analysed in New Brunswick [10, 11]. This analysis consisted in fitting the Annual Minimum Flow (AMF) to a Type III extremal distribution (3-parameter Weibull) function and calculating the discharge for different recurrence intervals. The low flow frequencies were calculated for 1-day, 7-day and 14-day durations and for different recurrence intervals (e.g., 2-year, 10-year, 20-year and 50-year). Following the single station analysis, a regression analysis was carried out in order to estimate low flows on a regional basis and for ungauged basins. Missing from these previous studies were the projections of future floods and low flows conditions under climate change. As such, this aspect will be addressed within the present study.

2. Data and Method

The present study analyses the climate change impacts on the discharge regimes of several catchments in New Brunswick. The hydrological response for given climate scenarios was simulated using artificial neural network models. Future climate data were extracted from the Canadian Coupled General Climate Model (CGCM) under different greenhouse gas emission scenarios defined by the Intergovernmental Panel on Climate Change [12]. Using an artificial neural network model, future river discharge was predicted for selected hydrometric stations. A frequency analysis was then carried out using the above described distribution functions, and the parame-

ters of the distribution were estimated using L-moments method [13].

2.1. Global Climate Model

General Circulation Models (GCMs) are based on mathematical representations of atmosphere, ocean, ice cap as well as land surface processes. These models are used for simulating the response of the global climate system to increasing "greenhouse gas" concentrations. Historically, gross CO_2 emissions have increased at an average rate of 1.7% per year since 1900 [13] and if this trend continues, global emissions would increase more than six fold by 2100. Different scenarios have been presented by the Intergovernmental Panel on Climate Change (IPCC) [12] which reflects a variety of CO_2 concentrations over the next 100 years. While the complexity of the global climate system is well captured by these models, they are unable to represent local scale features and processes due to their limited spatial resolution. Large geographic areas (e.g. 50,000 to 300,000 km^2) represent the basic unit of GCMs. In the present study the Canadian Global Coupled Model CGCM 3.1/T63, from CCCma, (Canadian Centre for Climate Modeling and Analysis), was used. This model has a geographic areas represented by a grid of approximately 2.81° latitude by 2.81° longitude (~60,000 km^2). In New Brunswick, a few grids points cover the entire province. At odds with GCM resolution is the fact that most researchers are focusing on the impacts of climate change primarily at the local and regional scales rather than focusing on large or global-scale changes. Therefore, downscaling of data is often used, and the delta change approach will also be used in the present study. The time-slice simulations follow the IPCC "observed 20th century" 20C3M scenario for years 1970-1999 and the Special Report on Emission Scenarios (SRES) B1 and A2 for years 2010-2100 [12].

2.2. Data Processing

Average temperatures in New Brunswick range from −10°C in January to 19°C in July. New Brunswick receives approximately 1100 mm of precipitation annually, with 20% to 33% falling in the form of snow [14]. Precipitation tends to be highest in southern parts of the province whereas the northern part of New Brunswick receives relatively higher amounts of precipitation in the form of snow due to colder winters.

Daily maximum and minimum air temperatures and total precipitation data from seven meteorological stations in New Brunswick were obtained from Environment Canada's National Climate Data Archive (**Table 1**, **Figure 1**). Air temperature data were available from dates going back as far as 1895 for some stations whereas precipitation data goes back to 1929 (Aroostook). Daily

Table 1. Meteorological stations.

Meteor. station	Latitude, Longitude	Temperature	Precipitation
Aroostook	46°48'N; 67°43'W	1913-1999	1929-2005
Charlo	47°59'N; 66°20'W	1945-1999	1966-2005
Chatham	47°01'N; 65°27'W	1895-1999	1943-2005
Doaktown	46°33'N; 66°09'W	1952-1999	1934-2005
Fredericton	45°52'N; 66°32'W	1895-1999	1951-2005
Moncton	46°06'N; 64°47'W	1895-1999	1939-2005
Saint John	45°19'N; 65°53'W	1895-1999	1946-2005

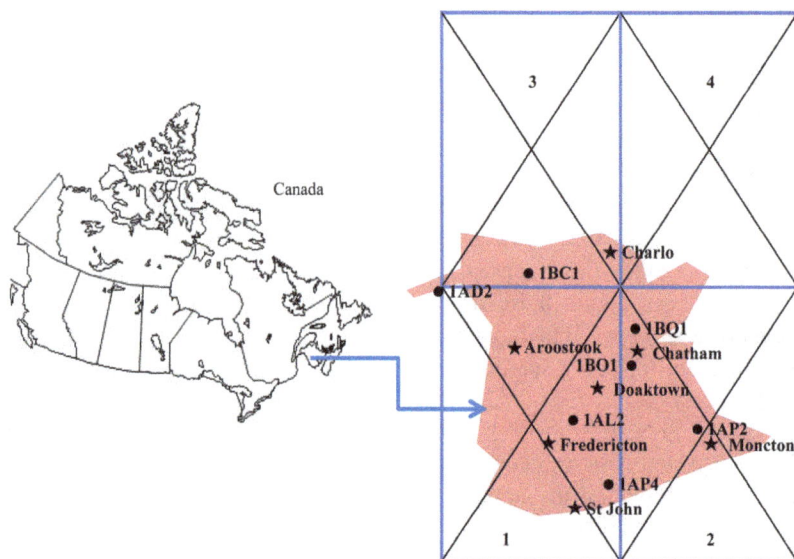

Figure 1. Subarea of 4 grid boxes corresponding to New Brunswick (box size ~200 × 300 km, • hydrometric station, ★ meteorological station).

discharge (m³/s) data from seven hydrometric stations in New Brunswick were obtained from Environment Canada's National Water Data Archive (HYDAT CD-ROM) (**Table 2**, **Figure 1**). Basin size of analysed stations ranged between 668 km² (Canaan River) and 14,700 km² (Saint John River).

Simulated monthly maximum and minimum air temperatures and total precipitation for the period of 1970-2100 was obtained from CCCma. The atmosphere model output was provided on a 128 × 64 Gaussian grid. **Figure 1** shows the sub-region of four tiles corresponding to New Brunswick.

When using simulated data from the CCCma model, an Inverse Distance Weighting (IDW) method was used to compute data corresponding to each meteorological site. This is a method for multivariate interpolation, a process of assigning values to unknown points by using values from usually scattered set of known points. The interpolated data were then downscaled using delta change approach [15,16]. This approach to downscaling of GCMs for hydrological modeling is one of the simpler

statistical downscaling techniques. In fact, Fowler *et al.* [17] noted that, if reproducing the mean characteristics are the main objectives, then simple statistical downscaling methods can perform as well as more sophisticated approaches. Changes in mean climate are applied as follows:

For air temperatures:

$$T_{new} = T_{obs} + T_{delta} \qquad (1)$$

For precipitation:

$$P_{new} = P_{obs} \times P_{fact} \qquad (2)$$

where T_{delta} is the difference between the GCM simulated mean temperature (from the future time period) and the historic mean temperature (1970-1999). P_{fact} is the ratio of the GCM simulated mean precipitation from the future time period relative to the historic period.

2.3. Hydrological Modelling Using Artificial Neural Networks

The average monthly discharge under climate change

Table 2. Hydrometric stations.

Hydrometric station	Latitude, Longitude	Drainage area (km^2)	Period of record
R. Saint John	47°15'N, 68°36'W	14,700	1927-2005
Nashwaak R.	46°08'N, 66°37'W	1450	1962-2005
Canaan R.	46°04'N, 65°22'W	668	1926-1940, 1963-2005
Kennebecasis R.	45°42'N, 65°36'W	1100	1961-2005
Restigouche R.	47°40'N, 67°29'W	3160	1963-2005
SW Miramichi R.	46°44'N, 65°50'W	5050	1919-1932, 1962-2005
NW Miramichi R.	47°06'N, 65°50'W	948	1962-2005

was simulated first using the Artificial Neural Networks (ANN) and then extreme event characteristics (floods and low flows) were predicted using a regression model based on historical events (ratio of daily to monthly values). This was done for the period 1970 to 1999. ANN approaches have been successfully used in water resources in a wide variety of applications [18,19]. Applications in river flow and groundwater level forecasting as well as sediment estimation are a few examples. Also, predicting conductivity and acidity of streams are some of the time series successfully modeled using this technique [20].

The following ANN model was used to predict mean discharge, Q_{ave}.

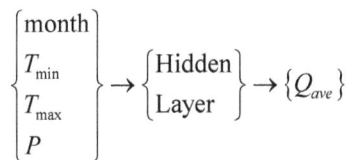

$$\begin{Bmatrix} month \\ T_{min} \\ T_{max} \\ P \end{Bmatrix} \rightarrow \begin{Bmatrix} Hidden \\ Layer \end{Bmatrix} \rightarrow \{Q_{ave}\}$$

where month is the month of year (1, 2, ⋯, 12), T is the air temperature (min or max) and P is the precipitation. The model is characterized by:

- Learning algorithm: back-propagation with Levenberg-Marquard method and early stopping;
- Training and validation data: 85% of data;
- Testing data: 15% of data;
- Hidden layer: 8 to 12 neurons.

It was observed that the extreme discharge events were better predicted using a $Ln(Q_{ave}) - Ln(Q_{max} \text{ or } Q_{min})$ regression model. Therefore, average monthly discharge, Q_{ave}, was simulated first using the ANN model. Once the Q_{ave} was predicted, then the regression equations were used to link maximum (Q_{max}) and minimum (Q_{min}) daily flows during each month.

3. Results

3.1. Method Evaluation

The ANN model showed a coefficient of determination, R^2, ranging between 0.69 and 0.79 in the prediction of

monthly flows (1970-1999). High flow predictions showed higher R^2 (0.85 - 0.92) than low flow predictions (0.66 - 0.82). Using the results from the Doaktown site (in close proximity SW Miramichi River) as an example, **Figure 2** shows how well the ANN model coupled with a Ln-Ln model simulated monthly Q_{ave}, Q_{max} and Q_{min} between 1985 and 1990 compared to observed discharge values when using historical input data (precipitation and temperature). Results here are typical of the whole study period. This figure shows that simulated discharges coincided fairly well with the observed discharges. It should be noted that input data consisted of local historical observations at the Doaktown station. As such, the ANN model and regression equations were effective in predicting the different flow components using historical data [16].

3.2. Future Climate Change Projections

Tables 3-5, in annual means, summarize the major changes in temperatures, precipitation and average flows throughout the province of New Brunswick. In terms of mean air temperature, the current air temperature in New Brunswick is expected to increase by 1.2°C (2010-2039), 2.2°C (2040-2069) and 2.9°C (2070-2099) under scenario B1 and 1.4°C (2010-2039), 3.2°C (2040-2069) and 5.2°C (2070-2099) under scenario A2 (**Table 3**). Precipitation (mean annual precipitation) is projected to increase significantly compared to current climate conditions (**Table 4**). At Doaktown, the precipitation will most likely increase from approximately 1140 mm annually to 1200 mm annually (B1 scenario) and to 1440 mm annually (A2 scenario) within the next hundred years. This represents an increase in precipitation of 5.1% (B1) and 14.6% (A2) compared to current conditions. Greatest increases are projected at the Moncton site (**Table 4**). Other sites show similar increases depending on the time slice and climate scenario. In general, the mean annual precipitation is expected to increase by 2% - 6% under the B1 scenarios and by 5% - 15% under the A2 scenario for the province of New Brunswick.

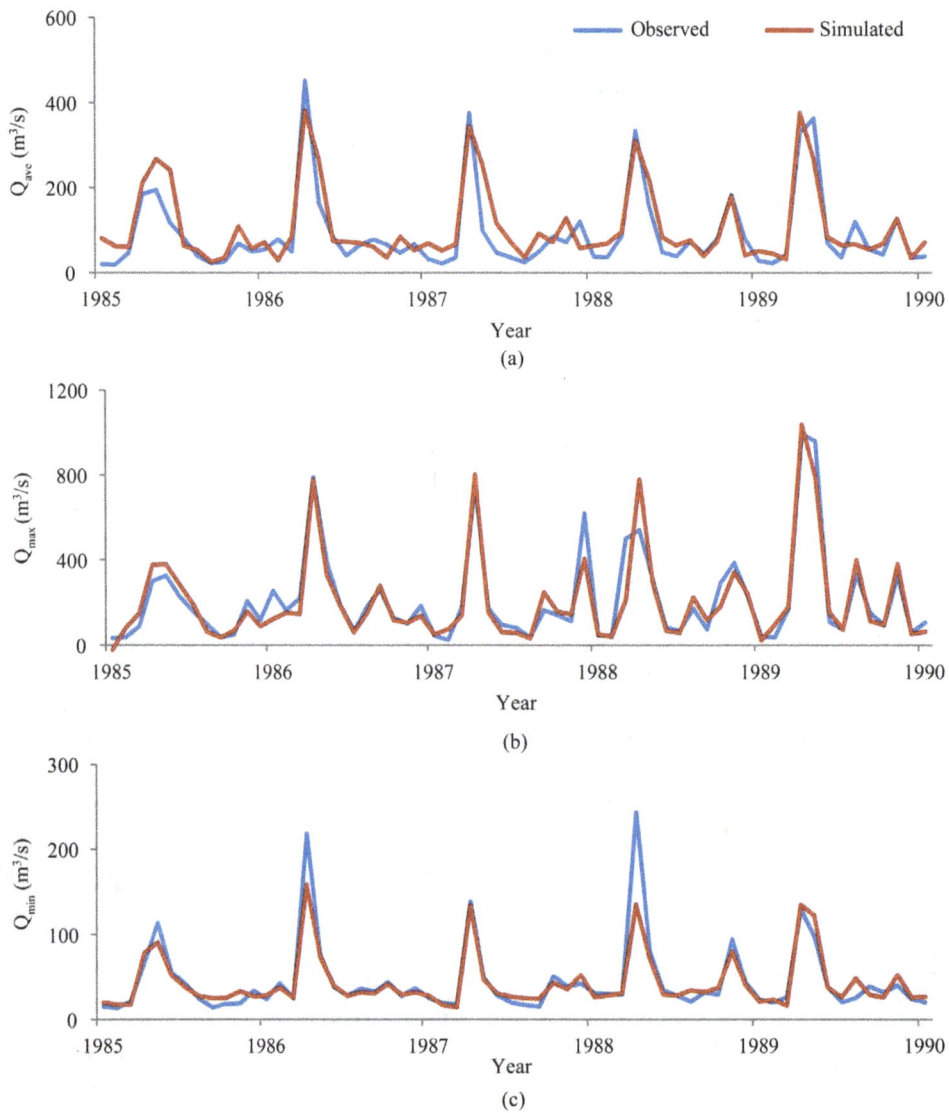

Figure 2. Observed and simulated monthly discharges for SW Miramichi River (a) mean discharge (b) maximum discharge (c) minimum discharge.

Table 3. Mean air temperature increase (°C).

	Scenario B1			Scenario A2		
	2010-2039	2040-2069	2070-2099	2010-2039	2040-2069	2070-2099
Aroostook	1.2	2.2	2.9	1.5	3.3	5.3
Charlo	1.2	2.2	2.9	1.4	3.2	5.2
Chatham	1.2	2.2	2.9	1.4	3.2	5.2
Doaktown	1.2	2.2	2.9	1.2	2.9	5.0
Fredericton	1.2	2.2	2.9	1.5	3.2	5.3
Moncton	1.2	2.2	2.9	1.5	3.2	5.3
Saint John	1.2	2.3	3.0	1.5	3.2	5.3

Table 4. Mean precipitation increase (%).

	Scenario B1			Scenario A2		
	2010-2039	2040-2069	2070-2099	2010-2039	2040-2069	2070-2099
Aroostook	2.2	6.3	5.6	4.2	8.0	15.4
Charlo	2.5	7.1	5.6	3.4	8.4	16.6
Chatham	2.3	7.1	5.6	4.5	8.4	15.2
Doaktown	2.1	6.5	5.1	4.4	8.1	14.6
Fredericton	2.3	6.6	5.3	4.8	7.7	14.7
Moncton	3.1	8.0	6.1	6.9	9.0	13.6
Saint John	2.8	7.0	5.6	5.1	7.9	14.7

Table 5. Mean flow increase (%).

	Scenario B1			Scenario A2		
	2010-2039	2040-2069	2070-2099	2010-2039	2040-2069	2070-2099
Saint John R.	3.8	8.8	11.3	4.6	13.4	26.6
Restigouche R.	3.8	6.0	6.7	3.9	7.2	11.0
NWMiramichi R.	0.2	1.7	1.0	0.7	1.9	4.0
SW Miramichi R.	4.7	10.2	11.4	5.6	13.6	22.6
Nashwaak River	1.3	3.2	2.9	2.2	3.9	8.4
Canaan River	4.4	10.4	11.9	8.1	14.1	24.4
Kennebecasis R.	3.5	7.6	7.3	5.5	9.4	15.5

3.3. Floods and Droughts under Future Climate Change

Using the projected maximum and minimum river discharge for each station, events of different recurrence intervals were computed using the generalized extreme value (GEV) distribution. The parameters of the GEV for both high and low flows were estimated by the method of L-moments and the goodness of fit was assessed by the Anderson-Darling statistics. As such, the discharge of different recurrence intervals (high and low flows) was based on the maximum (and minimum) daily flow recorded each year.

The frequency analysis show that for all sites under investigation, the intensity and frequency of discharges will most likely increase in severity for both climate scenarios. The increase in high flows for low return floods (e.g., 2-year) was generally higher than higher return floods (e.g. 100-year). Depending on the scenario and the time slice used, the increases for low return floods was about 30% whereas increases was about 15% for higher return floods. Low flows showed increases of about 10% for low return droughts and about 20% for higher return droughts. These results show that floods will most likely experienced a greater change for low recurrence intervals (e.g., 2-year) whereas the droughts will be most affected for higher recurrence intervals

3.4. Regional Climate Index (RCI)

An important part of the design process using frequency analysis is the estimation of future change in floods or droughts under climate scenarios at a given site for specific return periods or recurrence intervals. This analysis was carried out through the application of regional regression models linking floods and droughts to their frequencies under future climate scenarios (B1 and A2). To accomplish this, a so-called Regional Climate Index (RCI) was introduced. This index was calculated by dividing future flows over historical flows, while maintaining the same characteristics parameters. This RCI can be expressed as follows:

For floods:

$$RCI_F = Q_{F,T}^{x,ts,sc} / Q_{F,T}^{x,2010} \tag{3}$$

For droughts:

$$RCI_D = Q_{D,T}^{x,ts,sc} / Q_{D,T}^{x,2010} \tag{4}$$

where $Q_{F,T}^{x,ts,sc}$ and $Q_{D,T}^{x,ts,sc}$ are the high or low flow at a site x, during time slice ts (2010-2039, 2040-2069 or 2070-2099), under scenario sc (B1 or A2), and for return period T. $Q_{F,T}^{x,2010}$ and $Q_{D,T}^{x,2010}$ are the present time discharges at the same site x. RCI_F and RCI_D are the regional climate index equations for floods and droughts defined as the mean ratio of discharges from the future time period relative to the historic period.

The RCIs were obtained through regression analyses using historical and future simulated floods (and low flows). In all cases, the coefficient of determination, R^2, was about 0.99. These RCIs were developed for the whole province of New Brunswick, as future projections may contain some uncertainty at the site level. **Tables 6** and **7** show the regression equations while **Figures 3** and

4 show the RCI for floods and droughts for different return periods, for both scenarios B1 and A2 and for the different time periods. **Figure 3** show that low return floods will increase by a factor of 1.3 to 1.4 whereas higher return floods will only increase by 1.05 to 1.1 (scenario B1). Under the scenario A2, the increase will be more important (1.35 to 1.56 for low return flood and 1.16 to 1.21 for high return floods). **Figure 4** shows that droughts of low return periods will be less affected than higher return drought. However, in all cases, the severity of droughts will diminish, as an increase in flows is projected.

Table 6. Regional climate index equations for floods.

Time slice	Scenario B1	Scenario A2
2010-2039	$RCI_F(T) = -0.0727\mathrm{Ln}(T) + 1.3790$	$RCI_F(T) = -0.0756\mathrm{Ln}(T) + 1.4075$
2040-2069	$RCI_F(T) = -0.0759\mathrm{Ln}(T) + 1.4442$	$RCI_F(T) = -0.0795\mathrm{Ln}(T) + 1.4909$
2070-2099	$RCI_F(T) = -0.0794\mathrm{Ln}(T) + 1.4766$	$RCI_F(T) = -0.0907\mathrm{Ln}(T) + 1.6263$

Table 7. Regional climate index equations for droughts.

Time slice	Scenario B1	Scenario A2
2010-2039	$RCI_D(T) = 0.0100\mathrm{Ln}(T) + 1.0763$	$RCI_D(T) = 0.0191\mathrm{Ln}(T) + 1.0867$
2040-2069	$RCI_D(T) = 0.0194\mathrm{Ln}(T) + 1.1050$	$RCI_D(T) = 0.0195\mathrm{Ln}(T) + 1.1219$
2070-2099	$RCI_D(T) = 0.0193\mathrm{Ln}(T) + 1.1103$	$RCI_D(T) = 0.0194\mathrm{Ln}(T) + 1.7450$

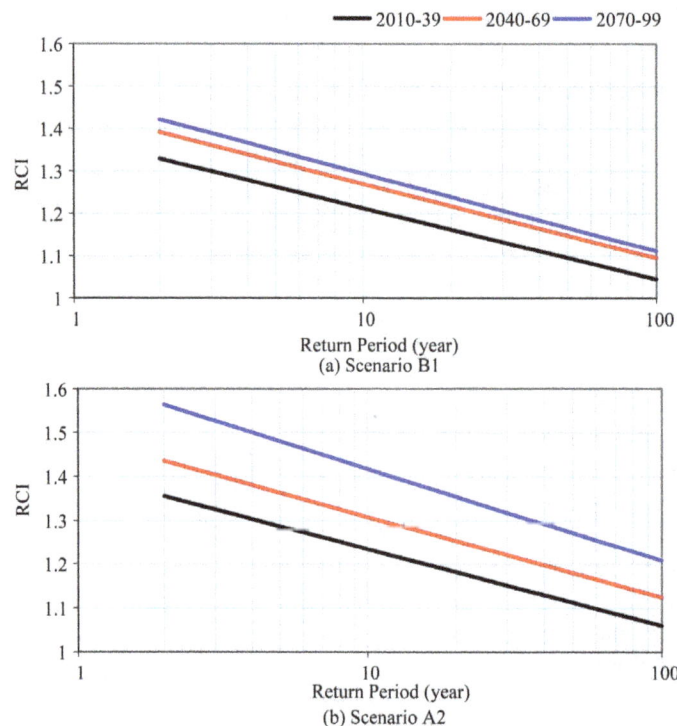

Figure 3. New Brunswick regional climate index curves for floods at time slices 2010-2039, 2040-2069, 2070-2099 under (a) scenario B1 (b) scenario A2.

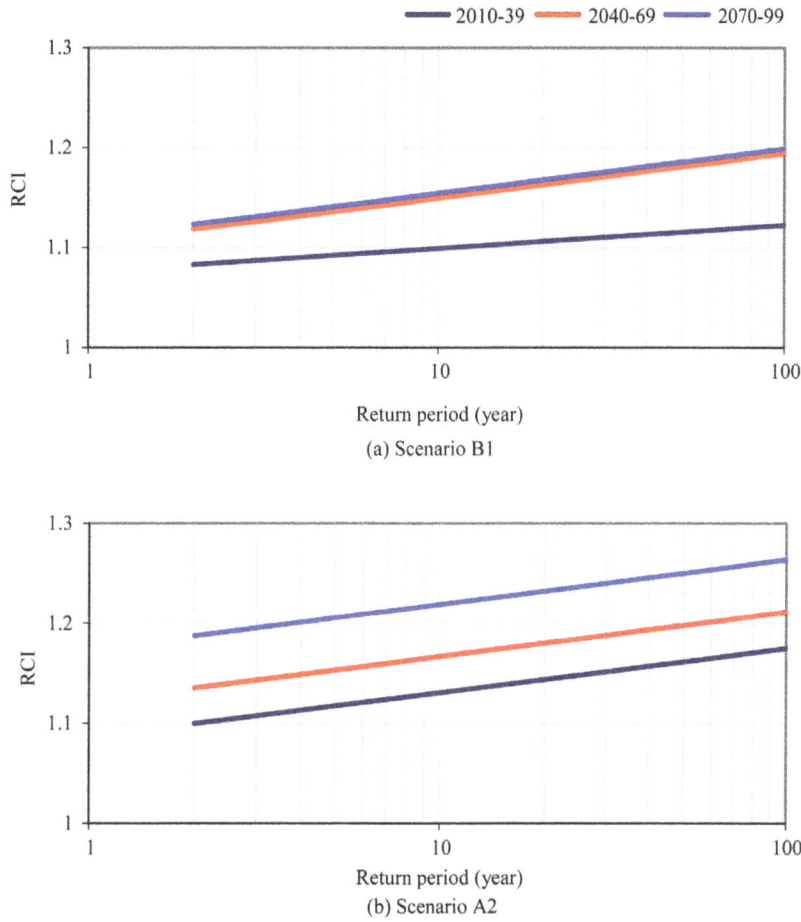

Figure 4. New Brunswick regional climate index curves for droughts at time slices 2010-2039, 2040-2069, 2070-2099 under a) scenario B1 b) scenario A2.

The future high and low flows in New Brunswick may therefore be estimated by:

For floods:

$$Q_{F,T}^{x,ts,sc} = Q_{F,T}^{x,2010} \times RCI_F(T) \qquad (5)$$

For droughts:

$$Q_{D,T}^{x,ts,sc} = Q_{D,T}^{x,2010} \times RCI_D(T) \qquad (6)$$

When expressed in percentage, it can be observed than future floods will most likely be higher than current floods by 30% - 40% (B1) or 35% - 55% (A2) for 2-year floods (**Figure 3**). Higher return floods (e.g. 100-year) are also expected to increase but not to the same extent as low return floods. For example, 100-year floods are projected to increase by 5% - 12% (B1) or 8% - 21% (A2). **Figure 4** shows that future low flows would most likely be higher than current low flows. For instance, a 2-year low flow could be 9% - 12% higher under the B1 scenario or 10% - 19% higher under the A2 scenario. Higher return (e.g. 100-year) low flows would most likely be less severe than current conditions. The increase in low flow discharge is expected to be in the range of 11% -

20% (B1) and 18% - 27% (A2).

4. Conclusion

Climate change will undoubtedly alter floods and droughts in New Brunswick. The success of industries (e.g., agriculture, forestry, fisheries and others) is intrinsically linked to climate and river flow, making New Brunswick particularly vulnerable to the impacts of climate change. Undoubtedly, these industries (among others) will be significantly affected by warmer and wetter climates with a potential shift in the availability of water during some seasons (e.g., summer and fall). Adaptation will be essential to maintaining the viability of some industries within the province. The hydrological response of seven catchments to two emission scenarios was simulated using artificial neural network (ANN). It was observed that air temperature in New Brunswick will increase by as much as 5.2°C by 2100. This rate of warming is much greater than that observed in the 20th century but is consistent with that predicted by Parks Canada [1] and Houghton *et al.*, [21] for the Atlantic Provinces.

As for precipitation, the mean annual precipitation showed an increase of 9% - 12% compared to current conditions.

The frequency analyses showed that flood magnitude would most likely increase by 11% - 21% towards the end of the century, depending on the emission scenario used. In term of low flows, the model is predicting a 20% - 26% increase towards the end of the century. This increase in low flow is most likely linked to the increase in precipitation in the future. Finally, future high and low flows were estimated by the introduction of a Regional Climate Index (RCI) for the province of New Brunswick. This RCI for floods and droughts was defined as the mean ratio of future discharge relative to the historic flow. As such, RCIs can used to estimate future floods and droughts conditions within the study area.

5. Acknowledgements

This study was funded by the New Brunswick Environmmental Trust Fund. The authors wish to thank anonymous reviewers for several helpful suggestions.

REFERENCES

[1]   Parks Canada, "Air Quality, Climate Change and Canada's National Parks," Parks Canada, Natural Resources Branch, Air Issues Bulletin No. 100, *Ottawa*, 1999.

[2]   H. G. Hengeveld, "Global Climate Change: Implication for Air Temperature and Water Supply in Canada," *Transactions of the American Fisheries Society*, Vol. 119, No. 2, 1990, pp. 176-182.

[3]   Natural Resources Canada, "Climate Change Impacts and Adaptation: A Canadian Perspective," Water Resources, Climate Change Impacts and Adaptation Directorate, Ottawa, 2002.

[4]   C. K. Minns, R. G. Randall, E. M. P. Chadwick, J. E. Moore and R. Green, "Potential Impact of Climate Change on the Habitat and Population Dynamics of Juvenile Atlantic Salmon (*Salmo salar*) in Eastern Canada. Climate Change and Northern Fish Population," *Canadian Special Publications of Fisheries and Aquatic Sciences*, 1995, pp. 699-708.

[5]   D. Caissie,"Hydrological Conditions for Atlantic Salmon Rivers in the Maritime Provinces in 1997," *Canadian Stock Assessment SecretariatResearch Document*, 1999.

[6]   D. Caissie, "Hydrological Conditions for Atlantic Salmon Rivers in the Maritime Provinces in 1998," *Canadian Stock Assessment Secretariat Research Document*, 1999.

[7]   D. Caissie, "Hydrological Conditions for Atlantic salmon Rivers in 1999," *Canadian Stock Assessment Secretariat Research Document*, 2000.

[8]   Environment Canada and New Brunswick Department of Municipal Affairs and Environment, "Flood Frequency Analyses, New Brunswick, a Guide to the Estimation of Flood Flows for New Brunswick Rivers and Streams," April 1987, p. 49.

[9]   F. Aucoin, D. Caissie, N. El-Jabi and N. Turkkan, "Flood Frequency Analyses for New Brunswick Rivers," Canadian Technical Report of Fisheries and Aquatic Sciences, 2011.

[10]   Environment Canada and New Brunswick Department of the Environment, "Low Flow Estimation Guidelines for New Brunswick," Inland Waters Directorate, Environment Canada, Dartmouth, NS and Water Resources Planning Branch, New Brunswick Department of the Environment, Fredericton, 1990.

[11]   D. Caissie, L. LeBlanc, J. Bourgeois, N. El-Jabi and N. Turkkan, "Low Flow Estimation for New Brunswick Rivers," Canadian Technical Report of Fisheries and Aquatic Sciences, 2011.

[12]   J. Alcamo, A. Bouwman, J. Edmonds, A. Grübler, T. Morita and A. Sugandhy, "An Evaluation of the IPCC IS92 Emission Scenarios," In J. T. Houghton, L. G. Meira Filho, J. Bruce, H. Lee, B. A. Callander, E. Haites, N. Harris and K. Maskell, Eds., *Radiative Forcing of Climate Change and an Evaluation of the IPCC IS92 Emission Scenarios*, Cambridge University Press, Cambridge, 1995, pp. 233-304.

[13]   N. Nakicenovic, A. Grubler, H. Ishitani, T. Johansson, G. Marland, J. R. Moreira and H.-H. Rogner, "Energy Primer in Climate Change 1995," In: R Watson, M. C. Zinyowera and R. Moss, Eds., *Impacts, Adaptations and Mitigation of Climate Change: Scientific Analysis*, Cambridge University Press, Cambridge, UK, 1996.

[14]   D. Caissie and S. Robichaud, "Towards a Better Understanding of the Natural Flow Regimes and Streamflow Characteristics of Rivers of the Maritime Provinces," Canadian Technical Report of Fisheries and Aquatic Sciences, 2009.

[15]   C. Prudhomme, N. Reynard and S. Crooks, "Downscaling from Global Climate Models for Flood Frequency Analysis: Where Are We Now?" *Hydrological Processes*, Vol. 16, No. 6, 2002, pp. 1137-1150.

[16]   N. Turkkan, N. El-Jabi and D. Caissie,"Floods and Droughts under Different Climate Change Scenarios in New Brunswick," Canadian Technical Report of Fisheries and Aquatic Sciences, 2011.

[17]   H. Fowler and C. Kilsby, "Using Regional Climate Model Data to Simulate Historical and Future River Flows in Northwest England," *Climatic Change*, Vol. 80, No. 3, 2007, pp. 337-367.

[18]   R. S. Govindaraju, "Artificial Neural Networks in Hydrology. I: Preliminary Concepts," *Journal of Hydrologic Engineering*, Vol. 5, No. 2, 2000, pp.115-123.

[19]   R. S. Govindaraju, "Artificial Neural Networks in Hydrology. II: Hydrologic Application," *Journal of Hydrologic Engineering*, Vol. 5, No. 2, 2000, pp. 124-137.

[20]   D. Bastarache, N. El-Jabi, N. Turkkan and T. A. Clair. "Predicting Conductivity and Acidity for Two Small Streams Using Learning Networks," *Canadian Journal of*

Civil Engineering, Vol. 24, 1997, pp. 1030-1039.

[21] J. T. Houghton, D. Ding, J. Griggs, M. Noguer, P. J. Z. van der Linden and D. Xiaosu, "Climate Change 2001: The Scientific Basis," Contribution of Working Group I to the Third Assessment Report of the Intergovernmental Panel on Climate Change (IPCC), Cambridge University Press, Cambridge, 2001.

Spatially Explicit Modeling of Long-Term Drought Impacts on Crop Production in Austria

Franziska Strauss[1,2], Elena Moltchanova[3], Erwin Schmid[2]
[1]Central Institute for Meteorology and Geodynamics, Vienna, Austria
[2]Institute for Sustainable Economic Development, University of Natural Resources and Life Sciences, Vienna, Austria
[3]Department of Mathematics and Statistics, University of Canterbury, Christchurch, New Zealand

ABSTRACT

Droughts have serious and widespread impacts on crop production with substantial economic losses. The frequency and severity of drought events may increase in the future due to climate change. We have developed three meteorological drought scenarios for Austria in the period 2008-2040. The scenarios are defined based on a dry day index which is combined with bootstrapping from an observed daily weather dataset of the period 1975-2007. The severity of long-term drought scenarios is characterized by lower annual and seasonal precipitation amounts as well as more significant temperature increases compared to the observations. The long-term impacts of the drought scenarios on Austrian crop production have been analyzed with the biophysical process model EPIC (Environmental Policy Integrated Climate). Our simulation outputs show that—for areas with historical mean annual precipitation sums below 850 mm—already slight increases in dryness result in significantly lower crop yields *i.e.* depending on the drought severity, between 0.6% and 0.9% decreases in mean annual dry matter crop yields per 1.0% decrease in mean annual precipitation sums. The EPIC results of more severe droughts show that spring and summer precipitation may become a limiting factor in crop production even in regions with historical abundant precipitation.

Keywords: Long-Term Drought Modeling; Dry Day Index; Biophysical Impacts; Spatial Variability; EPIC; Austria

1. Introduction

Drought is a natural phenomenon and can be defined as sustained and extensive occurrence of below average water availability [1]. A drought can be defined in several ways depending on the disciplinary perspective (e.g. soil moisture drought, hydrological drought). The most commonly used definition is the meteorological drought. It constitutes a deficit of precipitation (in comparison to climatological average values) in a given region and over a defined time span. In contrast, an agricultural drought results from lacking water supply for agricultural crops, leading to a reduction of crop yields in the affected region. The crop yields can be further affected by temperature increases, wind speed and humidity, which may lead to higher potential evapotranspiration rates and thus cause soil moisture deficits [e.g. 2]. An agricultural drought has been differently defined by several authors e.g. definition based on daily rainfall and crop water consumption [3], on the deficiency or absence of precipitation during the growing season or by long dry spells [4], on the number of consecutive days on which the actual evapotranspiration to potential evapotranspiration ratio remains below a certain threshold value [5].

Droughts might manifest themselves either as short but extreme single season drought (such as the extremely hot summer in Europe in 2003) or as a longer-term, multi-season drought, and they might be local or widespread in nature [6]. In Europe, the mean annual economic loss due to droughts amounts to € 5.3 billion, with a peak of € 8.7 billion in 2003 [1].

Precipitation deficits in many European regions have already become more severe [7]. On the European scale, some areas may be affected by a general increase in the number of drought occurrences and also by more extreme drought events with uncertain severity and spatial distribution [2,8]. For central and southern Europe, it has been projected that the total proportion of areas under water stress may increase from 19% in 2007 to 35% in 2070 [8]. The net global trend may result in an increase in the area of extreme drought from 1% to 30% by the end of this century [9]. An analysis of droughts in the Czech Republic in the period 1881-2006 confirms a statistically significant tendency to prolonged and more intensive dry episodes in terms of increased temperatures and decreased precipitation sums [2].

According to theories of physical geography and empirical field investigations, impacts of droughts on different landforms, soil types, and crop management systems may vary significantly. The soil types more often susceptible to droughts are mountain burozem, plain burozem, chestnut soil, chernozem, grassmarshland chernozem, and grassmarsh soil [4,10]. In the Czech Republic, persistently lower crop yields for spring barley, winter wheat, and forage crops were documented in years with drought episodes, in comparison to years without droughts [2]. The degree of damage to maize production has recently increased with global warming, and spatial distributions of crop yield losses caused by drought events are closely connected with the rainfall during growing season of the crop as well as the aridity index [10]. Furthermore, the degree of damage from droughts depends on frequency, duration, intensity, spatial extent (*i.e.* the area affected by drought), regional production level of the specific crop, and the timing of the drought related to crop phenology [11].

In summary, many studies have already assessed impacts of drought events on agricultural production [1,3-5, 11-14], but only few have investigated the impacts of increased frequency of extreme events on crop production [e.g. 15,16]. Consequently, it is important to quantitatively assess the impacts of increased drought occurrences on crop yields in geo-spatial contexts to account for the natural heterogeneity (with respect to temporal and spatial variations) in crop production. Such an analysis may provide a basis for the assessment of the potential damage by long-term droughts as well as of measures to mitigate the adverse impacts of drought risks (e.g. planning and management of water resources for irrigation to reduce the vulnerability in crop production).

A recent review presents common methodologies for future drought modeling [17], e.g. time series analysis, probability based modeling, spatio-temporal analysis, and the use of climate change scenarios of GCMs. However, the following weaknesses were identified: 1) the assumption of linearity between predictor and predictand in a regression model, 2) the modeling of drought events with ARIMA for up to two months in advance but no longer, 3) the application of Markov chains in probability models (firstly used by [18] having the property that the value of the next state depends only on the current state and not on the entire past), and 4) the development of future drought events based on precipitation anomalies derived from GCMs. Based on such GCM outputs, an increase in the frequency of long-duration droughts was found for catchment areas in southern Europe and less

frequent droughts for a catchment in northern England, but the model used in the analysis is plagued by problems of correctly reproducing mean monthly precipitation sums [19]. The usefulness of GCMs to perform a proper analysis of dry spells is therefore limited because of the arbitrariness in the generation of day-to-day precipitation data as well as the spatio-temporal and physical inconsistency between other weather parameters [17].

In the present study, we analyze both the meteorological and agricultural droughts. We use an alternative statistical approach to develop meteorological drought scenarios for Austria. Our approach is based on a climatological study for Austria, which provides daily weather data for the period 1975-2007 on a 1 km grid resolution [20]. We have divided the original datasets into blocks of seven and eight days (the blocks were not allowed to span more than one month) and calculated the dry day index which is the proportion of total area dry during those days. We have then applied a block bootstrap method [21] to generate hypothetical more extreme meteorological drought scenarios for the future. We have entered our developed drought scenarios as well as data on soil, topography and crop management into the biophysical process model EPIC—Environmental Policy Integrated Climate [22-23], which is applied on Austrian cropland area at 1 km grid resolution. EPIC provides outputs inter alia on dry matter crop yields, soil moisture, evapotranspiration, and stress factors of crop growth. Hence, the impacts of increased meteorological drought occurrences on crop production can be demonstrated in a spatially explicit and consistent manner. Furthermore, regions can be identified which are most prone to drought risk. A 31 years period is investigated allowing for the analysis of both long-term drought impacts and interannual variability in agricultural crop production.

This article is organized as follows. Section 2 presents the data and the methodology for the development of drought scenarios in Austria. In Section 3, we show the consequences and impacts of increased long-term drought occurrence on both simulated crop yields and evapotranspiration followed by discussion and conclusion in Section 4.

2. Data, Methods and Impact Modeling

2.1. Dry Day Index

We have used the climate dataset for Austria [20], which includes daily time series of solar radiation, maximum and minimum temperatures, precipitation, relative humidity and wind speed for the period 1975-2007. This dataset is based on both spatially interpolated climatologies of mean annual precipitation sums and mean annual temperatures on a 1 km grid [24] including high-quality homogenized daily time series [25]. The spatially interpolated climatologies have been combined into clusters with homogenous climate characteristics (*i.e.* climate clusters), and for each climate cluster the daily time series represent the long-term (inter-annual) and short-term (daily) variabilities. In total, 60 climate clusters have been derived which represent the small-scale climate patterns in Austria [cp. 20]. They are named with numbers including the climatological characteristics: e.g., climate cluster 0509 represents regions with mean annual precipitation sums smaller than or equal to 500 mm and mean annual temperatures ranging between 8.6°C and 9.5°C; climate cluster 0901 represents regions with mean annual precipitation sums ranging between 801 mm and 900 mm and mean annual temperatures between 0.1°C and 2.5°C; and climate cluster 2006 represents regions with mean annual precipitation sums greater than 1500 mm and mean annual temperatures ranging between 5.6°C and 6.5°C.

The highest annual precipitation sums are in the mountainous regions of the west and the south as well as in the northern foothills of the Austrian Alps. The lowest annual precipitation sums are in the flat areas of the east, the north-east and the south-east of Austria. We are particularly interested in these regions considering the long-term impacts and the inter-annual variability of increased meteorological drought occurrences on crop production. For this purpose, we model meteorological drought scenarios based on the following dry day index DI_B:

$$DI_B = \frac{\sum_c \sum_{d \in B} I\{prec_{cd} = 0\} A_c}{l_b \sum_c A_c} \qquad (1)$$

where $I\{prec_{cd} = 0\}$ equals to 1 if there was no precipitation recorded for the cluster c on day d and equals to 0 otherwise, l_b is the block length (seven or eight days) and A_c is the area of the respective climate cluster c. Note that $\sum_c A_c$ is the total area of Austria. The block bootstrap was used rather than the simple (one-day-at-a-time) bootstrap in order to account for temporal autocorrelation. Each month was thus divided into four blocks. The first three blocks were of eight days each and the last block was always incomplete (for example, for a month of 30 days, the last block is of length $30 - 8 \times 3 = 6$ days). However, for convenience, we refer to an "eight-day block" in this text. Thus, the dry day index DI_B reflects the average proportion of the total territory which is dry on any random day. Note, that $DI_B = 1$ for absolutely dry and $DI_B = 0$ for daily rainfall in the entire country (**Figure 1**).

In order to take into account seasonal variation in other weather parameters, our block bootstrap set-up was also month specific. Thus, for example, blocks for June 2040 were sampled from the set of blocks from any past June.

We have chosen to analyze the impacts of three hypothetical meteorological drought scenarios for Austria. The first drought scenario reflects a baseline scenario (S0), *i.e.* the distribution of the dry day index remains nearly the

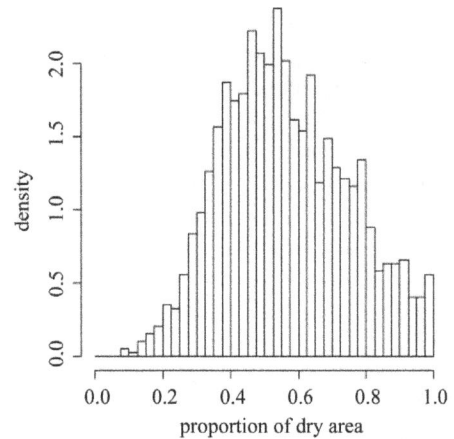

Figure 1. The empirical distribution of the dry day index DI_B in the period 1975-2007.

same as for the past period 1975-2007. The other two drought scenarios (S1 and S2) project higher proportion of dry days, and are modeled by sampling more frequently from the drier eight-day blocks. We use a weighted resampling scheme where the probability P_B of sampling block B is evaluated by:

$$P_B = \frac{DI_B^a}{\sum DI_B^a} \qquad (2)$$

where a is the resampling weight. Note that if $a = 0$, we get S0 so that each block has an equal probability of being resampled, and if $a = 1$ (respectively 2) we obtain S1 (respectively S2). In S1, the situation where almost the entire country is wet in an eight-day block occurs less often than in S0, reflecting an increase in drought events. S2 describes an even more extreme increase in the frequency of drought events as the fraction of days within an eight-day block—during which many parts of the country are wet—decreases even more (**Table 1**). Most pronounced decreases in precipitation sums with S2 are in the range of 40% and occur in autumn and winter. By comparison, the climate projections with the Climate Local Model (CLM) and the A1B emissions scenario (out of the Special Report on Emissions Scenarios—SRES—by IPCC) show highest precipitation decreases of 41% in the south-western Alps followed by 25% in the north-eastern Alps in summer; however, these projections until the end of the 21st century depend on the GCMs and regions selected. On average, precipitation decreases between 1% and 11% are simulated for the Alpine region [26].

The procedure for obtaining time series of the set of meteorological parameters is described in the following: 1) We have calculated DI_B (Equation (1)) for each eight-day block from the period 1975-2007. 2) The blocks have then been bootstrapped based on DI_B from the past and on month-specific blocks with the probability evaluated in Equation (2). The spatial autocorrelation is taken into

Table 1. Probabilities of a given percentage of dry areas in Austria for S0, S1 and S2 based on one of the 30 re-allocations for each of the drought scenarios.

% of dry area	S0 $P_B = 1$	S1 $P_B = DI_B / \sum DI_B$	S2 $P_B = DI_B^2 / \sum DI_B^2$
0 - 20	0.02	0.01	0.00
21 - 40	0.20	0.12	0.07
41 - 60	0.41	0.38	0.33
61 - 80	0.26	0.35	0.38
81 - 100	0.12	0.15	0.21

account as well since we sample the weather parameters (precipitation, maximum and minimum temperatures, solar radiation, relative humidity and wind speed) for the entire country at a time. 3) 30 bootstrapped samples have been produced for each drought scenario in order to assess the uncertainty of our estimates.

Other statistical approaches (e.g. LARS WG) exist as well see [27-29] which are capable of generating local-scale and daily climate scenarios based on the output from GCMs. The motivation of our statistical approach can be summarized by the weaknesses of GCMs and Regional Climate Models (RCMs) to consistently project near climate futures (defined as the upcoming 30 years period) such as: 1) uncertainties in initialization and parameterization; 2) coarse resolution of GCMs (and often also RCMs) in representing the complex terrain in Austria; 3) climate change signals arising from greenhouse forcing, which is still small for the upcoming 30 years period compared to the model differences and model internal long-term variability; and 4) the spatio-temporal and physical inconsistency of multiple weather parameters. Furthermore, the scenario members generated by GCMs (RCMs) might not be independent from each other. However, the main limitation of our approach comes from the assumptions made in generating the climate scenarios without taking into account any possible other developments (*i.e.* economy, population, technology) as described in SRES. But such developments may gain in importance and show significant impacts in the second half of the 21st century.

2.2. Biophysical Impact Modeling with EPIC

The impact assessment of increased drought occurrences on crop production has been performed by analyzing annual simulation outputs of the biophysical process model EPIC, which has been already validated for some Austrian regions [e.g. 30-33]. In previous studies, EPIC crop yields were found to be more sensitive to temperature changes compared to precipitation changes [32]; however, in the present analysis, precipitation is expected to change by more units than temperature, and thus resulting in higher sensitivities of crop yields to precipitation.

Biophysical processes simulated with EPIC include among others crop growth, nutrient leaching, nitrification, mineralization, wind and water erosion, and soil carbon respiration [22,23]. These processes are simulated at daily time steps or smaller. EPIC contains algorithms that allow for a complete description of the hydrological balance at field to small watershed scale (up to 100 ha) including snowmelt, surface runoff, infiltration, soil water content, percolation, lateral flow, water table dynamics, and evapotranspiration. EPIC offers five equations to compute potential evapotranspiration (PET) including Penman-Monteith [34]—which has been used in the present analysis, and [35-38].

EPIC uses the concept of radiation-use efficiency by which a fraction of daily photosynthetically active radiation is intercepted by the plant canopy and converted into plant biomass. The leaf area index is simulated as a function of heat units, crop stresses and development stages. Daily accumulation of plant biomass is affected by vapor pressure deficits and atmospheric CO_2 concentration [39]. Crop yield is a function of the harvest index and aboveground biomass. Stress indices for water, temperature, nitrogen, phosphorus, aluminum toxicity and aeration are calculated daily using the most limiting value to reduce actual plant growth and crop yield. Similarly, stress factors such as soil bulk density, temperature, and aluminum toxicity are used to adjust potential root growth [40].

The soil water balance depending on the potential water use, the root zone depth, and the water use distribution parameter is applied in a water use function in which any water deficit can be overcome if a soil layer that is encountered has adequate water storage. The potential water use is reduced when the soil water storage is less than 25% of plant-available soil water by using dependencies on the soil water contents at field capacity and wilting point.

The site conditions in Austria comprise topographical (slope and elevation) and soil data which are classified into Homogenous Response Units (HRUs) cp. [33]. A total of 443 HRUs has been delineated for Austria and comprises six elevation classes (from sea level to above 2100 m), seven slope classes (from smaller 5% to above 100%), and 15 soil types which are merged with data from 60 climate clusters (cp. Section 2.1). Soil data are extracted from [41] and contain soil layer specific contents of silt, sand and clay, humus, pH, calcium carbonate, and coarse fragments. Up to 25 crop rotation systems per municipality have been derived with the CropRota model [42] using historical land use data of 22 crops [32]. These crops, respectively crop rotations, cover about 89% of total arable land in Austria. In EPIC, annual planting and harvesting dates are automatically adjusted to account for changing seasonal growing conditions (*i.e.* planting and harvesting dates are triggered when certain fractions of total heat units per crop are attained by daily heat unit accumulation from planting to maturity). Fertilizer application rates (N, P, K) are computed crop and regional specific [33] and refer to the guidelines of good agricultural practices in Austria. Irrigation is omitted to isolate the drought impacts on crop production.

The biophysical impacts of drought scenarios on crop yields and evapotranspiration are simulated for the future period and presented next.

3. Results

3.1. Description of Meteorological Drought in Austria

On average, the mean annual precipitation sums in S2 from the period 2008-2040 are lower than in S0 and S1. In addition, we obtain significant temperature increases in S1 and S2, predominantly in spring, summer and autumn. Mean annual maximum temperatures are rising most in S2. In **Figure 2**, we show for the climate clusters 0509, 0901 and 2006 the development of monthly precipitation sums in a randomly selected year (2040) compared to the mean monthly precipitation sums of 1975-2007. These climate clusters have been chosen, because they span typical climates in Austria (lowlands and Alpine regions). **Figure 2** depicts the deficiency of precipitation in several months when compared to the respective mean monthly sums of the past period. The occurrence of meteorological drought is thus obvious even if monthly precipitation

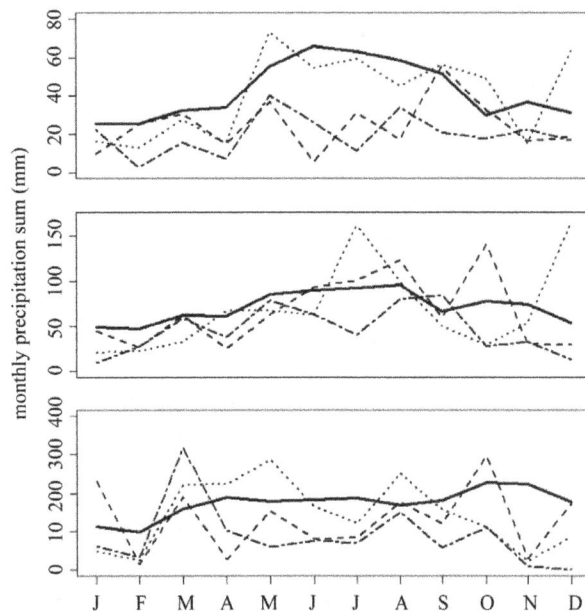

Figure 2. Monthly precipitation sums (mm) for the climate clusters 0509 (top), 0901 (middle), and 2006 (bottom) based on one bootstrapped re-allocation in 2040 for S0 (dotted), S1 (dashed), S2 (dashed-dotted) and mean monthly precipitation sums of 1975-2007 (continuous line).

might be above the mean value of 1975-2007 in certain months. In the case that the highest past precipitation sums occur in summer (autumn), we also receive the highest future precipitation sums in summer (autumn), on average. **Table 2** summarizes in which season the decrease in precipitation sums is most pronounced, as it may have decisive effects on particular crops.

The largest decreases in precipitation sums occur for climate cluster 0509 with 43% in autumn (September, October, November), and for climate clusters 0901 and 2006 with 41% in winter (December, January, February). The precipitation decreases are also important in spring and summer (ranging between 20% and 34%) with S2. Even in regions with abundant precipitation (e.g. climate cluster 2006), the seasonal precipitation sums, particularly in the spring and summer seasons, can be so low that water may become a limiting factor in crop production.

Figure 3 shows the aridity index *AI* which is defined as the relation between annual precipitation and potential evapotranspiration [43]. Its spatial distribution is shown for S0, S1 and S2, and the number of years (in total 31 years * 30 bootstrapped re-allocations) where $AI \leq 0.5$ (specifying a semi-arid climate) increases significantly in S1 and S2, predominantly in the eastern parts of Austria.

3.2. Spatial Analysis of Long-Term Drought Impacts on Crop Yields and Evapotranspiration

We have analyzed the impacts on annual crop yields for S1 or S2 and S0 (**Figure 4**). Therefore, we have averaged the sum of dry matter crop yields of the respective crop rotation for each 1 km grid and over the period 2010-2040 as well as the 30 bootstrapped re-allocations. S1 leads to lower dry matter crop yields of ~2 t/ha in comparison to S0 for many parts of the country, especially in the eastern lowlands. In contrast, simulated dry matter crop yields are higher by up to 2 t/ha in the foothills of the Alps and in the south-east of Austria. In these regions, enough water seems to be available for crop production and the increase in temperature leads to additional crop yields due to less temperature stress on average.

The adverse effects on crop yields are more pronounced in S2. The extent of areas where the lower precipitation becomes a limiting factor for crop production increases: Crop yields may decrease between 2 t/ha and 4 t/ha, most severely in the eastern lowlands and the northern foothills of the Alps. Furthermore, small increases in crop yields in S1 turn into crop yield losses of ~2 t/ha in S2 in many parts of the north and south-east of Austria.

In comparing the results of S1 with S0, we reveal that the highest decreases of average dry matter crop yields (about 2 t/ha) are linked with the highest decreases in evapotranspiration rates (between 60 mm and 80 mm; **Figure 5**) and important increases of water stress (up to

Table 2. Seasonal average temperatures (tave in °C) and precipitation sums (prec in mm) for the climate clusters 0509, 0901, and 2006 for the past, S0, S1, and S2. In parentheses, anomalies are presented in °C for tave and % for prec.

	Spring		Summer		Autumn		Winter	
	tave	prec	tave	prec	tave	prec	tave	prec
Climate cluster 0509								
Past	10.0	121	19.0	188	9.8	118	0.0	82
S0	9.8 (−0.2)	124 (2)	19.1 (0.1)	184 (−2)	10.0 (0.2)	116 (−2)	0.0 (0.0)	80 (−2)
S1	10.1 (0.1)	102 (−16)	19.1 (0.1)	151 (−20)	10.0 (0.2)	87 (−26)	−0.2 (−0.2)	67 (−18)
S2	10.0 (0.0)	97 (−20)	19.5 (0.5)	140 (−25)	10.1 (0.3)	67 (−43)	−0.3 (−0.3)	48 (−41)
Climate cluster 0901								
Past	0.6	208	9.5	277	3.0	217	−5.4	150
S0	0.3 (−0.3)	202 (−3)	9.7 (0.2)	275 (−1)	3.2 (0.2)	201 (−7)	−5.2 (0.2)	156 (4)
S1	0.8 (0.2)	203 (−2)	9.9 (0.4)	246 (−11)	3.7 (0.7)	175 (−19)	−5.1 (0.3)	119 (−21)
S2	0.9 (0.3)	141 (−32)	10.3 (0.8)	222 (−20)	4.5 (1.5)	134 (−38)	−4.8 (0.6)	88 (−41)
Climate cluster 2006								
Past	6.0	526	14.9	539	7.4	634	−1.3	386
S0	5.8 (−0.2)	600 (14)	15.0 (0.1)	518 (−4)	7.5 (0.1)	629 (−1)	−1.2 (0.1)	354 (−8)
S1	6.2 (0.2)	463 (−12)	15.1 (0.2)	427 (−21)	7.8 (0.4)	461 (−27)	−1.2 (0.1)	306 (−21)
S2	6.2 (0.2)	403 (−23)	15.5 (0.6)	353 (−34)	8.1 (0.7)	417 (−34)	−1.4 (−0.1)	228 (−41)

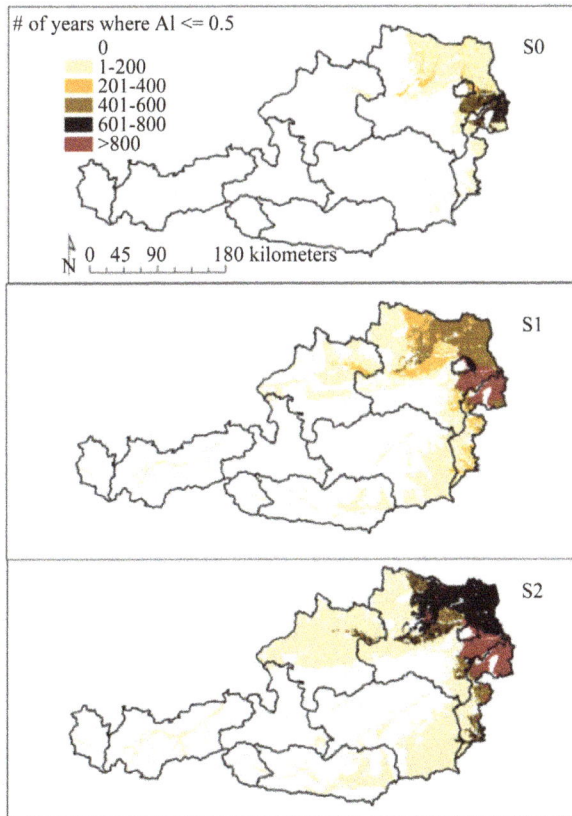

Figure 3. Occurrence of a semi-arid climate ($AI \leq 0.5$) on Austrian cropland by scenarios (S0, S1, S2).

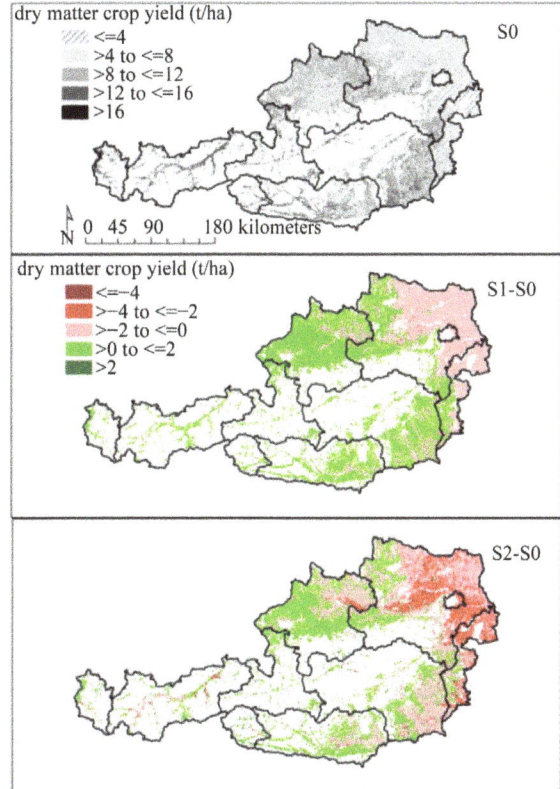

Figure 4. Simulated average annual dry matter crop yields (t/ha) on Austrian cropland for S0, and respective differences (middle: S1-S0, bottom: S2-S0).

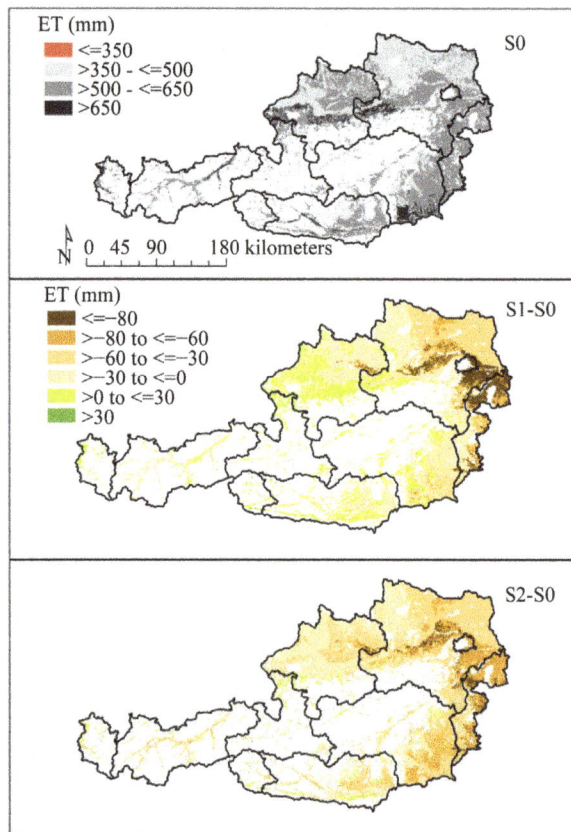

Figure 5. Simulated average annual evapotranspiration rates (mm) on Austrian cropland for S0, and respective differences (middle: S1-S0, bottom: S2-S0).

50% more days with water stress compared to S0 in the 31 years period and for the 30 bootstrapped re-allocations), predominantly in the north-eastern parts of Austria. Severely limiting precipitation and higher temperatures can lead to lower crop transpiration rates and biomass accumulation as well as soil evaporation. In contrast, it can also lead to higher evapotranspiration rates by up to 30 mm in regions, where precipitation is still sufficiently available. However, slight crop yield increases may also induce decreases of evapotranspiration rates (in the comparison with S1 in the range of 0 mm to 30 mm; in the comparison with S2 in the range of 0 mm to 80 mm). In these regions, the crop transpiration slightly increases due to higher temperatures, but soil evaporation decreases due to deficient precipitation amounts. The highest absolute dry matter crop yield losses simulated with S2 range between 2 t/ha and 4 t/ha, which are associated with decreases of annual evapotranspiration rates between 60 mm and 80 mm and an increase of water stress days by 75%.

To account for long-term drought impacts of different soil types, elevation and slope classes on dry matter crop yields, we have averaged the EPIC outputs over the respective area (out of the total cropland area of about 1.3 million ha). Most pronounced dry matter crop yield dif-

ferences between S1 and S0 occur on chernozems (−1 t/ha) and alluvial soils (−0.2 t/ha). The comparison between S2 and S0 leads to substantial decreases in dry matter crop yields of about 2 t/ha on chernozems, 0.8 t/ha on alluvial soils, 0.4 t/ha on gley and pseudogley, and 0.2 t/ha on brown earth. Brown earth, chernozems and pseudogley cover about 40%, 25% and 9% of total cropland in Austria, respectively.

The comparison of simulated average crop yields between elevation classes shows that average crop yield declines are most significant at low elevations. The dry matter crop yield reductions are 0.8 t/ha in S1 and 1.6 t/ha in S2 for elevations below 300 m. For elevations above 2100 m, there is a crop yield increase of 0.9 t/ha in S1 and of 1.6 t/ha in S2. Mainly forage crops are grown at this elevation.

Similarly, average dry matter crop yield decreases are higher on flatter slopes (respective decreases of 0.3 t/ha and 0.8 t/ha in S1 and S2 for slopes <5%), and turning into increases at steeper slopes *i.e.* 0.4 t/ha in S1 and S2 for slopes >100% due to less erosion.

The average decreases in evapotranspiration rates of S1 (S2) with respect to S0 are about 70 mm (130 mm) on chernozems, followed by 30 mm (70 mm) on alluvial soils and 20 mm (60 mm) on pseudogley. For different elevation classes, the respective decreases in evapotranspiration rates of S1 and S2 are 60 mm and 120 mm at elevations <300 m, and turning into increases of 25 mm and 40 mm at elevations >2100 m, respectively. For different slope classes, decreases in evapotranspiration rates of S1 (S2) are 30 mm (70 mm) at slopes <5% and again turning into increases of ~10 mm for both S1 and S2 at slopes >100%.

3.3. Inter-Annual Variability

The inter-annual variability is described by the spread consisting of 2790 simulated crop yield values for each 1 km grid point, which result from 31 years of simulations (2010-2040), three drought scenarios (S0, S1 and S2), and 30 bootstrapped re-allocations per drought scenario. As standard measure, we have investigated the standard deviations. The development of simulated crop yields under drier conditions is more "certain" in the eastern and north-eastern parts of the country. Standard deviations range between ±1 t/ha and ±3 t/ha in the east and north-east of Austria and between +5 t/ha and +7 t/ha in other parts of Austria.

The smaller inter-annual variability of the simulated crop yields in the eastern and north-eastern parts of Austria can be explained by the rather persistent small precipitation amounts observed in the past period 1975-2007, which become even smaller in our drought scenarios. In the other parts of Austria, our drought scenarios may lead to very dry conditions in particular years, however, years with "usual" precipitation sums (similar to the past observations) may also occur and therefore lead to minor or no crop yield impacts (predominantly in the southern and western parts of Austria as well as for some parts in the foothills of the Alps).

3.4. Elasticities

Figure 6 shows the percent changes in mean annual dry matter crop yields and mean annual precipitation sums of S1 on the top (S2 on the bottom) compared to S0 of one bootstrapped re-allocation out of the 30 re-allocations. The correlation plots for other bootstrapped re-allocations look similar and are therefore not presented. The comparison with S1 shows that even small increases in dryness will result in significantly lower dry matter crop yields, particularly in areas with annual precipitation sums below 850 mm in S0: Simulated dry matter crop yields decrease by 0.6% per 1.0% decrease in annual precipitation sums.

In areas with annual precipitation sums above 850 mm in S0, simulated dry matter crop yields decrease by 0.4% per 1.0% decrease in annual precipitation sums. The comparison with S2 shows an even more pronounced sensi-

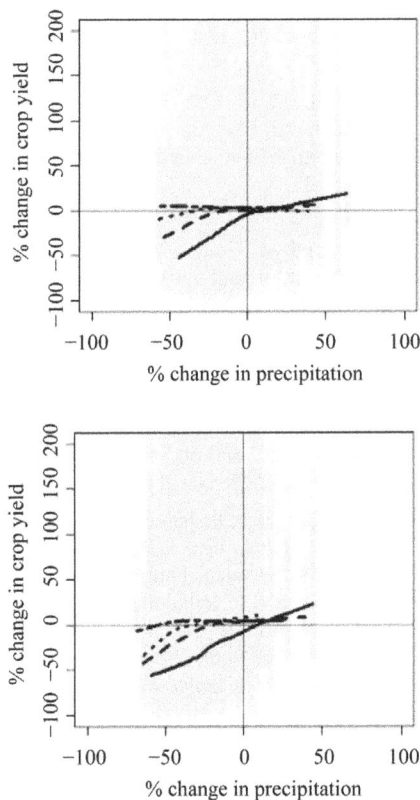

Figure 6. Elasticities between percent changes in simulated dry matter crop yields and annual precipitation sums (P): 400 mm < P < 670 mm (continuous line); 671 mm < P < 850 mm (dashed); 851 mm < P < 1050 mm (dotted); P > 1051 mm (dashed-dotted).

tivity: With up to 50% less precipitation, dry matter crop yields still only slightly change in regions with mean annual precipitation sums above 1050 mm in S0, but the crop yields become more sensitive to decreases in annual precipitation sums smaller than 1050 mm. Dry matter crop yields may decrease by about 20% with up to 50% less precipitation in regions with mean annual precipitation sums between 851 mm and 1050 mm in S0 (elasticity of +0.6), followed by decreases of about 35% in regions with mean annual precipitation sums between 671 mm and 850 mm (elasticity of +0.8) and decreases of about 55% in regions with mean annual precipitation sums between 400 mm and 670 mm (elasticity of nearly +1.0).

4. Discussion and Conclusions

The block bootstrap, which has been applied in the present analysis for generating drought scenarios, is a straightforward method of re-sampling time series while relating to the inherent autocorrelation. It does not require explicit modeling of the autocorrelation structure which in our case would be further compounded by the spatial autocorrelation. The choice of block length naturally plays a role. If the blocks are too short, the autocorrelation is weak; if they are too long, they become unmanageable and permit very little variation. We have chosen the block of eight days because it is convenient for building up the length of a month, and because preliminary simulations showed that the autocorrelation structure for the weather parameters was retained to the sufficient lag.

The analysis shows the usefulness of our developed dry day index to analyze in a spatial consistent way the long-term impacts of increased drought occurrences on crop production. Thus, we have identified the regions in Austria that would be more at risk when the meteorological drought occurrences increase. These regions are predominantly located in the eastern and north-eastern parts of the country, as expected, due to their already lower annual precipitation sums compared to other regions in Austria. Therefore, future research should assess the agricultural adaptation potential—especially in regions with lower annual precipitation sums—e.g. switching to more drought tolerant cultivars with slow leaf dehydration rates. The development of such cultivars (e.g. plant breeding of genotypes which are less sensitive to drought stress by a better developed root system) may come to the fore to avoid overexploitation of ground water resources. In addition, investments in irrigation systems, accompanied by well-defined diversification in crops and crop management practices will gain in importance to mitigate drought risk [31,44].

Austrian regions with historically abundant precipitation amounts (e.g. in the southern and western parts of the country) may also suffer from significant crop yield losses in particular years when confronted with extreme drought events. However, years with precipitation sums similar to the past observations may also occur and therefore may lead to minor crop yield impacts, although temperature is increasing and leading to higher evapotranspiration rates. Moreover, the simulated crop yields and evapotranspiration rates tend to increase along the elevation slope gradients due to higher precipitation amounts in mountainous terrains which, thus, tend to be less vulnerable to droughts. However, only a small fraction of cropland is located in such terrain.

Finally, it will be also interesting to investigate the impacts of significantly higher precipitation amounts on crop production and environment (e.g. flooding, soil water erosion). In any case, our approach can also be used to generate increased occurrences of heavy rainfall and flood events—which may also occur in a warmer climate—by mirroring the dry day index distribution.

5. Acknowledgements

This study has been supported by the ACRP projects CAFEE (grant number B068725)—Climate change in agriculture and forestry: an integrated assessment of mitigation and adaptation measures in Austria, and CC2BBE (grant number B286285)—Vulnerability of a bio-based economy to global climate change impacts as well as by the project MACSUR (grant number 100875)—Modeling European Agriculture with Climate Change for Food Security, a FACCE JPI knowledge hub supported by the Federal Ministry of Agriculture, Forestry, Environment and Water Management of Austria.

REFERENCES

[1] Commission of the European Communities, "Addressing the Challenge of Water Scarcity and Droughts in the European Union," Communication from the Commission to the European Parliament and the Council, 414 Final, Brussels, 2007.

[2] R. Brázdil, M. Trnka, P. Dobrovolný, K. Chromá, P. Hlavinka and Z. Žalud, "Variability of Droughts in the Czech Republic, 1881-2006," *Theoretical and Applied Climatology*, Vol. 97, No. 3-4, 2009, pp. 297-315.

[3] K. Yurekli and A. Kurunc, "Simulating Agricultural Drought Periods Based on Daily Rainfall and Crop Water Consumption," *Journal of Arid Environments*, Vol. 67, No. 4, 2006, pp. 629-640.

[4] J. Q. Zhang, "Risk Assessment of Drought Disaster in the Maize-Growing Region of Songliao Plain, China," *Agriculture, Ecosystems and Environment*, Vol. 102, No. 2, 2004, pp. 133-153.

[5] G. M. Richter and M. A. Semenov, "Modelling Impacts of Climate Change on Wheat Yields in England and Wales: Assessing Drought Risks," *Agricultural Systems*, Vol. 84, No. 1, 2005, pp. 77-97.

[6] S. J. Eisenreich, "Climate Change and European Droughts," In: *Climate Change and the European Water Dimension*, European Commission, Luxemburg, 2005, pp. 121-135.

[7] A. K. Mishra and V. P. Singh, "A Review of Drought Concepts," *Journal of Hydrology*, Vol. 391, No. 1-2, 2010, pp. 202-216.

[8] Intergovernmental Panel on Climate Change IPCC, Synthesis Report. Contribution of Working Groups I, II and III to the 4th Assessment Report of the Intergovernmental Panel on Climate Change, 2007.

[9] E. J. Burke, S. J. Brown and N. Christidis, "Modeling the Recent Evolution of Global Drought and Projections for the Twenty-First Century with the Hadley Centre Climate Model," *Journal of Hydrometeorology*, Vol. 7, No. 5, 2006, pp. 1113-1125.

[10] J. Q. Zhang, "A Study on Damage Degree and Risk Assessment and Regional Classification of Meteorological Disasters: Case Studies of Yamaguchi Prefecture in Japan and Songliao Plain in China," Ph.D. Thesis, Tottori University, Tottori, 2000.

[11] M. H. Yang and J. Q. Zhang, "Assessment and Regionalization of Meteorological Disaster to Maize in Jilin Province, China," In: E. P. Wan and X. R. Xu, Eds., *Prediction and Monitoring of Maize Yield by Remote Sensing in China*, Chinese Science and Technology Press, Beijing, 1996, pp. 196-218.

[12] V. Kumar and U. Panu, "Predictive Assessment of Severity of Agricultural Droughts Based on Agro-Climatic Factors," *Journal of the American Water Resources Association*, Vol. 33, No. 6, 1997, pp. 1255-1264.

[13] J. E. Olesen and M. Bindi, "Consequences of Climate Change for European Agricultural Productivity, Land Use and Policy," *European Journal of Agronomy*, Vol. 16, No. 4, 2002, pp. 239-262.

[14] J. Eitzinger, C. Kersebaum and H. Formayer, "Landwirtschaft im Klimawandel—Auswirkungen und Anpassungsstrategien für die Land-und Forstwirtschaft in Mitteleuropa," AgriMedia, Clenze, 2009.

[15] F. Ewert, D. Rodriguez, P. Jamieson, M. A. Semenov, R. A. C Mitchell, J. Goudriaan, J. R. Porter, B. A. Kimball, P. J. Pinter Jr., R. Manderscheid, H. J. Weigel, A. Fangmeier, E. Fereres and F. Villalobos, "Effects of Elevated CO_2 and Drought on Wheat: Testing Crop Simulation Models for Different Experimental and Climatic Conditions," *Agriculture, Ecosystems and Environment*, Vol. 93, No. 1-3, 2002, pp. 249-266.

[16] Q. Zhang, P. Sun, V. P. Singh and X. Chen, "Spatial-Temporal Precipitation Changes (1956-2000) and Their Implications for Agriculture in China," *Global and Planetary Change*, Vol. 82-83, 2012, pp. 86-95.

[17] A. K. Mishra and V. P. Singh, "Drought Modeling—A

Review," *Journal of Hydrology*, Vol. 403, No. 1-2, 2011, pp. 157-175.

[18] K. R. Gabriel and J. Neumann, "A Markov Chain Model for Daily Rainfall Occurrences at Tel Aviv," *Quarterly Journal of the Royal Meteorological Society*, Vol. 88, No. 375, 1962, pp. 90-95.

[19] S. Blenkinsop and H. J. Fowler, "Changes in Drought Characteristics for Europe Projected by the PRUDENCE Regional Climate Models," *International Journal of Climatology*, Vol. 27, No. 12, 2007, pp. 1595-1610.

[20] F. Strauss, H. Formayer and E. Schmid, "High Resolution Climate Data for Austria in the Period from 2008 to 2040 from a Statistical Climate Change Model," *International Journal of Climatology*, Vol. 33, No. 2, 2013, pp. 430-443.

[21] M. Mudelsee, "Climate Time Series Analysis," Springer, Dordrecht, 2010.

[22] J. R. Williams, "The EPIC Model," In: V. P. Singh, Ed., *Computational Models of Watershed Hydrology*, Water Resources Publications, Highlands Ranch, Colorado, 1995, pp. 909-1000.

[23] R. C. Izaurralde, J. R. Williams, W. B. McGill, N. J. Rosenberg and M. C. Quiroga, "Simulating Soil C Dynamics with EPIC: Model Description and Testing against Long-Term Data," *Ecological Modelling*, Vol. 192, No. 3-4, 2006, pp. 362-384.

[24] I. Auer, R. Böhm, H. Mohnl, R. Potzmann and W. Schöner, "ÖKLIM—A Digital Climatology of Austria 1961-1990," *Proceedings of the 3rd European Conference on Applied Climatology*, Pisa, 16-20 October 2000, CD Rom.

[25] W. Schöner, I. Auer, R. Böhm and S. Thaler, "Qualitätskontrolle und Statistische Eigenschaften Ausgewählter Klimaparameter auf Tageswertbasis im Hinblick auf Extremwertanalysen," Subproject of StartClim: Erste Analysen extremer Wetterereignisse und ihrer Auswirkungen in Österreich, Vienna, 2003. http://www.austroclim.at/index.php?id=startclim2003

[26] European Environment Agency EEA, "Regional Climate Change and Adaptation—The Alps Facing the Challenge of Changing Water Resources," EEA Report No. 8/2009, ISSN 1725-9177, Copenhagen, 2009.

[27] P. Racsko, L. Szeidl and M. Semenov, "A Serial Approach to Local Stochastic Weather Models," *Ecological Modelling*, Vol. 57, No. 1-2, 1991, pp. 27-41.

[28] M. A. Semenov and E. M. Barrow, "Use of a Stochastic Weather Generator in the Development of Climate Change Scenarios," *Climatic Change*, Vol. 35, No. 4, 1997, pp. 397-414.

[29] M. A. Semenov and P. Stratonovitch, "The Use of Multi-Model Ensembles from Global Climate Models for Impact Assessments of Climate Change," *Climate Research*, Vol. 41, No. 1, 2010, pp. 1-14.

[30] E. Schmid, F. Sinabell and P. Liebhard, "Effects of Reduced Tillage Systems and Cover Crops on Sugar Beet Yield and Quality, Ground Water Recharge and Nitrogen Leaching in the Pannonic Region Marchfeld, Austria," *Pflanzenbauwissenschaften*, Vol. 8, No. 1, 2004, pp. 1-9.

[31] C. Heumesser, S. Fuss, J. Szolgayová, F. Strauss and E. Schmid, "Investment in Irrigation Systems under Precipitation Uncertainty," *Water Resources Management*, Vol. 26, No. 11, 2012, pp. 3113-3137.

[32] F. Strauss, E. Schmid, E. Moltchanova, H. Formayer and X. Wang, "Modeling Climate Change and Biophysical Impacts of Crop Production in the Austrian Marchfeld Region," *Climatic Change*, Vol. 111, No. 3, 2012, pp. 641-664.

[33] B. Stürmer, J. Schmidt, E. Schmid and F. Sinabell, "Implications of Agricultural Bioenergy Crop Production in a Land Constrained Economy—The Example of Austria," *Land Use Policy*, Vol. 30, No. 1, 2013, pp. 570-581.

[34] J. L. Monteith, "Evaporation and Environment," *Symposia of the Society for Experimental Biology*, Vol. 19, 1965, pp. 205-234.

[35] H. L. Penman, "Natural Evaporation from Open, Bare Soil and Grass," *Proceedings of the Royal Society of London*, Series A, Vol. 193, No. 1032, 1948, pp. 120-145.

[36] C. H. B. Priestly and R. J. Taylor, "On the Assessment of Surface Heat Flux and Evaporation Using Large Scale Parameters," *Monthly Weather Review*, Vol. 100, No. 2, 1972, pp. 81-92.

[37] G. H. Hargreaves and Z. A. Samani, "Reference Crop Evapotranspiration from Temperature," *Applied Engineering in Agriculture*, Vol. 1, No. 2, 1985, pp. 96-99.

[38] W. Bair and G. W. Robertson, "Estimation of Latent Evaporation from Simple Weather Observations," *Canadian Journal of Plant Science*, Vol. 45, No. 3, 1965, pp. 276-284.

[39] C. O. Stockle, J. R. Williams, N. J. Rosenberg and C. A. Jones, "A Method for Estimating the Direct and Climatic Effects of Rising Atmospheric Carbon Dioxide on Growth and Yield of Crops: Part I-Modification of the EPIC Model for Climate Change Analysis," *Agricultural Systems*, Vol. 38, No. 3, 1992, pp. 225-238.

[40] C. A. Jones, P. T. Dyke, J. R. Williams, J. R. Kiniry, V. W. Benson and R. H. Griggs, "EPIC: An Operational Model for Evaluation of Agricultural Sustainability," *Agricultural Systems*, Vol. 37, No. 4, 1991, pp. 341-350.

[41] Bundesforschungszentrum für Wald BFW, "Digital Soil Map for Austria," 2009. http://gis.lebensministerium.at/eBOD

[42] M. Schönhart, E. Schmid and U. Schneider, "Crop Rota—A Crop Rotation Model to Support Integrated Land Use Assessments," *European Journal of Agronomy*, Vol. 34, No. 4, 2011, pp. 263-277.

[43] United Nations Environment Programme UNEP, "World

Atlas of Desertification," Second Edition, 1997.

[44] Organization for Economic Co-Operation and Development OECD, "Scoping Paper on Crop Insurance and Farmer Incentives to Adapt to Climate Change," Joint Working Party on Agriculture and the Environment, 2010.

4

Coral Reef Populations in the Caribbean: Is There a Case for Better Protection against Climate Change?

Michael James C. Crabbe

Institute for Biomedical and Environmental Science and Technology, Faculty of Creative Arts, Technologies and Science, University of Bedfordshire, Luton, UK

ABSTRACT

Knowledge of factors that are important in coral reef growth help us to understand how reef ecosystems react following major environmental disturbances due to climate change and other anthropogenic effects. This study shows that despite a range of anthropogenic stressors, corals on the fringing reefs south of Kingston harbour, as well as corals on fringing reefs on the north coast of Jamaica near Discovery Bay can survive and grow. Skewness values for *Sidastrea siderea* and *Porites astreoides* were positive (0.85 - 1.64) for all sites, implying more small colonies than large colonies. Coral growth rates are part of a demographic approach to monitoring coral reef health in times of climate change, and linear extension rates (mm·yr^{-1}) of *Acropora palmata* branching corals at Dairy Bull, Rio Bueno, and Pear Tree Bottom on the north coast of Jamaica were c. 50 - 90 mm·year^{-1} from 2005-2012. The range of small-scale rugosities at the Port Royal cay sites studied was lower than that at the Discovery Bay sites; for example Rio Bueno was 1.05 ± 0.15 and Dairy Bull the most rugose at 2.3 ± 0.16. Diary Bull reef has for several years been the fringing reef with the most coral cover, with a benthic community similar to that of the 1970s. We discuss whether Jamaica can learn from methods used in other Caribbean countries to better protect its coral reefs against climate change. Establishing and maintaining fully-protected marine parks in Jamaica and elsewhere in the Caribbean is one tool to help the future of the fishing industry in developing countries. Developing MPAs as part of an overall climate change policy for a country may be the best way of integrating climate change into MPA planning, management, and evaluation.

Keywords: Demographics; Belize; Jamaica; MPAs; GDP; Hurricanes; Fishing; Bleaching; Climate Change; Global Warming

1. Introduction

Coral reefs throughout the world are under severe challenges from environmental factors including overfishing, destructive fishing practices, coral bleaching, ocean acidification, sea-level rise, algal blooms, agricultural run-off, coastal and resort development, marine pollution, increasing coral diseases, invasive species, ocean acidification, rising sea level, changing circulation patterns, increasing severity of storms, changing freshwater influxes and hurricane/cyclone damage (e.g. [1-4]). The fringing reefs around Jamaica constitute one of the best documented areas of reef decline in the Caribbean, where significant loss of corals and macroalgal domination has been due to hurricanes [5-7], overfishing [8,9], die-off of the long-spined sea urchin *Diadema antillarum* in 1983-1984 [10], and coral disease [11]. Nutrient enrichment does not appear to have been a causal factor in the development of these reef macroalgal communities [12].

Warming ocean (sea surface) temperatures due to climate change are considered to be an important cause of the degradation of the world's coral reefs. Marine protected areas (MPAs) have been proposed as one tool to increase coral reef ecosystem resistance and resilience (*i.e.* recovery) to the negative effects of climate change. However, few studies have evaluated their efficacy in achieving these goals.

While healthy reefs usually have high numbers of coral recruits and juvenile corals, degraded systems typically have limited numbers of young colonies [13,14]. To manage coral reefs it is important to have an understanding of coral population demography-structure and dynamics. Ideally, this involves the quantification of numbers of individual colonies of different size classes-the population structure-through time, in addition to quantifying coral growth rates, recruitment and survival.

Knowledge of coral population structure helps in understanding how reefs react following disturbance, and provides us with an early warning system for predicting future reef health.

Here we studied recent non-branching coral population structure at sites near Discovery Bay on the north coast of Jamaica, and at sites near Kingston Harbour, on the south coast of Jamaica.

Kingston harbour is a major trans-shipment post for the Caribbean, and in 2002 a major ship channel (East Channel) was constructed by Rackham's Cay, near Port Royal to accommodate vessels with draft of up to 14.5 metres and beams of 42 metres [15]. In addition, the harbour is highly polluted, mostly due to an excessive input of sewage [16-19]. This study shows that despite these and other environmental stressors, corals can survive on the fringing reefs south of the harbour.

MPAs provide *in-situ*-based management of marine ecosystems through various degrees and methods of protective actions. As impacts of climate change strengthen they may exacerbate effects of existing stressors and require new or modified management approaches; for example MPA networks may be an improvement over individual MPAs in addressing multiple threats to the marine environment. We discuss whether Jamaica can learn from methods used in other Caribbean countries to better protect its coral reefs in times of climate change.

2. Materials and Methods

2.1. Reef Sites

GPS coordinates were determined using a hand-held receiver (Garmin Ltd.).

In Discovery Bay, Jamaica, surface areas of non-branching corals between 5 - 9 m depth, were measured using SCUBA at four sites [Rio Bueno (18°28.805'N; 77°21.625'W), M1 (18°28.337'N; 77°24.525'W), Dancing Ladies (18°28.369'N; 77°24.802'W), Dairy Bull (18°28.083'N; 77°23.302'W) and Pear Tree Bottom (18°27.829'N; 77°21.403'W)] along the fringing reefs surrounding Discovery Bay, Jamaica. Overall, surface areas of 209 non-branching corals were measured in 2012 (**Table 1(a)**).

This work was conducted at Discovery Bay during August 8-August 10 in 2012.

For sites near Port Royal, Jamaica, surface areas of non-branching corals were measured using SCUBA at seven sites south of Port Royal. Six of these were fringing reefs around the cays South of Port Royal: SE Barrier Cay (17°53.714'N, 76°48.226'W); Lime Cay (17°54.948'N, 76°49.134'W); Gun Cay (17°55.901'N, 76°50.141'W); Drunkenman's Cay (17°54.128'N, 76°50.736'W), the face of the ship channel made in 2002 along the reef of Rackham's Cay (17°55.571'N, 76°50.307'W) and Maiden Cay (17°54.506'N, 76°48.728'W); a seventh site was the

wreck of the ship *Edina*, south of the barrier reef (17°49.525'N, 76°50.723'W). Depth of samples at six of the sites was between 5 - 12 m, at the *Edina* wreck it was between 22 - 28 m.

Overall, surface areas of 347 non-branching corals were measured in 2010 (**Table 1(b)**) and of 451 non-branching corals in 2012 (**Table 1(c)**).

This work was conducted in April 14-16 in 2010 and July 30-August 2 in 2012. SE Barrier Cay was only studied in 2010, owing to time limitations caused by tropical storm Ernesto in 2012.

2.2. Sampling

Details of data sampling have been described for North Jamaica and South Jamaica [6,20]. In summary, corals 2m either side of transect lines were photographed for archive information, and surface areas measured with flexible tape as described previously using SCUBA. For non-branching corals, this was done by measuring the widest diameter of the coral and the diameter at 90° to that.

In all cases except that of the Edina wreck near Port Royal, which was 22 - 28 m, depth of samples was between 5 - 12 m, to minimise variation in growth rates due to depth [21]. To increase accuracy, surface areas rather than diameters of live non-branching corals were measured [6]. Sampling was over as wide a range of sizes as possible. Colonies that were close together (<50 mm) or touching were avoided to minimise age discontinuities through fission and altered growth rates [22,23]. In this study we ignored *Montastrea annularis* colonies, because their surface area does not reflect their age [22], and because hurricanes can increase their asexual reproduction through physical damage [23].

Computer digital image analysis for coral linear extension rates was undertaken using the UTHSCSA (University of Texas Health Science Center, San Antonio, Texas, USA) Image Tool software (see [6]). One-factor ANOVA was used; ±error values represent standard errors of the data.

Skewness [24] (sk) was used to estimate the distribution of small and large colonies in the coral populations. Negative skewness implies more large colonies than small colonies, while conversely positive skewness implies more small colonies than large colonies.

2.3. Rugosity

Rugosity (R) was determined according to the formula:

$$R = Sr/Sg$$

where Sr = real surface distance between two points, and Sg = straight line geometric distance between two points. This was calculated over a 20 m distance, performed in

Table 1. Coral species and numbers studied at sites around Port Royal in 2010 and 2012, and around Discovery Bay in 2012. (a) Corals at sites around Discovery Bay in 2012; (b) Corals at sites around Port Royal in 2010; (c) Corals at sites around Port Royal in 2012.

(a)

Species	2012				
	Dairy Bull	M1	Pear Tree	Rio Bueno	Total
ss	37	12	2	7	58
pa	19	10	12	9	50
mean mean	6	25	4	1	36
dip strig	2			8	10
ag ag	5		2	3	10
mont cav	6			2	8
dip laby	4	4	3		11
eusmilia	2		1		3
manicina					0
muss ang					0
mycetophyllia	10				10
col natans	4	5	1	3	13
Total	95	56	25	33	209

(b)

Species	2010							
	Rackhams	Edina wreck	Drunkenmans	SE barrier	Maiden	Lime	Gun	Total
ss	78	20	12	21	5	23	10	169
pa	10	6	28	15		2	7	68
mean mean	9	3	6	4	3	4	1	30
dip streg	1	16	8	12		4		41
ag ag		1						1
mont cav		8	4	2		2	10	26
dip laby			2	2	2	3	1	10
eusmilia						1		1
col natans							1	1
Total	98	54	60	56	10	39	30	347

(c)

Species	2012						
	Rackhams	Edina wreck	Drunkenmans	Maiden	Lime	Gun	Total
ss	35	17	22	42	43	33	192
pa	6	9	10	27	20	9	81
mean mean		1	8	1	7	5	22
dip strig		5	8	1	4	4	22
ag ag		8	1	8	7	1	25
mont cav	1	8	11	10	7	8	45
dip laby	1	2	1	3			7
eusmilia				4	1		5
manicina						3	3
muss ang			1				1
mycetophyllia	1	2	7	14	12		36
col natans		1		1	6	4	12
Total	44	53	69	111	107	67	451

Legend. Coral species: ss, *Sidastrea siderea* (Ellis, 1786); pa, *Porites astreoides* (Lamarck, 1816); meanmean, *Meandrina meandrites* (Linnaeus, 1758); Dip strig, *Diploria strigosa* (Dana, 1848); agag, *Agaricia agaricites* (Linnaeus, 1758); mont cav, *Montastrea cavernosa* (Linnaeus, 1758); dip laby, *Diploria labyrinthiformis* (Linnaeus, 1758); eusmilia, *Eusmilia fastigiata* (Pallas, 1766); col natans, *Colpophyllia natans* (Houttuyn, 1772); manicina, *Manicina areolata* (Linnaeus, 1758); muss ang, *Mussa angulosa* (Pallas, 1766); mycetophyllia, *Mycetophyllia lamarckiana* (Milne Edwards 1848). GPS coordinates of all sites are given in the text. Gaps in the tables indicate 0.

triplicate, at each site, using photographic image analysis verified by the chain method, as described previously for Discovery Bay sites [25].

3. Results

The reefs fringing the cays around Port Royal in Jamaica did not exhibit extensive three-dimensional complexity; this is exemplified by their rugosities: Rackhams cay, 1.42 ± 0.15; Edina wreck, 1.37 ± 0.17; Drunkenman's cay 1.41 ± 0.16; SE Barrier cay, 1.39 ± 0.15; Maiden cay, 1.1 ± 0.1; Lime cay, 1.16 ± 0.12; and Gun cay, 1.17 ± 0.2.

Despite their low three-dimensional complexity, there was a high proportion of small size classes of non-branching corals that included new recruits and juveniles on these reefs. This is illustrated for both 2010 and 2012 in **Figures 1(a)** and **(b)** for *Sidastrea siderea* and in **Figures 1(c)** and **(d)** for *Porites astreoides*. These patterns are typical of other species of corals, where the numbers of corals are smaller. Skewness values for both species were positive (for example 1.36 for *Sidastrea siderea* at the Edina site in 2010, and 0.85 for *Porites astreoides* at Drunkenman's Cay in 2012), implying more small colonies than large colonies.

There was also a high proportion of small size classes of non-branching corals on the fringing reefs around Discovery Bay on the Jamaican north coast in 2012. This is illustrated for *Sidastrea siderea* and *Porites astreoides* in **Figures 2(a)** and **(b)**. Once again, this pattern was typical of other coral species. Skewness values for both species were positive, with Dairy Bull having the highest values (for example 1.64 for *Sidastrea siderea* and 1.12 for *Porites astreoides* in 2012 at Dairy Bull), implying more small colonies than large colonies.

Coral growth rates are part of a demographic approach to monitoring coral reef health in times of climate change, and **Figure 3** presents linear extension rates ($mm \cdot yr^{-1}$) of *Acropora palmata* branching corals (n = 4) at Dairy Bull,

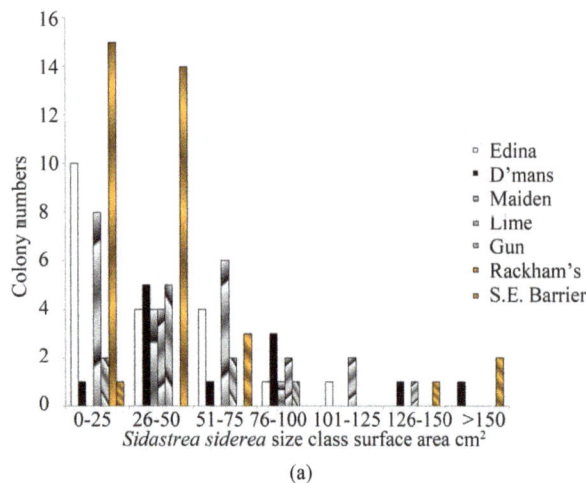

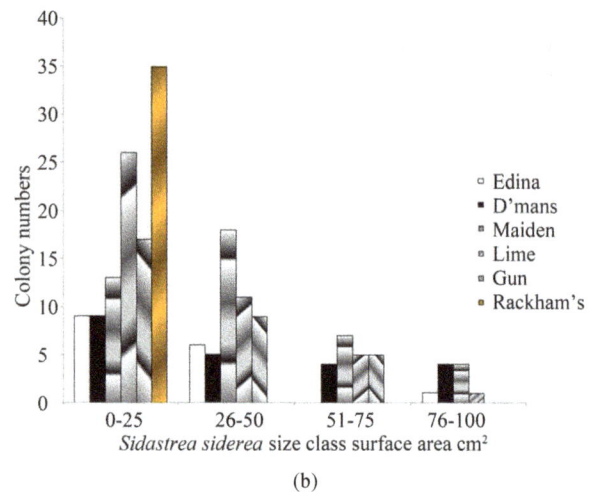

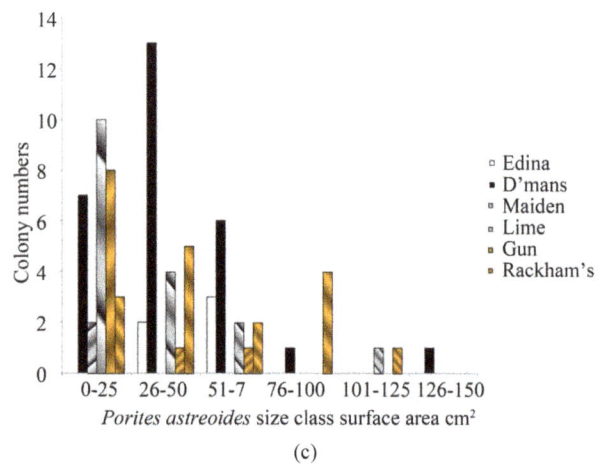

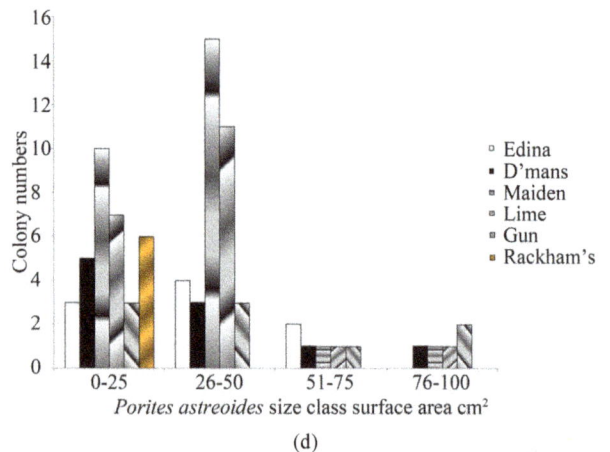

(b)

(c)

(d)

Figure 1. Size classes of non-branching corals on the fringing reefs on the Port Royal Cays. (a) *Sidastrea siderea* **in 2010; (b)** *Sidastrea siderea* **in 2012; (c)** *Porites astreoides* **in 2010; (d)** *Porites astreoides* **in 2012.**

Rio Bueno, and Pear Tree Bottom on the north coast of Jamaica from 2005-2008, and from 2009-2012. Growth rates are similar to those reported previously [21]. There were no significant differences between the sites, or across the time periods of the study. Where growth rates

(a)

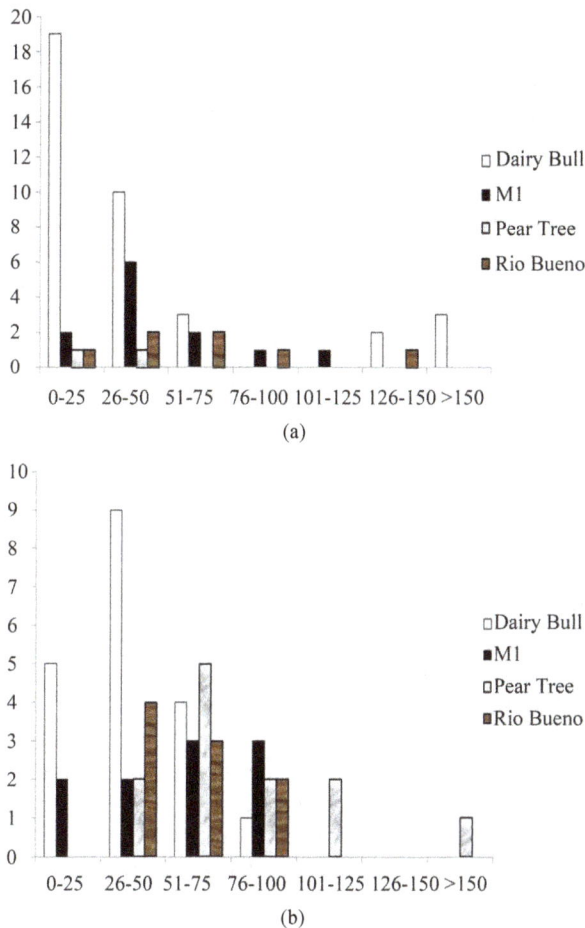

Figure 2. Size classes of non-branching corals on the fringing reefs near Discovery Bay, Jamaica. (a) *Sidastrea siderea* in 2012; (b) *Porites astreoides* in 2012.

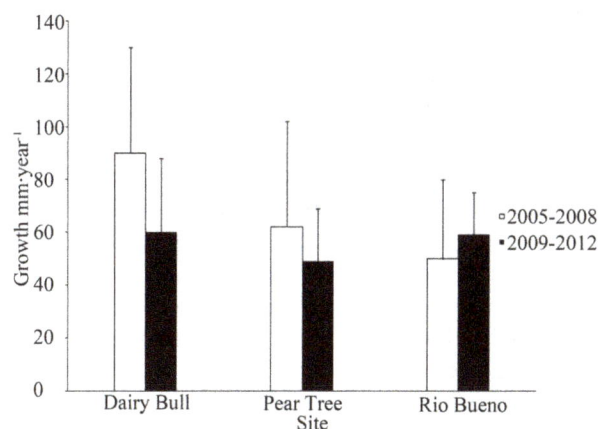

Figure 3. Linear extension rates of *Acropora palmata* at sites near Discovery Bay, Jamaica, from 2005-2008 and 2009-2012. Error bars represent standard errors of the data (n = 4).

were higher, they tended to be higher at Dairy Bull reef. With the increase of *Diadema antillarum* at Rio Bueno in recent years, clearing the macroalgae, many healthy *A.*

palmata colonies have appeared at the Rio Bueno site from 2006.

4. Discussion

Mesoscale rugosity (larger scale) has been found important for predicting intra-habitat variation in coral reef fish assemblages, and explicitly in predicting the impacts of coral mortality on ecosystem process and services. This particularly relates to large, tall (>50 cm) corals [26]. Loss of architectural complexity in Caribbean reefs due to climate change appears to be linked to physical impacts, such as hurricanes and bio-erosion [3].

The range of small-scale rugosities at the Port Royal cay sites studied was lower than that at the Discovery Bay sites [25]; for example Rio Bueno was 1.05 ± 0.15 and Dairy Bull the most rugose at 2.3 ± 0.16. Diary Bull reef has for several years been the fringing reef with the most coral cover, with a benthic community similar to that of the 1970s [21], and it was the subject of the study which suggested a rapid phase-shift reversal [27].

In addition to climate change, Jamaican reefs are subject to a number of acute and chronic stressors, the last including overfishing and continuing coastal development, including the much-publicised development on land adjacent to Pear Tree Bottom reef and the resurfacing of the North Jamaican coastal highway [28]. Faster rates of macroalgal growth, higher rates of algal recruitment iron enrichment from Aeolian dust, lack of acroporid corals, lower herbivore biomass and missing groups of herbivores all predispose the Caribbean to low resilience, relative to the Indo-Pacific region [29].

The Kingston harbour area has been impacted for many decades [16,19] and there have been efforts at mitigation, particularly in association with the ship channel [30]. In addition, some coral communities may adapt to chronic stressors, for example sedimentation, over long periods of time [31].

After the 2005 bleaching event there was a major loss of live coral cover [32,33], and it is encouraging that the population size studies show that there are numbers of both small and large colonies at both Discovery Bay and Port Royal sites. Also, the linear extension rates of *A. palmata* branching corals at Dairy Bull, Rio Bueno, and Pear Tree Bottom on the north coast of Jamaica were maintained from 2005-2012.

The influence of *M. annularis* colonies on the reef, acting as structural refugia [27], with maintenance of the biological legacies, may have facilitated this recovery. In addition, there have been no major hurricanes or bleaching events due to climate change since 2005 until this study which have impacted Jamaican fringing reefs.

Unlike other areas in the Caribbean, Jamaica has few Marine Protected Areas (MPAs) or Marine Reserves.

MPAs have been suggested as means for enhancing local resilience and population growth of marine species [e.g. 34,35]. However, some studies have highlighted continued climatic and other impacts in regions of MPAs [36-38].

Belize has the highest annual capture production-the annual volume of aquatic species caught by country for all commercial, industrial, recreational and subsistence purposes, in 2010, for countries in the Caribbean for which data is available [39]. The Belize continental shelf also includes the MesoAmerican Barrier reef; this World Heritage Site has levered the adoption of Marine Protected Areas (MPAs) for Belize. In 2010, Belize had 11.86% of its territorial waters as MPAs. This compares with Jamaica, 4.2%; Trinidad and Tobago, 2.8%; Barbados, 0.1%; St. Kitts and Nevis, 0.5%; St. Lucia, 0.1%; St. Vincent and the Grenadines, 0.6%. **Figure 4** shows the relationship between percentage of a country's territorial waters as MPAs with volume of catch ($r^2 = 0.88$); there is a similar correlation ($r^2 = 0.66$) with fisheries as a percentage of GDP [40] for the country concerned. Catch is a function of effort, and capture production may reflect this in addition to the influence of MPAs.

Recognition of the importance of fisheries in a country's GDP may be a factor in empowering conservation policy as well as local stakeholder action to conserve coral reefs. Having a large percentage of its territorial waters as MPAs, and in a coordinated network of MPAs, may reflect in the value of a country's fishing industry.

The 2012 IUCN report [41] on Caribbean coral reefs points to the decline of Caribbean reefs from c. 50% cover in the 1970 to just 8% today. They call for strictly enforced local action to improve the health of corals, including limiting fishing through catch quotas, an extension of MPAs, a halt to nutrient run-off from the land, and a reduction on the global resilience on fossil fuels.

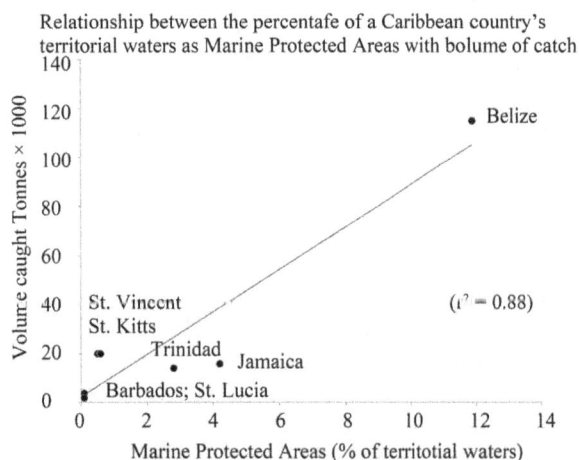

Relationship between the percentafe of a Caribbean country's territorial waters as Marine Protected Areas with bolume of catch

Figure 4. Relationship between the percentage of a country's territorial waters as Marine Protected Areas (MPAs) with volume of catch ($r^2 = 0.88$).

Through the IUCN-coordinated Global Coral Reef Monitoring Network (e.g. [42]) there are moves to strengthen the data available at a worldwide level.

Protection using *fully protected* MPAs is one of the few management tools that governments and local communities can use to combat large scale environmental impacts. In Bajja California, such an MPA provided protection to marine populations both within and outside the protected area [43]. The degree of protection is critical [44]. At Glover's Reef Marine Reserve, a no-take policy has not resulted in increases in grazing fish abundance, although commercially important fish, such as black grouper (*Mycteroperca bonaci*) and lane snapper (*Lutjanus synagris*) and invertebrates have increased [45]. This could be for many reasons, not least if they are poorly managed and maintained over long periods, particularly when there are stressful environmental events such as hurricanes of high sea surface temperatures (SSTs).

In an environment where carbon dioxide concentrations predicted to occur at the end of this century would significantly reduce coral settlement and crustose coralline algae cover, and would so reduce successful coral recruitment and larval settlement [46,47], establishing and maintaining fully-protected marine parks in Jamaica and elsewhere in the Caribbean is one tool to help the future of the fishing industry in developing countries.

Where protection in MPAs was not found to reduce the effect of warm temperature anomalies on coral cover declines [48], then shortcomings in MPA design, including size and placement, may have contributed to the lack of an MPA effect. It may be that the benefits from single MPAs may not be great enough to offset the magnitude of losses from acute thermal stress events. Although MPAs are important conservation tools, their limitations in mitigating coral loss from acute thermal stress events suggest that they need to be complemented with policies aimed at reducing the activities responsible for climate change. One way forward is to have networks of MPAs [49], and they could be more effective in conjunction with other management strategies, such as fisheries regulations and reductions of nutrients and other forms of land-based pollution. Developing MPAs as part of an overall climate change policy for a country [50] may be the best way of integrating climate change into MPA planning, management, and evaluation.

5. Acknowledgements

I thank Mr. Peter Gayle and the staff of the Discovery Bay Marine Laboratory Jamaica, Mr. Hugh Small and colleagues at Port Royal Marine Station, Jamaica for their invaluable help and assistance. I also thank Ms. Marcia Creary and staff of the University of the West

Indies, Mona campus, for their logistical help, and Margaret O'Rorke for invaluable conversations.

REFERENCES

[1] T. A. Gardner, I. M. Côté, J. A. Gill, A. Grant and A.R. Watkinson, "Long-Term Region-Wide Declines in Caribbean Corals," *Science*, Vol. 301, No. 5635, 2003, pp. 958-960.

[2] C. M. Eakin, J. A. Morgan, S. F. Heron, T. B. Smith, G. Liu, L. Alvarez-Filip, B. Baca, E. Bartels, C. Bastidas, C. Bouchon, M. Brandt, A. Bruckner, L. Bunkley-Williams, A. Cameron, B. D. Causey, M. Chiappone, T. R. L. Christensen, M. J. C. Crabbe, O. Day, E. de la Guardia, G. Díaz-Pulido, D. DiResta, D. L. Gil-Agudelo, D. Gilliam, R. Ginsburg, S. Gore, H. M. Guzman, J. C. Hendee, E. A. Hernández-Delgado, E. Husain, C. F. G. Jeffrey, R. J. Jones, E. Jordán-Dahlgren, L. Kaufman, D. I. Kline, P. Kramer, J. C. Lang, D. Lirman, J. Mallela, C. Manfrino, J.-P. Maréchal, K. Marks, J. Mihaly, W. J. Miller, E. M. Mueller, E. Muller, C. A. Orozco Toro, H. A. Oxenford, D. Ponce-Taylor, N. Quinn, K. B. Ritchie, S. Rodríguez, A. Rodríguez Ramírez, S. Romano, J. F. Samhouri, J. A. Sánchez, G. P. Schmahl, B. Shank, W. J. Skirving, S. C. C. Steiner, E. Villamizar, S. M. Walsh, C. Walter, E. Weil, E. H. Williams, Woody K. Roberson and Y. Yusuf, "Caribbean Corals in Crisis: Record Thermal Stress, Bleaching, and Mortality in 2005," *PLOS ONE*, Vol. 5, No. 11, 2010, Article ID: e13969.

[3] L. Alvarez-Filip, J. A. Gill, N. K. Dulvy, A. L. Perry, A. R. Watkinson and I. M. Coté, "Drivers of Region-Wide Declines in Architectural Complexity on Caribbean Reefs," *Coral Reefs*, Vol. 30, No. 4, 2011, pp. 1051-1060.

[4] R. Rodolfo-Metalpa, F. Houlbréque, E. Tambutté, F. Biosson, C. Baggini, F. P. Patti, R. Jefree, M. Fine, A. Foggo, J-P. Gattuso and J. M. Hall-Spencer, "Coral and Mollusc Resistance to Ocean Acidification Adversely Affected by Warming," *Nature Climate Change*, Vol. 1, 2011, pp. 308-312.

[5] J. D. Woodley, E. A. Chornesky, P. A. Clifford, J. B. C. Jackson, L. S. Kaufman, N. Knowlton, J. C. Lang, M. P. Pearson, J. W. Porter, M. C. Rooney, K. W. Rylaarsdam, V. J. Tunnicliffe, C. M. Wahle, J. L. Wulff, A. S. G. Curtis, M. D. Dallmeyer, B. P. Jupp, M. A. R. Koehl, J. Neigel and E. M. Sides, "Hurricane Allen's Impact on Jamaican Coral Reefs," *Science*, Vol. 214, No. 4522, 1981, pp. 749-755.

[6] M. J. C. Crabbe, J. M. Mendes and G. F. Warner, Lack of Recruitment of Non-Branching Corals in Discovery Bay Is Linked to Severe Storms," *Bulletin of Marine Science*, Vol. 70, 2002, pp. 939-945.

[7] J. Mallela and M. J. C. Crabbe, "Hurricanes and Coral Bleaching Linked to Changes in Coral Recruitment in Tobago," *Marine Environmental Research*, Vol. 68, No. 4, 2009, pp. 158-162.

[8] J. B. C. Jackson, "Reefs since Columbus," *Proceedings of the 8th International Coral Reef Symposium*, Vol. 1, 1997, pp. 97-106.

[9] J. P. Hawkins and C. M. Roberts, "Effects of Artisanal Fishing on Caribbean Coral Reefs," *Conservation Biology*, Vol. 18, No. 1, *Conservation Biology*, Vol. 18, 2004, pp. 215-226.

[10] T. P. Hughes, "Catastrophes, Phase Shifts and Large-Scale Degradation of a Caribbean Coral Reef," *Science*, Vol. 265, No. 5178, 1994, pp. 1547-1551.

[11] R. B. Aronson and W. F. Precht, "Evolutionary Paleoecology of Caribbean Coral Reefs," In: W. D. Allmon and D. J. Bottjer, Eds., *Evolutionary Paleoecology: The Ecological Context of Macroevolutionary Change*, Columbia University Press, New York, 2001, pp. 171-233.

[12] A. M. Greenaway and D.-A. Gordon-Smith, "The Effects of Rainfall on the Distribution of Inorganic Nitrogen and Phosphorus in Discovery Bay, Jamaica," *Limnology and Oceanography*, Vol. 51, 2006, pp. 2206-2220.

[13] E. H. I. Meesters, M. Hilterman, E. Kardinaal, M. Keetman, M. de Vries and R. P. M. Bak, "Colony Size-Frequency Distributions of Scleractinian Coral Populations: Spatial and Interspecific Variation," *Marine Ecology Progress Series*, Vol. 209, 2001, pp. 43-54.

[14] L. D. Smith, M. Devlin, D. Haynes and J. Gilmour, "A Demographic Approach to Monitoring the Health of Coral Reefs," *Marine Pollution Bulletin*, Vol. 51, No. 1-4, 2005, pp. 399-407.

[15] Monitoring Report, "Final Monitoring Report, Dredging and Reclamation Programme in Kingston Harbour," National Environmental Protection Agency, Jamaica, 29 November 2002.
http://www.nrca.org/publications/coastal/monitoring_rep/reports/MONITORING_REPORTFnl.pdf

[16] B. A. Wade, L. Antonio and R. Mahon, "Increasing Organic Pollution in Kingston Harbour, Jamaica," *Marine Pollution Bulletin*, Vol. 3, No. 7, 1972, pp. 106-111.

[17] A. Mansingh and A. Wilson, "Insecticide Contamination of Jamaican Environment III. Baseline Studies on the Status of Insecticidal Pollution of Kingston Harbour," *Marine Pollution Bulletin*, Vol. 30, No. 10, 1995, pp. 640-645.

[18] J. E. Andrews, A. M. Greenaway, G. R. Bigg, D. F. Webber, P. F. Dennis and G. A. Guthrie, "Pollution History of a Tropical Estuary Revealed by Combined Hydrodynamic Modelling and Sediment Geochemistry," *Journal of Marine Systems*, Vol. 18, No. 4, 1999, pp. 333-343.

[19] A. M. Maxam and D. F. Webber, "Using the Distribution of Physicochemical Variables to Portray Reefal Bay Waters," *Journal of Coastal Research*, Vol. 25, No. 6, 2009, pp. 1210-1221.

[20] M. J. C. Crabbe, "The Influence of Extreme Climate Events on Models of Coral Colony Recruitment and Survival in the Caribbean," *American Journal of Climate Change*, Vol. 1, 2012, pp. 33-40.

[21] M. Huston, "Variation of Coral Growth Rates with Depth at Discovery Bay, Jamaica," *Coral Reefs*, Vol. 4, No. 1, 1985, pp. 19-25.

[22] T. P. Hughes and J. B. C. Jackson, "Do Corals Lie about Their Age? Some Demographic Consequences of Partial Mortality, Fission, and Fusion," *Science*, Vol. 209, No. 4457, 1980, pp. 713-715.

[23] N. L. Foster, I. B. Baums and P. J. Mumby, "Sexual vs Asexual Reproduction in an Ecosystem Engineer: The Massive Coral *Montastrea annularis*," *Journal of Animal Ecology*, Vol. 76, No. 2, 2007, pp. 384-391.

[24] Z. H. Zar, "Biostatistical Analysis," 4th Edition, Prentice-Hall, Upper Saddle River, 1999, p. 663.

[25] M. J. C. Crabbe, "Topography and Spatial Arrangement of Reef-Building Corals on the Fringing Reefs of North Jamaica May Influence Their Response to Disturbance from Bleaching," *Marine Environmental Research*, Vol. 69, No. 3, 2010, pp. 158-162.

[26] A. R. Harborne, P. J. Mumby and R. Ferrari, "The Effectiveness of Different Meso-Scale Rugosity Metrics for Predicting Intra-Habitat Variation in Coral Reef Assemblages," *Environmental Biology of Fishes* Vol. 94, 2012, pp. 431-442.

[27] J. A. Idjadi, S. C. Lee, J. F. Bruno, W. F. Precht, L. Allen-Requa and P. J. Edmunds, "Rapid Phase-Shift Reversal on a Jamaican Coral Reef," *Coral Reefs*, Vol. 25, 2006, pp. 209-211.

[28] I. Westfield, S. Dworkin, R. Bonem and E. Lane, "Identification of Sediment Sources Using Geochemical Fingerprinting at Pear Tree Bottom Reef, Runaway Bay, Jamaica," *Abstracts of the* 11*th International Coral Reef Society*, 2008, p. 137.

[29] G. Roff and P. J. Mumby, "Global Disparity in the Resilience of Coral Reefs," *Trends in Ecology & Evolution*, Vol. 27, No. 7, 2012, pp. 404-413.

[30] P. M. H. Gayle, P. Wilson-Kelly and S. Green, V. S. Flood, J. M. Pitt and S. R. Smith, "Transplantation of Benthic Species to Mitigate Impacts of Coastal Developments in Jamaica," *Revista de Biologia Tropical*, Vol. 53,2005, pp. 105-115.

[31] V. S. Flood, J. M. Pitt and S. R. Smith, "Historical and Ecological Analysis of Coral Communities in Castle Harbour (Bermuda) after More than a Century of Environmental Perturbation," *Marine Pollution Bulletin*, Vol. 51, No. 5-7, 2005, pp. 545-557.

[32] N. J. Quinn and B. L. Kojis, "The Recent Collapse of a Rapid Phase-Shift Reversal on a Jamaican North Coast Reef after the 2005 Bleaching Event," *International Journal of Tropical Biology*, Vol. 56, Suppl. 1, 2008, pp. 149-159.

[33] M. J. C. Crabbe, "Scleractinian Coral Population Size Structures and Growth Rates Indicate Coral Resilience on

the Fringing Reefs of North Jamaica," *Marine Environmental Research*, Vol. 67, No. 4-5, 2009, pp. 189-198.

[34] F. R. Gell and C. M. Roberts, "Benefits beyond Boundaries: The Fishery Effects of Marine Reserves," *Trends in Ecology and Evolution*, Vol. 18, 2003, pp. 448-455.

[35] S. Macia, M. P. Robinson and A. Nalevanko, "Experimental Dispersal of Recovering *Diadema antillarum* increases Grazing Intensity and Reduces Macroalgal Abundance on a Coral Reef," *Marine Ecology-Progress Series*, Vol. 348, 2007, pp. 173-182.

[36] I. M. Côté and E. S. Darling, "Rethinking Ecosystem Resilience in the Face of Climate Change," *PLoS Biology*, Vol. 8, No. 7, 2010, Article ID: e1000438.

[37] T. P. Hughes, N. A. J. Graham, J. B. C. Jackson, P. J. Mumby and R. S. Stenek, "Rising to the Challenge of Sustaining Coral Reef Resilience," *Trends in Ecology and Evolution*, Vol. 25, No. 11, 2010, pp. 633-642.

[38] C. Mora and P. F. Sale, "Ongoing Global Biodiversity Loss and the Need to Move beyond Protected Areas: A Review of the Technical and Practical Shortcomings of Protected Areas on Land and Sea," *Marine Ecology Progress Series*, Vol. 434, 2011, pp. 251-266.

[39] FAO, "The State of World Fisheries and Aquaculture," FAO, Rome, 2012. http://www.fao.org/corp/statistics/en/

[40] World Bank, "*Global Financial Development Report* 2013: *Rethinking the Role of the State in Finance*," World Bank, New York, 2012. http://data.worldbank.org/

[41] IUCN (International Union for the Conservation of Nature), "2012 Report on Caribbean Coral Reefs," IUCN.

[42] L. Jones, P. M. Alcolado, Y. Cala, D. Cobián, V. Coelho, A. Hernández, R. Jones, J. Mallela and C. Manfrino, "The Effects of Coral Bleaching in the Northern Caribbean and Western Atlantic," In: C. Wilkinson and D. Souter, Eds., *Status of Caribbean Coral Reefs after Bleaching and Hurricanes in* 2005, Global Coral Reef Monitoring Network, and Reef and Rainforest Research Centre, Townsville, 2008, pp. 73-83.

[43] F. Micheli, A. Saenz-Arroyo, A. Greenley, L. Vazquez, J. A. E. Montes, M. Rossetto and G. A. Leo, "Evidence that Marine Reserves Enhance Resilience to Climatic Impacts," *PLoS ONE*, Vol. 7, No. 7, 2012, Article ID: e-40832

[44] Z. Sary, H. A. Oxenford and J. D. Woodley, "Effects of an Increase in Trap Mesh Size on an Over-Exploited Coral Reef Fishery at Discovery Bay, Jamaica," *Marine Ecology Progress Series*, Vol. 154, 1997, pp. 107-120.

[45] B. E. Huntington, M. Karnauskas and D. Lirman, "Corals Fail to Recover at a Caribbean Marine Reserve Despite Ten Years of Reserve Designation," *Coral Reefs*, Vol. 30, No. 4, 2012, pp. 1077-1085.

[46] C. Doropoulos, S. Ward, G. Diaz-Pulido, O. Hoegh-

Guldberg and P. J. Mumby, "Ocean Acidification Reduces Coral Recruitment by Disrupting Intimate Larval-Algal Settlement Interactions," *Ecology Letters*, Vol. 15, No. 4, 2012, pp. 338-346.

[47] K. Frieler, M. Meinhausen, A. Golly, M. Mengel, K. Lebek, S. D. Donner and O. Hoegh-Guldberg, "Limiting Global Warming to 2˚C Is Unlikely to Save Most Coral Reefs," *Nature Climate Change*, Vol. 3, 2013, pp. 165-170.

[48] E. R. Selig, K. S. Casey and J. F. Bruno, "Temperature-Driven Coral Decline: The Role of Marine Protected Areas," *Global Change Biology*, Vol. 18, No. 5, 2012, pp. 1561-1570.

[49] B. D. Keller, D. F. Gleason, E. McLeod, C. M. Woodley, S. Airame, B. D. Causey, A. M. Friedlander, R. Grober-Dunsmore, J. E. Johnson, S. L. Miller and R. S. Steneck, "Climate Change, Coral Reef Ecosystems, and Management Options for Marine Protected Areas," *Environmental Management*, Vol. 44, No. 6, 2009, pp. 1069-1088.

[50] P. Söderholm, "Modeling the Economic Costs of Climate Policy: An Overview," *American Journal of Climate Change*, Vol. 1 No. 1, 2012, pp. 14-32.

5

The Influences of Climate Change on the Runoff of Gharehsoo River Watershed

Kazem Javan, Farzin Nasiri Saleh, Hamid Taheri Shahraiyni

Faculty of Civil and Environmental Engineering, Tarbiat Modares University, Tehran, Iran

ABSTRACT

The purpose of this study is to survey the impact of climate change on the runoff of Gharehsoo River in northwest of Iran. In this research the outputs of monthly precipitation and temperature data of PRECIS model, which is a regional climate model with 50 × 50 km resolution on the basis of B2 scenario, have been utilized for base (1961-1990) and future (2071-2100) periods. The output results of PRECIS model show that the average temperature of watershed increased up to 2°C - 5°C. In addition, future precipitation is more than the base precipitation on January, February, March, September and December. The observed data of 1996-2002 used for calibration of HSPF model and the data of 2003-2004 were used for HSPF validation. The present monthly patterns for precipitation and temperature were estimated using the geostatistical techniques and the future monthly patterns were retrieved by the combination of future monthly PRECIS data and monthly patterns of precipitation and temperature. Then, the base and future precipitation and temperature patterns were introduced to validate HSPF model for the simulation of monthly runoff in the base and future periods. The results show that in the future, the discharge of Gharehsoo River watershed decreases in all of the months. In addition, the peak discharge in the future period happens one month earlier, in April, because of increase of temperature and earlier beginning of snow melting season. Finally the sensitivity analysis was performed on the monthly runoff. The results showed that monthly discharge increases 0.3% - 35.6% and decreases 0.3% - 32.6% due to 20% increase and decrease of precipitation, respectively. In addition, 1°C and 2°C increase of temperature leads to 0% - 8% and 0.1% - 15% decrease of average monthly discharge, respectively.

Keywords: Climate Change; HSPF Model; Regional Climate Model; Gharehsoo River

1. Introduction

Climate change can influence the ecosystems, environment and water resources. One of the most important impacts of climate change is the changes of regional and local available water. Different studies have been performed on the impact of climate change on the water resources [1-5]. Recently, some studies have been performed on the impacts of climate change on water resources in Asia. Chen *et al.* [6] analyzed the climate change in the Danjiangkou reservoir that is a source of water in China. The results for period 2021-2050 showed that runoff and precipitation of Danjiangkou reservoir will increase in all of the seasons. Sensitivity analysis in their study revealed that 1°C and 2°C increases in temperature reduce the mean annual runoff about 3.5% and 7%, respectively and 10% and 20% decrease/increase of mean monthly precipitation decreases/increases the mean

annual runoff about 15% and 30%, respectively. Akhtar *et al.* [7] showed that estimates of runoff changes in three river basins in the Hindukush-Karakorum-Himalaya region are related to the climate change. In this study, PRECIS Regional Climate Model was utilized for the simulation of future climate. The results showed that the temperature and precipitation will increase at the end of 21st century. Vicuna *et al.* [8] studied on the impacts of climate change scenarios in the north-central Chile in the first half of the 21st century. Their results showed an increase in temperature of about 3°C - 4°C and a reduction in precipitation of 10% - 30% during the first half of 21st century. Zarghami *et al.* [9] used General Circulation Models (GCM) to predict the climate change. They used the three scenarios (A1B, A2 and B1) with the horizons 2020, 2055 and 2090. Their study revealed that average annual temperature will increase 2.3°C and annual precipitation will decrease about 30% in the middle of this century.

In this study, the impacts of climate change on the Gharehsoo River runoff were investigated. A new method was developed for reasonable prediction of spatial patterns of precipitation and temperature. This method uses the results of a Regional Climate Model (PRECIS model) coupled with the appropriate spatial modeling techniques. HSPF model was used to simulate the future runoff of Gharehsoo River. Different studies demonstrated the ability of HSPF for runoff simulation [10-14].

2. The Study Area

The present study was conducted for the watershed of Ardebil province in North-western Iran, which lies between latitude 37° to 38°N and longitude 47° to 48°E (**Figure 1**). The geographical information and the mean observed climate data for the main synoptic stations of the province for the baseline years between 1996 and 2004 are presented in **Table 1**. The mean annual precipitation in this watershed (stations are presented in **Table 1**) is

very little in comparison with world average of 800 mm. In recent years, the water shortage in Ardebil city (the capital of the province) for the reason that used in excess of water resource in agricultural province and industry consumptions has become change into a serious problem for this province.

There are very strict conflicts on using its recharge sources and new water transfers are limited. The water providing to the cities is now more vulnerable, and the Ardebil Regional Water Company needs to notice the future trend lines of the climate and their impacts on the water resources. This data aids them to understand the extents of the uncertainties and the real threats they will face in future years. The purpose of this research is therefore to predict the climate change and its impacts on the water resources in this regional.

3. Methodology

This algorithm of this study is presented in **Figure 2**. It has two important steps: First, it prepares the future me-

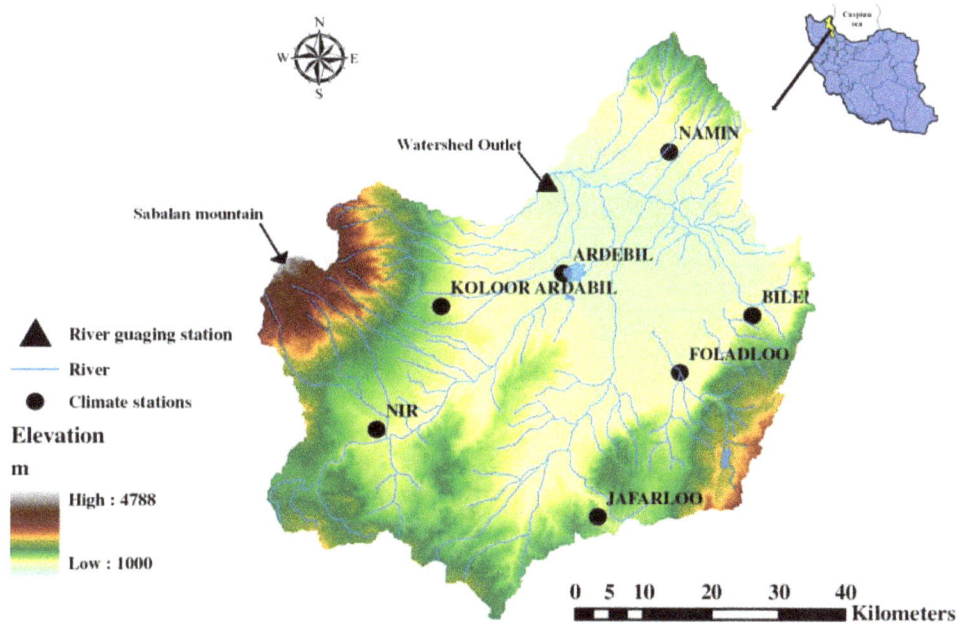

Figure 1. Gharehsoo River watershed and its location in Iran with its topography, drainage network and meteorological stations.

Table 1. The positions and the averages of the temperature and precipitation of seven synoptic stations.

	Stations						
	Ardebil	Bile	Foladloo	Jafarloo	Namin	Nir	Koloor Ardebil
Latitude (°E)	38.25	38.02	38.12	37.92	38.42	38.03	38.20
Longitude (°N)	48.28	48.60	48.48	48.35	48.45	47.98	48.08
Elavation (m)	1332	1680	1490	1680	1500	1450	1581
Available data (years)	1951-2007	1975-2007	1994-2007	1969-2007	1960-2007	1960-2007	1975-2007
Mean precipitation (mm)	445	480	334	359	360	376	458
Mean temperature (°C)	10.3	8.4	6.3	7.6	8.3	7.4	8.4

teorological data for region under a scenario, and second, it assesses the impacts of climate change on Gharehsoo River watershed by using the HSPF model.

3.1. Climate Change Data by PRECIS and Preparation

Despite the important increase in the resolution of General Circulation Models (GCMs), they cannot yet predict meteorological outputs for small scales. Different dynamic and statistical models have developed to downscale the GCM outputs. The PRECIS (Providing Regional Climates for Impacts Studies) model is a RCM (Regional Climate Model) that it was developed by the Hadley center on the basis of the atmospheric of HadCM3 [15] to generate high resolation climate change scenarios as described in Jones *et al.* 2004 [16]. The PRECIS simulated region with a horizontal resolution of 50 × 50 km. The base climate (1961-1990) and future climate SRES B2 scenario (2071-2100), have been selected. Comparison between observed data and PRECIS Model simulated data of the base period (1961-1990) demonstrated that there is an appropriate similarity between these two data

series; so that, the base data series of PRECIS model could be used for the runoff simulation using HSPF during the base period. Statistical analysis of precipitation and temperature data series (observed and output data of PRECIS model) shows that these two time series have approximately the same mean and standard deviation. In the study, we have applied a new method for preparing future data that the algorithm of calculation is as below. In the study, we have applied a new method for preparing future data that the algorithm of calculation is as below.

First, it's necessary to have a series of precipitation and temperature unit patterns in producing of these patterns in monthly periods in future. This series of maps are generated using the precipitation and temperature patterns of present data. For this work, we use of interpolation methods for preparation these patterns. Utilizing interpolation methods to estimate hydrological parameters can increase the accuracy of rainfall-runoff calculations [17]. These methods are including of Inverse Distance Weighting (IDW), Global Polynomial, Local Polynomial, Radial Basis Functions (RBF), Ordinary Kriging and Simple Kriging. The cross validation technique is utilized for identification of the best interpolation technique for each month. Then, precipitation and temperature unit patterns by the algorithm of calculation are as follows. **Figure 3** shows use of the new approach for preparation precipitation unit pattern in future.

Using of algorithm **Figure 3**, the appropriate present unit patterns (maps) are determined for each month and in the other word, 12 monthly present unit patterns (maps) are generated. Then, the future patterns are calculated using the following formula:

$$fmp_i = pmp_i \times (fh_i/ph_i) \qquad (1)$$

$$fmt_i = pmt_i \times (ft_i/pt_i) \qquad (2)$$

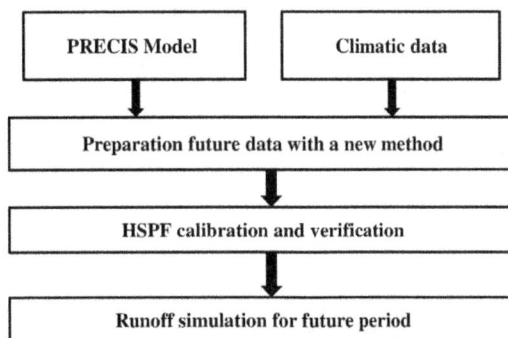

Figure 2. The algorithm of study.

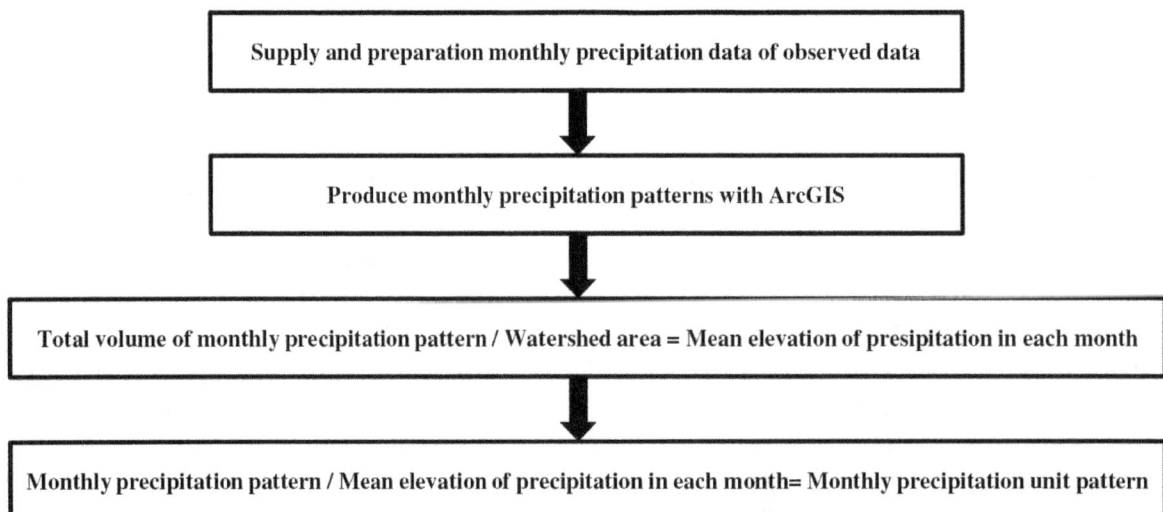

Figure 3. Algorithm of the new approach for preparation precipitation unit pattern in future.

Where, fmp_i and fmt_i are future patterns of precipitation (mm) and temperature (°C) in month i-th ($i = 1 \cdots 12$), respectively. pmp_i and pmt_i are present unit patterns of precipitation (mm) and temperature (°C) in month i-th, respectively and are calculated using interpolation techniques as explained above. fh_i and ft_i are future mean precipitation (mm) and temperature (°C), respectively and are calculated using the PRECIS model. ph_i and pt_i are present mean precipitation (mm) and temperature (°C), and are calculated by averaging of pmp_i and pmt_i patterns, respectively.

3.2. HSPF Hydrological Model

In this study, we use of Hydrological Simulation Program FORTRAN (HSPF) for simulation outlet discharge of Gharehsoo River watershed. HSPF is a set of computer codes, developed by the US Environmental Protection Agency. It is based on the Stanford Watershed Model IV [18]. HSPF has been generated by the combination of Stanford Watershed Model IV with Agricultural Runoff Management Model (ARM) [19], Non-point Source Runoff Model (NPS) [20], and Hydrological Simulation Program (HSP) [21-23]. This model can simulate the hydrologic processes on permeable and impermeable land surfaces and streams [24]. It has been widely used in Asian and other parts of the world in the climate change studies [13,14,25].

HSPF is a semi distributed deterministic, continuous and physically based model. The PERLND, IMPLND, and RCHRES modules are three main modules of HSPF which help to simulate permeable land segments, impermeable land segments, and free-flow reaches, respectively. Detailed information about these modules can be found in the literatures [20,24,26,27]. HSPF model uses a Storage Routing technique to route water in each reach. Infiltration in permeable land is calculated based on Richard's equation [24]. Actual evapotranspiration (ET) is calculated by Penman or Jensen formulas. **Table 2** shows key HSPF parameters. These parameters should be calibrated during the calibration process. LZSN is the most important parameter in infiltration capacity which is called

in HSPF with the INFILT parameter. AGWRC is depended on topography, climate, soil properties and land use. UZSN is influenced of LZSN [11]. Other parameters that they have not presented in **Table 2** are estimated using the BASINS software based on topographic, soil properties and land use data. Then the estimated parameters are introduced to HSPF. The data from 1996 to 2002 were utilized for HSPF model calibration and the data from 2003-2004 were used as validation dataset.

4. Results and Discussion

4.1. Calibration and Validation of HSPF Model

7-year daily average discharge data of Gharehsoo River of 1996-2002 are used for calibration of HSPF model in the simulation of daily discharge in the hydrometric outlet station of Samian. Two years (2003-2004) are used for the model validation. **Table 3** shows the values of calibrated parameters in this study. For example, LZSN in **Table 3** is an average value 38.1 mm/h that has been estimated according to the Linsley equation [29]. Linsley equation for the LZSN estimation is LZSN = $100 + 0.25 \times$ (Yearly mean precipitation). For estimation of the other parameters, BASINS Technical Note 6 [28] has been utilized.

Figures 4 and **5** show the observed and simulated hydrographs for calibration and validation periods, respectively. These figures present good agreement between observed and simulated daily runoff in the calibration and validation periods. The correlation coefficients for calibration and validation periods are 0.814 and 0.806, respectively. It implies that HSPF simulation is acceptable. Moreover, Nash-Sutcliff coefficient (model efficiency) is 0.87 in calibration period and 0.76 in validation period. Nash-Sutcliffe efficiency coefficient value less than 0.5 are considered as unacceptable, while values greater than 0.6 are considered as good and greater than 0.8 are considered excellent results. Therefore, HSPF has been presented good daily runoff simulation. Results show that HSPF simulation of watershed discharge is acceptable in calibration period and can be used in this research.

Table 2. The parameters of HSPF model in simulation process [28].

Parameter	Definition	Units	Possible range	
			MIN	MAX
INFILT	Index of infiltration capacity	mm/h	0.25	12.7
AGWRC	Base groundwater recession	dimensionless	0.85	0.999
LZSN	Lower zone nominal soil moisture storage	mm	50.8	381
UZSN	Upper zone nominal soil moisture storage	mm	1.27	50.4
DEEPFR	Fraction of groundwater inflow to deep recharge	dimensionless	0	0.5
INTFW	Interflow inflow parameter	dimensionless	1	10
IRC	Interflow recession parameter	dimensionless	0.3	0.85
BASETP	Fraction of remaining ET from base flow	dimensionless	0	0.2
LZETP	Lower zone ET parameter	dimensionless	0.1	0.9

4.2. Future Changes of Temperature and Precipitation

Figure 6 shows the average of 30-years monthly temperature in Ardabil station for the base (solid line) and future (dash line) periods. As it is shown, this scenario forecasts that temperature will increase in Ardabil station in all of the seasons. Temperature increases 2°C - 4°C in winter, 2°C - 5°C in spring, 3°C - 5°C in summer and 2°C - 4°C in autumn. Maximum temperature will happen in July and the Minimum temperature will take place in January. **Figure 7** shows the average of 30-year monthly precipitation in Ardabil station for the base and future

periods. Future precipitation is more than the base precipitation on January, February, March, September and December.

PRECIS model forecasts that maximum precipitation happens on February and the minimum on July. It is concluded that climate change impacts on climate variables such as temperature and precipitation of Gharehsoo river watershed in the future; although according to **Figures 6** and **7**, the impact of climate change on temperature would be more than precipitation. Comparison between observed data and PRECIS Model simulated data of the base period (1961-1990) demonstrated that there is an appropriate similarity between these two data series; so that, the base data series of PRECIS model could be used for the runoff simulation using HSPF during the base period. Statistical analysis of precipitation and temperature data series (observed and output data of PRECIS model) shows that these two time series have approximately the same mean and standard deviation.

In order to prepare base monthly precipitation and temperature patterns, different geostatistical methods are compared to each other using cross validation technique. **Tables 4** and **5**, show RMSE (Root Mean Square Error) values of the six interpolation methods for precipitation and temperature data, respectively. Results show that RBF and IDW methods can be utilized for preparation of

Table 3. Values of parameters, used in simulation.

Parameter	Value
INFILT	0.35 mm/h
AGWRC	0.977
LZSN	38.1 mm
UZSN	22.86 mm
DEEPFR	0.2
INTFW	2
IRC	0.9
BASETP	0.1
LZETP	0.7

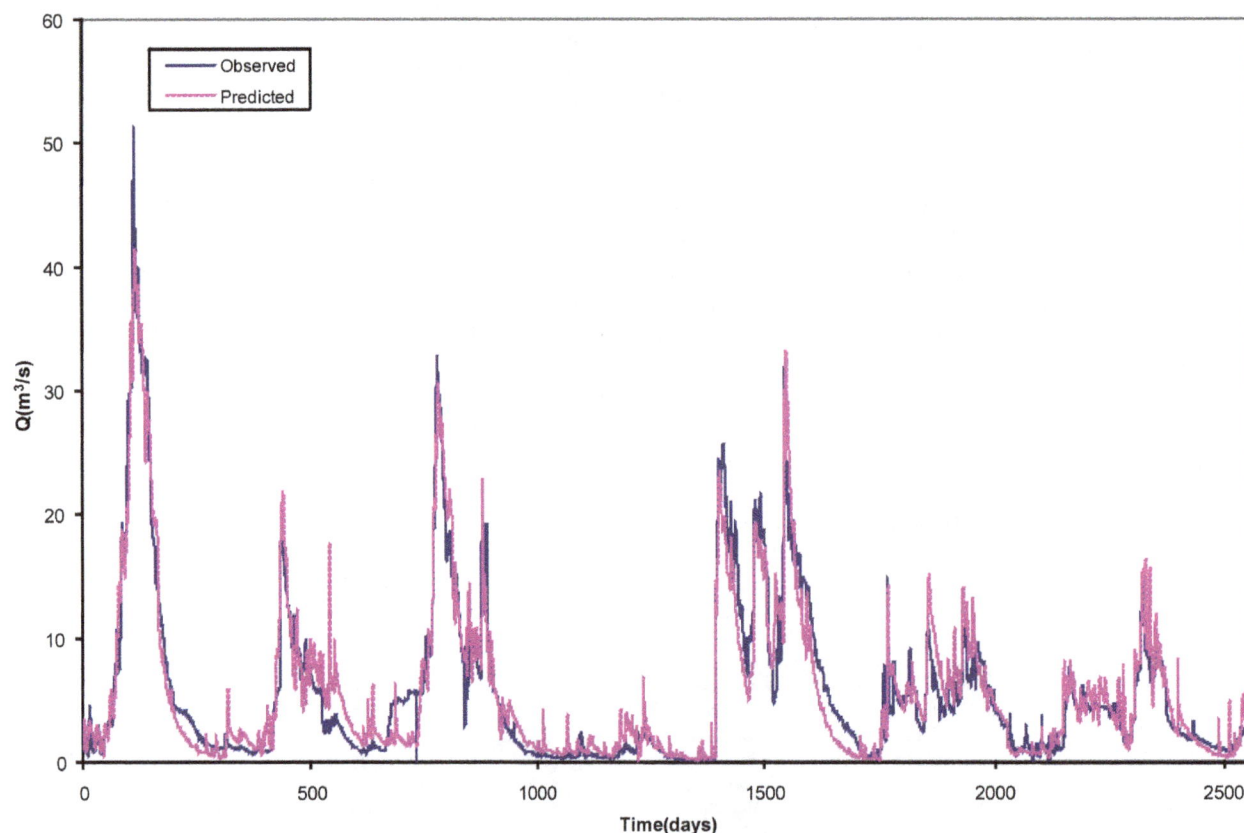

Figure 4. Simulated and observed hydrographs for calibration period during 1996-2002.

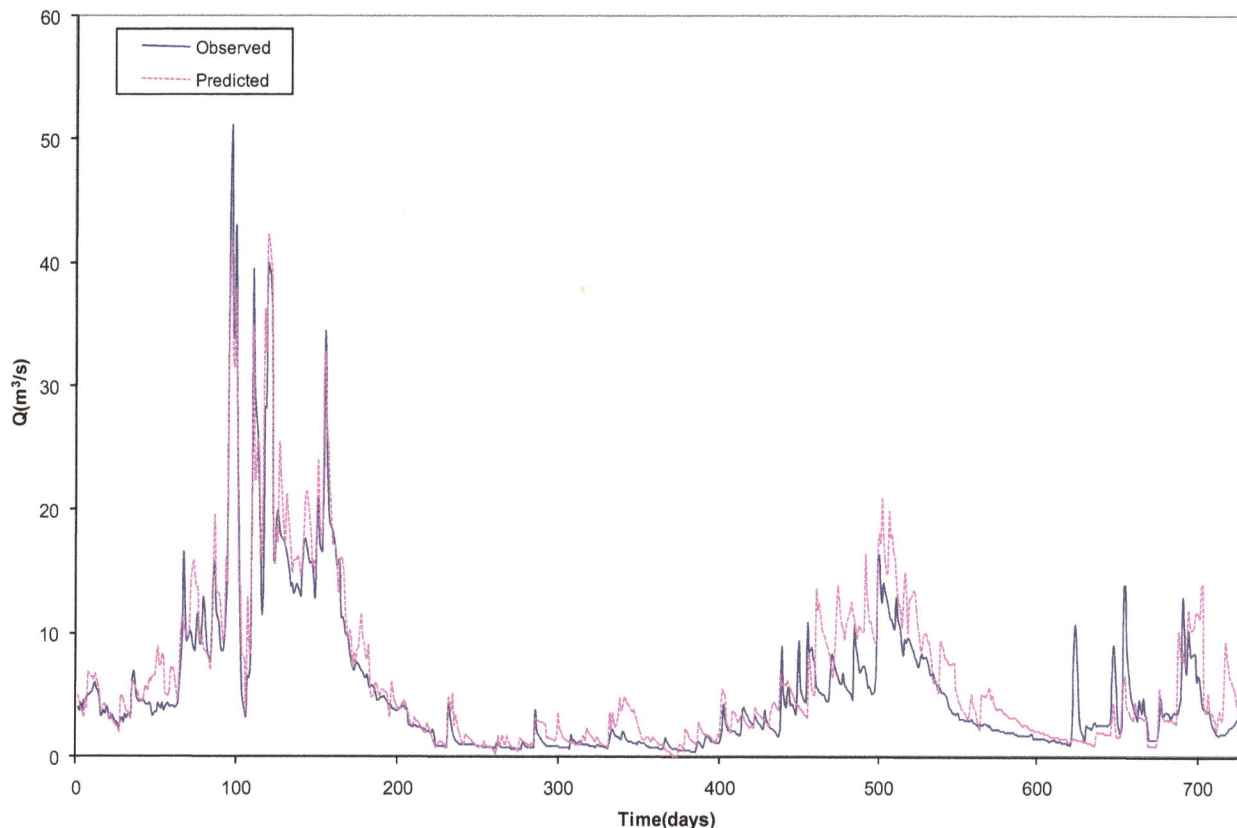

Figure 5. Simulated and observed hydrographs for validation period during 2003-2004.

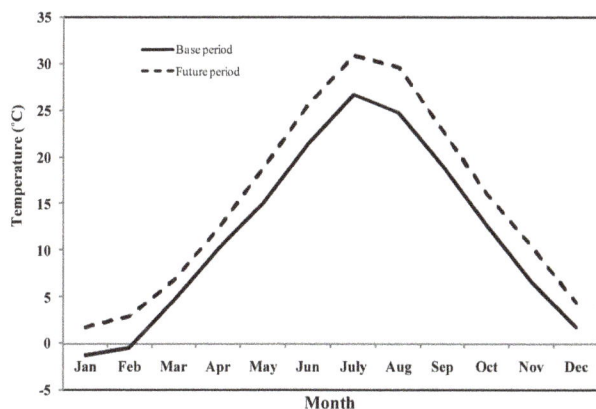

Figure 6. Mean of temperature in Ardabil station for the base (1961-1990) and future (2070-2100) period.

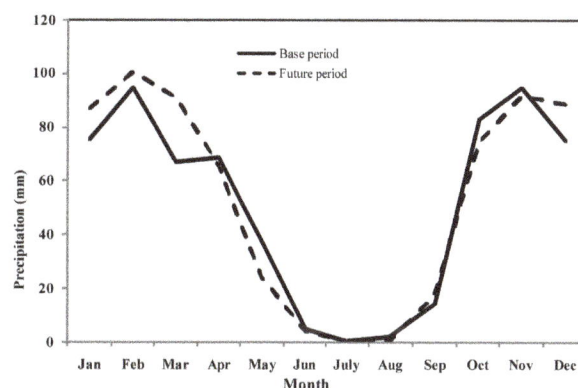

Figure 7. Mean of precipitation in Ardabil station for the base (1961-1990) and future (2070-2100) period.

precipitation and temperature patterns. Therefore, precipitation and temperature patterns are prepared for all months using these methods. Future monthly precipitation and temperature patterns are retrieved using the explained method in Section 3.1.3. The samples of precipitation and temperature patterns in January of 2100 are shown in **Figures 8** and **9**.

4.3. Future Discharge Results

After calibration and validation of HSPF model to the

watershed, PRECIS model base and future data series is used as the input to HSPF model. **Table 6** shows the difference between monthly discharges for base (1960-1990) and future (2070-2100) periods. In spite of April, monthly discharges of future period decrease in all of the months. The differences between monthly discharge of base and future periods in the warm months are more than the other months; this is because of increase of future temperature and evapotranspiration and decrease of future precipitation in the warm months in comparison with base data. The least discharge difference is 28%, which

Table 4. RMSE values of the six interpolation methods for precipitation data.

Methods	Jan	Feb	Mar	Apr	May	Jun	July	Aug	Sep	Oct	Nov	Dec
IDW	21.44	33.38	21.20	23.65	34.44	16.35	18.95	2.63	2.82	27.03	24.25	22.24
Global polynomial	21.73	34.33	22.52	25.55	36.11	16.76	19.45	2.69	3.15	24.07	27.50	21.35
Local polynomial	21.42	34.22	23.66	24.65	40.22	18.28	19.33	2.84	3.28	21.88	28.90	21.46
RBF	**20.85**	**31.01**	**20.59**	**19.63**	**32.18**	**16.08**	**18.62**	**2.55**	**2.05**	23.78	24.77	**21.12**
Ordinary Kriging	22.37	33.60	21.67	23.32	33.37	17.04	18.66	2.62	2.89	**21.73**	**23.82**	21.14
Simple Kriging	21.44	33.06	21.54	23.35	33.60	16.09	19.64	2.983	2.63	21.87	24.06	21.84

Table 5. RMSE values of the six interpolation methods for temperature data.

Method	Jan	Feb	Mar	Apr	May	Jun	July	Aug	Sep	Oct	Nov	Dec
IDW	**4.88**	2.81	4.22	2.34	2.56	**3.39**	**3.49**	**4.72**	**4.37**	6.13	6.33	**3.95**
Global polynomial	6.16	3.64	5.91	2.36	3.35	4.18	4.50	5.82	5.56	7.97	9.04	5.54
Local polynomial	6.45	3.88	6.26	2.34	3.16	4.30	4.59	5.96	5.72	8.16	9.45	5.88
RBF	5.15	2.78	4.13	2.24	2.42	3.51	3.62	4.99	4.63	**5.96**	**6.07**	4.08
Ordinary Kriging	4.96	**2.62**	**3.78**	**2.20**	**2.32**	3.50	3.76	4.81	4.99	6.45	6.55	4.23
Simple Kriging	5.07	2.93	3.86	2.45	2.38	3.64	3.82	4.92	4.86	6.21	6.30	3.38

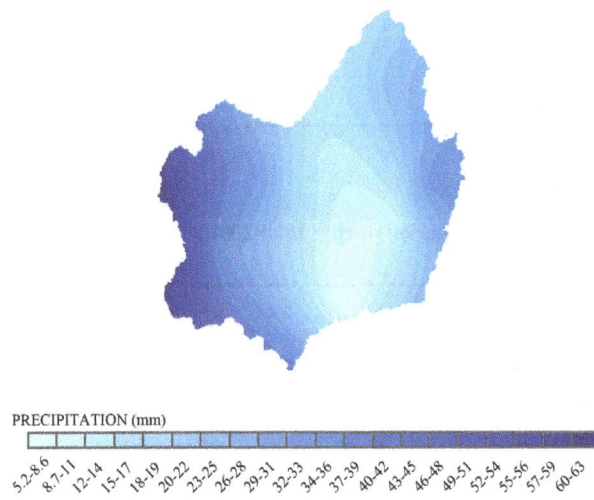

PRECIPITATION (mm)

Figure 8. Precipitation pattern on January of 2100.

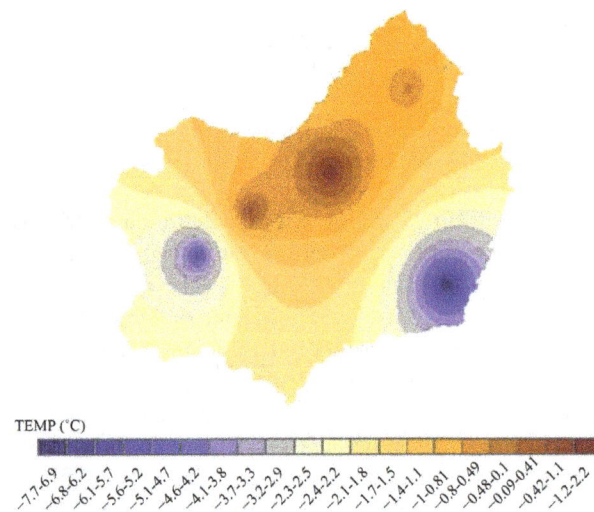

TEMP (°C)

Figure 9. Temperature pattern on January of 2100.

takes place on April (**Table 6**).

Figure 10 shows monthly discharge in the future and base periods. This figure demonstrates that the peak discharge in the future period would happen one month earlier. It is because of increasing temperature and earlier beginning of snow melting. Generally, results show that in the future, the discharge of Gharehsoo River watershed would decrease in the all of months. It might make problem for agriculture of studied region; because, Gharehsoo watershed is one of the most important regions for production of crops in Iran and plays an important role in economic growth and food of this country.

4.4. Sensitivity Analysis

In this section, the sensitivity of precipitation and temperature to runoff is investigated. Sensitivity analysis is performed in four hypothetical scenarios for future climate (**Table 7**). In two hypothetical scenarios, the precipitation is increased and decreased 20 percent. In the other scenarios, the temperature is increased 1°C and 2°C. Results of sensitivity analysis are exhibited in **Figure 11**. It is obvious that 1°C and 2°C increase of temperature lead to 0% - 8% and 0.1% - 15% decrease of average monthly discharge, respectively. In addition, **Figure 11** exhibits that monthly discharge increases 0.3 - 35.6 percent due to 20% increase of precipitation. Similarly, monthly discharge decreases 0.3 - 32.6 percent due to 20% decrease of precipitation.

5. Conclusions

HSPF was utilized in this study as a hydrological model. The results of calibration and validation of this model demonstrated its ability for runoff simulation in the

Table 6. Difference between monthly discharges in the base and future periods.

Period	Jan	Feb	Mar	Apr	May	Jun	Jul	Aug	Sep	Oct	Nov	Dec
Base period	22.78	11.16	19.36	53.42	82.35	77.83	37.64	16.39	8.12	6.47	32.35	78.46
Future period	21.76	7.32	10.29	53.70	49.67	27.50	15.06	5.34	2.67	5.05	18.78	41.46
Changes	−1.03	−3.84	−9.07	0.28	−32.69	−50.32	−22.58	−11.04	−5.45	−1.42	−13.57	−37.00
Percent of changes	−4.50	−34.41	−46.84	0.53	−39.69	−64.66	−60.00	−67.40	−67.08	−21.97	−41.95	−47.16

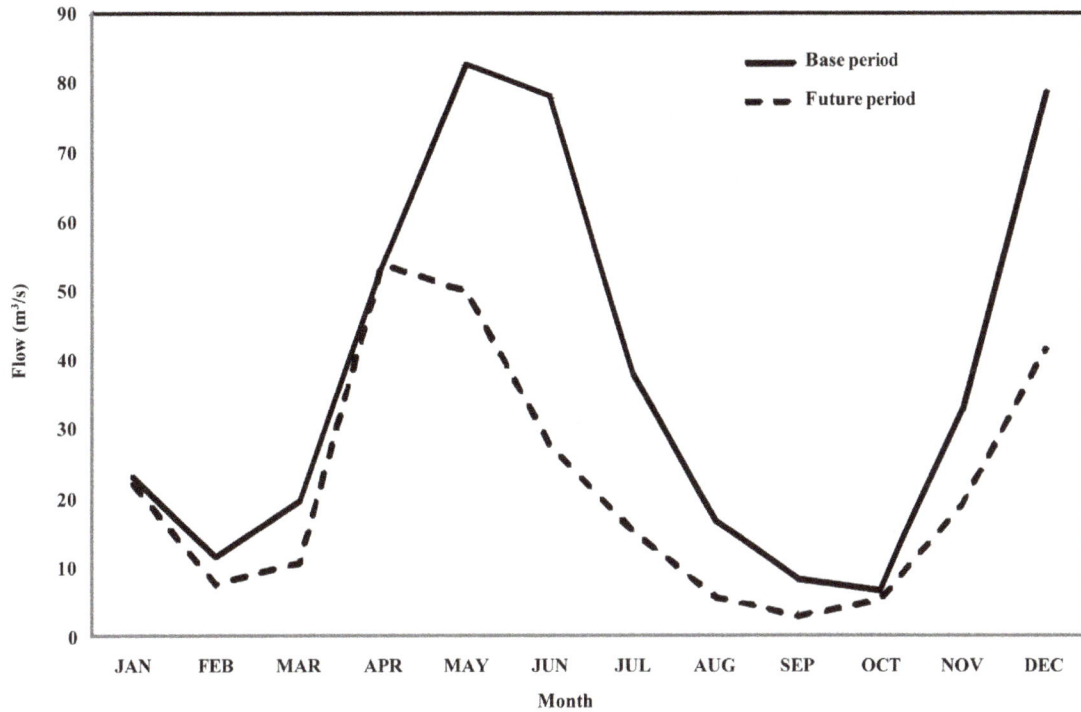

Figure 10. Monthly discharge future and base period.

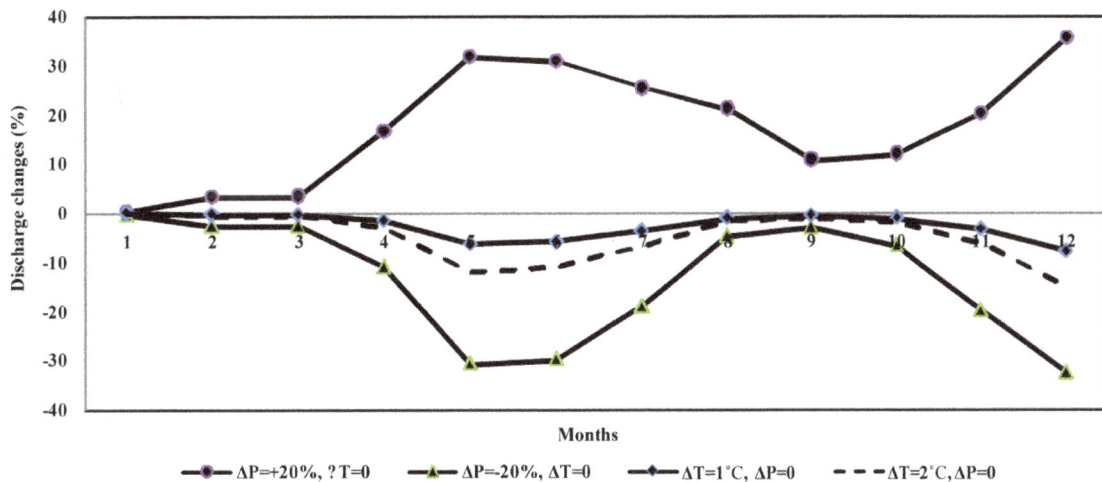

Figure 11. Results of sensitivity analysis of mean monthly runoff to the precipitation and temperature in the Gharehsoo River watershed.

Gharehsoo River. In this study, the different geostatistical methods were utilized for the estimation of present monthly patterns of precipitation and temperature. The cross validation technique was utilized for evaluation of different geostatistical methods. The results showed that the best methods for extraction of precipitation and temperature pattern are RBF and IDW methods, respectively. PRECIS as a regional climate model was utilized to pro-

Table 7. Hypothetical scenarios of future climate.

		Precipitation scenarios		
		$\Delta P = -20\%$	$\Delta P = +20\%$	$\Delta P = 0$
	$\Delta T = 0$	☑	☑	-
Temperature scenarios	$\Delta T = 1°C$	-	-	☑
	$\Delta T = 2°C$	-	-	☑

duce climate data of base (1960-1990) and future (2070-2100) periods based on B2 scenario with 50 × 50 km resolution. In addition, by combination of base and future precipitation and temperature data with the extracted precent patterns for precipitation and temperature, the future patterns for monthly precipitation and temperature were extracted. The comparison between base and future monthly precipitation and temperature showed that future precipitation is more than the base precipitation on January, February, March, September and December and future temperature increases 2°C - 4°C in winter, 2°C - 5°C in spring, 3°C - 5°C in summer and 2°C - 4°C in autumn.

The base and future precipitation and temperature patterns were introduced to validated HSPF model for the simulation of monthly runoff in the base and future periods. The results show that in the future, the discharge of Gharehsoo River watershed decreases in all of the months. In addition, the peak discharge in the future period happens one month earlier, because of increase of temperature and earlier beginning of snow melting season. Finally, the sensitivity of precipitation and temperature to runoff was investigated and the results showed that 1°C and 2°C increase of temperature leads to 0% - 8% and 0.1% - 15% decrease of average monthly discharge, respectively. In addition, monthly discharge increases 0.3% - 35.6% and decreases 0.3% - 32.6% due to 20% increase and decrease of precipitation, respectively.

REFERENCES

[1]   L. Nash and P. Gleick, "The Sensitivity of Streamflow in the Colorado Basin to Climatic Changes," *Journal of Hydrology*, Vol. 125, No. 3, 1991, pp. 221-241.

[2]   R. L. Wilby, L. E. Hay and G. H. Leavesley, "A Comparison of Downscaled and Raw GCM Output: Implications for Climate Change Scenarios in the San Juan River Basin, Colorado," *Journal of Hydrology*, Vol. 225, No. 1, 1999, pp. 67-91.

[3]   W. Arnell, "Climate Change and Global Water Resources," *Journal of Global Environmental Change*, Vol. 9, No. 1, 1999, pp. 31-49.

[4]   P. S. Yu, T. C. Yang and C. K. Wu, "Impact of Climate Change on Water Resources in Southern Taiwan," *Journal of Hydrology*, Vol. 260, No. 1, 2002, pp. 161-175.

[5]   T. G. Huntington, "Climate Warming Could Reduce Run-
off Significantly in New England, USA," *Journal of Agricultural and Forest Meteorology*, Vol. 117, No. 3-4, 2003, pp. 193-201.

[6]   H. Chen, S. Guo, C. Y. Xu and V. P. Singh, "Historical Temporal Trends of Hydro-Climatic Variables and Runoff Response to Climate Variability and Their Relevance in Water Resource Management in the Hanjiang Basin," *Journal of Hydrology*, Vol. 344, No. 3-4, 2007, pp. 171-184.

[7]   M. Akhtar, N. Ahmad and M. J. Booij, "The Impact of Climate Change on the Water Resources of Hindukush-Karakorum-Himalaya Region under Different Glacier Coverage Scenarios," *Journal of Hydrology*, Vol. 355, No. 1, 2008, pp. 148-163.

[8]   S. Vicune, R. D. Garreaud and J. McPhee, "Climate Change Impacts on the Hydrology of a Snowmelt Driven Basin in Semiarid Chile," *Journal of Climatic Change*, Vol. 105, No. 3, 2011, pp. 469-488.

[9]   M. Zarghami, A. Abdi, I. Babaeian, Y. Hassanzadeh and R. Kanani, "Impacts of Climate Change on Runoffs in East Azerbaijan, Iran," *Journal of Global and Planetary Change*, Vol. 78, No. 3-4, 2011, pp. 137-146.

[10] V. M. F. Jacomino and E. D. Fields, "A Critical Approach to the Calibration of a Watershed Model," *Journal of American Water Resource Association*, Vol. 33, No. 1, 1997, pp. 143-154.

[11] M. Albek, U. Ogutveren and E. Albek, "Hydrological Modeling of Seydi Suyu Watershed (Turkey) with HSPF," *Journal of Hydrology*, Vol. 285, No. 1, 2004, pp. 260-271.

[12] J. C. Imhoff, J. L. Kittle, M. R. Gray and T. E. Johnson, "Using the Climate Assessment Tool (CAT) in US EPA BASINS Integrated Modeling System to Assess Watershed Vulnerability to Climate Change," *Journal of Water Science & Technology*, Vol. 56, No. 8, 2007, pp. 49-56.

[13] N. Al-Abed and M. Al-Sharif, "Hydrological Modeling of Zarqa River Basin—Jordan Using the Hydrological Simulation Program—FORTRAN (HSPF) Model," *Water Resources Management*, Vol. 22, No. 9, 2008, pp. 1203-1220.

[14] F. Abdulla, T. Eshtawi and H. Assaf, "Assessment of the Impact of Potential Climate Change on the Water Balance of a Semi-Arid Watershed," *Water Resources Management*, Vol. 23, No. 10, 2009, pp. 2051-2068.

[15] C. C. Gordon, R. Cooper, C. A. Senior, H. Banks, J. M. Gregory, T. C. Johns, J. F. B. Mitchell and R. A. Wood, "The Simulation of SST, Sea Ice Extents and Ocean Heat Transports in a Version of the Hadley Centre Coupled Model without Flux Adjustments," *Climate Dynamics*, Vol. 16, No. 2-3, 2000, pp. 147-168.

[16] R. G. Jones, M. Noguer, D. C. Hassell, D. Hudson, S. S.

Wilson, G. J. Jenkins and J. F. B. Mitchell, "Generating High Resolution Climate Change Scenarios Using PRECIS," Met Office Hadley Centre, Exeter, 2004.

[17] K. Johnston, J. M. VerHoef, K. Krivoruchko and N. Lucas, "Using ArcGIS Geostatistical Analyst," ESRI Press, Redlands, 2001.

[18] H. H. Crawford and R. K. Linsley, "Simulation in Hydrology: Stanford Watershed Model IV," Technical Report No. 39, Stanford University, Stanford, 1966.

[19] A. S. Donigian and H. H. Davis, "User's Manual for Agricultural Runoff Management (ARM) Model," USEPA, Athens, 1978.

[20] A. S. Donigian and N. H. Crawford, "Modelling Nonpoint Pollution from the Land Surface," Environmental Research Laboratory, Athens, 1976.

[21] A. S. Donigian and W. C. Huber, "Modeling of Nonpoint Source Water Quality in Urban and Non-Urban Areas," USEPA, Athens, 1991.

[22] A. S. Donigian, B. R. Bicknell and J. C. Imhoff, "Hydrological Simulation Program—FORTRAN (HSPF)," In: V. P. Singh, Ed., *Computer Models of Watershed Hydrology*, Water Resources Pubs, Highlands Ranch, 1995, pp. 395-442.

[23] Hydrocomp Inc., "Hydrocomp Water Quality Operations Manual," Hydrocomp, Inc., Palo Alto, 1997.

[24] B. R. Bicknell, J. C. Imhoff, J. L. Kittle, T. H. Jobes and A. S. Donigian, "Hydrological Simulation Program Fortran: User's Manual for Release 12.2. US EPA Ecosystem Research Division, Athens, GA and US," 2005.

[25] E. S. Chung, K. Park and K. S. Lee, "The Relative Impacts of Climate Change and Urbanization on the Hydrological Response of a Korean Urban Watershed," *Journal of Hydrological Processes*, Vol. 25, No. 4, 2011, pp. 544-560.

[26] B. R. Bicknell, J. C. Imhoff, J. L. Kittle, R. C. Johanson and A. S. Donigian, "Hydrological Simulation Program-Fortran User's Manual for Release 10," Environmental Research Laboratory Office of Research and Development US Environmental Protection Agency, Athens, 1993.

[27] A. S. Donigian, B. R. Bicknell, J. C. Imhoff and J. L. Kittle, "Application Guide for Hydrological Simulation Program-Fortran (HSPF)," Prepared for US EPA, EPA-600/3-84-065, Environmental Research Laboratory, Athens, GA, 1984.

[28] EPA, "BASINS Technical Note 6," Estimating Hydrology and Hydraulic Parameters for HSPF, 2001.

[29] R. K. Linsley, M. A. Kohler and J. L. H. Paulhus, "Hydrology for Engineers," McGraw-Hill, New York, 1988.

6

Global River Basin Modeling and Contaminant Transport

Rakesh Bahadur, Christopher Ziemniak, David E. Amstutz, William B. Samuels
Center for Water Science and Engineering, Science Applications International Corporation, McLean, USA

ABSTRACT

Using geographic information system techniques, elevation derived datasets such as flow accumulation, flow direction, hillsope and flow length were used to delineate river basin boundaries and networks. These datasets included both HYDRO1K (based on 1 km resolution DEM) and HydroSHEDs (based on 100 meter Shuttle Radar Topography Mission). Additional spatial data processing of global landuse and soil type data were used to derive grids representing soil depth, texture, hydraulic conductivity, water holding capacity, and curve number. These grids were input to the Geospatial Stream Flow model to calculate overland flow (both travel time and velocity). The model was applied to river basins across several continents to calculate river discharge and velocity based on the use of satellite derived rainfall estimates, numerical weather forecast fields, and geographic data sets describing the land surface. Model output was compared to historical stream gauge observations as a validation step. The stream networks with associated discharge and velocity are used as input to a riverine water contamination model.

Keywords: Hydrology; Geographic Information Systems; Surface Water

1. Introduction

Modeling the fate and transport of waterborne contaminants in rivers and watersheds requires fundamental knowledge of the hydrologic cycle. The processes are well known and hydrologic models have been developed. The limiting factor in applying these models is the underlying data. Data sources for rivers and watersheds in the United States have been integrated with models [1-5] to simulate both deliberate and accidental releases. However, for applications outside the US, little or no waterborne modeling has been done for chemical, biological or radiological constituents.

The physical processes involved in watershed analysis start with the deposition of water on the earth's surface as rain or snow. The liquid water (including snow melt) then moves over the surface forced by gravity to seek the lowest point in the terrain. As the liquid flows over the surface, some of it percolates into the soil. The fraction going into the soil depends on the land cover, soil texture and saturation, which in turn depends on the rate at which the soil dries out due to evapotranspiration.

The application of transport models is dependent on the availability of data to implement the modeling and to verify model fidelity. To apply complex models to watersheds, simplification and adaptation are necessary to address the complexity of each individual modeling domain. For a given setting, some terms in the governing equations are less important than others, allowing simplification and a more efficient implementation. However, over-simplification can result in simulation models that are far removed from the physical, chemical, or biological characteristics of water bodies.

Global river flows are an important input (boundary condition) to estuarine, coastal and oceanic models. Real-time river flow [6] is also a critical input to river models used to portray transport and dispersion of toxic contaminants released deliberately or accidentally onshore. In the absence of a network of real-time river gages, as is available in the US, alternative means are required for calculating the flow of drainage streams and rivers.

Two models (GeoSFM and ICWater) were used, respectively, to create drainage networks (with associated flows and velocities) and to perform contaminant transport based on these networks. The GeoSFM processes and datasets are described in section 2 below. The application of ICWater for downstream contaminant transport and dispersion is discussed in Section 4.

2. Hydrologic Data Processing

In this study, the first step in the process was to assemble hydrologic and terrain data sets from remotely sensed data. This includes: digital elevation model, land use, soils, catchment boundaries, stream network, precipitation, and evapotranspiration. Parameterization of the basins' hydrologic properties is accomplished through the use of three data sets describing the surface topography, land cover, and soils. In addition, literature searches are performed to gather additional hydrologic data to fill data gaps (this may yield additional local data that can be used to enhance the digital datasets and to perform model calibration). Once this process is complete, river discharge is calculated using hydrologic modeling techniques within the GeoSpatial Stream Flow Model (Geo-SFM) (**Figure 1**).

The USGS [7,8] developed GeoSFM, which makes use of terrain, vegetative cover, soil absorption characteristics, precipitation and evapotranspiration data to calculate river flow. Much of the input data for GeoSFM are derived from satellite observations. The GeoSFM has been applied successfully in a portion of Africa [9,10] using HYDRO1K [11]. In recent applications [12], HYDRO1K has been replaced by HydroSHEDS, a higher resolution stream network [13].

HydroSHEDS is derived from elevation data of the Shuttle Radar Topography Mission (SRTM) at 3 arc-second resolution. The original SRTM data have been hydrologically conditioned using a sequence of automated procedures. Existing methods of data enhancement and newly developed algorithms have been applied, including void-filling, filtering, stream burning, and up-scaling techniques. Manual corrections were made where necessary. Preliminary quality assessments indicate that the resolution of HydroSHEDS significantly exceeds that of existing global watershed and river maps [13]. HydroSHEDS vector and raster datasets include: stream networks, watershed boundaries, drainage directions, and ancillary data layers such as flow accumulations, distances, and river topology information.

The Digital Soil Map of the World [14] was derived from an original compilation at 1:5,000,000 scale. Attributes for the soil associations are used to set hydraulic parameters that govern interflow, soil moisture content, and deep percolation to the ground water table. Rates at which subsurface layers release water to the stream network also depend on these physical soil attributes. Global Land Use/Land Cover was provided by the USGS [15].

Daily precipitation is obtained from the National Oceanic and Atmospheric and Administration (NOAA) Climate Prediction Center Morphing technique (CMORPH) [16]. CMORPH produces global precipitation analyses at very high spatial and temporal resolutions. This technique uses precipitation estimates that have been derived exclusively from low Earth orbit satellite microwave observations, and whose features are transported via spatial propagation information that is obtained entirely from geostationary satellite IR data. CMORPH is not a precipitation estimation algorithm but a means by which estimates from existing microwave rainfall algorithms can be combined. Therefore, this method is extremely flexible such that any precipitation estimates from any microwave satellite source can be incorporated. CMORPH data is available in GIS format on a 1/4 × 1/4 degree grid.

Daily net precipitation and evapotranspiration (PET) data is also obtained from the USGS Global Data Assimilation System. PET is the maximum extraction rate from soil and is based on air temperature, atmospheric pressure, wind speed, relative humidity, and solar radiation (long wave, short wave, outgoing and incoming). The daily PET is calculated on a spatial basis using the Penman-Monteith equation and the formulation of Shuttleworth [17]. GeoSFM converts PET to actual daily evapotranspiration based on antecedent soil moisture conditions. PET is available on a 1 × 1 degree grid [18].

3. River Basin Modeling Results

For this project, six river basins (Danube, Dneister, Kura-Araks, Yangtze, Hwang he and Mekong) were selected as case studies (**Figure 2**). The river basin boundaries are based on two datasets: a revised version of the Major Watersheds of the World dataset-distributed through the International Water Management Institute [19] and the EROS Data Center HYDRO1K basin boundaries developed at the US Geological Survey [11]. River basin boundaries were digitally derived using ETOPO5, 5-minute gridded elevation data, and known locations of

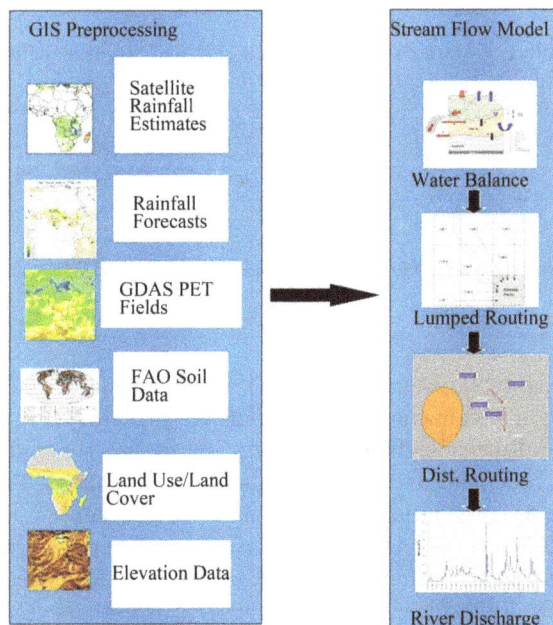

GIS Preprocessing

Satellite Rainfall Estimates

Rainfall Forecasts

GDAS PET Fields

FAO Soil Data

Land Use/Land Cover

Elevation Data

Stream Flow Model

Water Balance

Lumped Routing

Dist. Routing

River Discharge

Figure 1. GeoSFM data sets and processing steps [7].

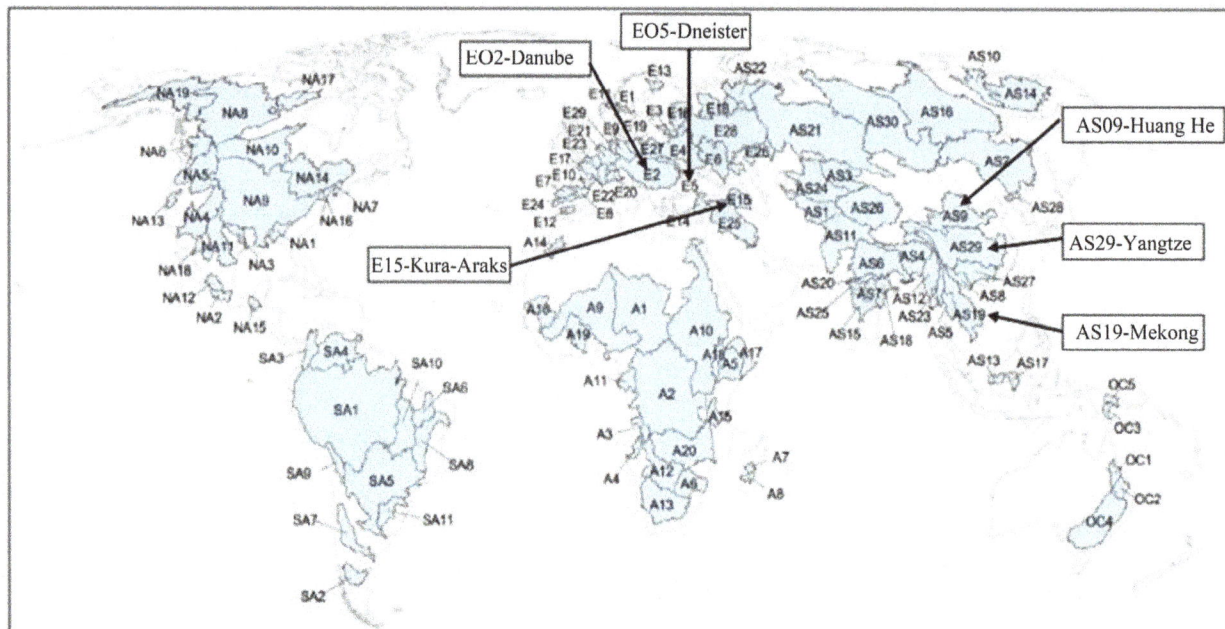

Figure 2. Major river basins of the world and the six case study basins [21].

rivers. The HYDRO1K is a geographic database derived from the USGS' 30 arc-second digital elevation model of the world, GTOPO30. The results of the stream delineation for the Dneister River Basin are shown in **Figure 3**. Observed [21] and simulated peak and annual average flows at the mouth of the Dneister are in good agreement as shown in **Table 1**. Similar results [22] are presented for the Kura-Araks basin (**Figure 4** and **Table 2**).

The basin map for the Mekong River is shown in **Figure 5**. The average flow from nine gauging stations was obtained from the Mekong River Commission [23] for use in comparing observed and simulated flows. This comparison of averaged observed and simulated flows is tabulated in **Table 3**. Although not shown in this paper, similar river delineations and flow comparisons were made for the Danube, Yangtze and Hwang He basins.

4. Riverine Contamination Modeling

The output of GeoSFM can serve as input to the Incident Command Tool for Drinking Water protection (ICWater). ICWater was developed with the RiverSpill modeling tool [24] as the hydrological engine. The RiverSpill system allows the user to track a chemical or biological agent, under real-time flow conditions, from the point of introduction to downstream water supply intakes. It determines the concentration and decay rate of an agent as it is dispersed within the water and identifies the population served by the water system that may be at risk. ICWater integrates multiple sources of information to give decision makers concise summaries of current conditions and forecasts of future consequences of contami-

Figure 3. Map of the Dneister River Basin.

Table 1. Comparison of observed [21] and simulated flows for the Dneister River Basin.

Parameter	Observed [16] (@ the mouth of the Dniester)	Simulated (2006-2010)
Peak Flow	2600 m^3/s	2175 m^3/s
Annual Average Flow	10.7 Billion m^3	10.6 Billion m^3
Total Flow	9.1 km^3	10.6 k m^3

nated public water supplies. The time-dependent distribution of contaminant concentrations, simulated by modeled dispersion, dilution and substance decay, is reported for contaminants arriving at drinking water intakes. **Figure 6** shows the current functionality in ICWater.

ICWater calculates the downstream concentration using the dispersion equation to create the downstream trace. Runoff is incorporated into the downstream calculation based on deposition from an atmospheric transport/dispersion model or user input. Runoff from atmos-

Figure 4. Map of the Kura-Araks River basin.

Table 2. Comparison of observed [22] and simulated flows for the Kura-Araks River Basin.

Station Name	River	Location	Observed Q_{av} [22]	Simulated Q_{av} (2006-2010)
Khertvisi	Kura	Khertvisi (Georgia, downstream of the border with Turkey): latitude: 41 29'; longitude: 43 17'	33	35.50
Tbilisi City	Kura	Tbilisi city (Georgia): latitude: 41 44'; longitude: 44 47'	204	148.43
Kyragkesaman	Kura	Kyragkesaman (Azerbaijan, on the border with Georgia): latitude: 41 00'; longitude: 46 10'	270	284.01
Agrichai	Alazani	Discharge characteristics at the Agrichai gauging station (Azerbaijan) latitude: 41 16'; longitude: 46 43'	110	52.54
Skhvilisi	Potskhovi	Discharge characteristics at the gauging station "Skhvilisi" in Georgia (10 km upstream of the river mouth):latitude: 41 38'; longitude: 42 56'	21.3	18.35
Red Bridge	Ktsia-Khrami	Discharge characteristics at the transboundary gauging station "Red bridge": latitude: 41 20'; longitude: 45 06'	51.7	58.91
Arenji	Arpa	Discharge characteristics of the Arpa River at the Areni gauging station (Armenia) upstream of the border with Azerbaijan	23.2	11.99
Vorotan	Vorotan	Discharge characteristics of the Vorotan River at the Vorotan gauging station (Armenia) upstream of the border with Azerbaijan	21.8	16.44
Kapan	Voghji	Discharge characteristics of the Voghji River at the Kapan gauging station (Armenia) upstream of the border with Azerbaijan	11.6	3.11
Sadaghlo	Debet	Discharge characteristics at the Sadaghlo gauging station at the Georgian-Armenian border	29.2	28.48
Airum	Debet	Discharge characteristics at the Airum gauging station (Armenia) upstream of the border with Georgia	38.1	29.26
Idshevan	Agstev	Discharge characteristics of the Agstev River at the Idshevan gauging station (Armenia) upstream of the border with Azerbaijan	9.07	37.47

Table 3. Comparison of observed [23] and simulated flows for the Mekong River Basin.

River	Station	Country	Range	Observed Q_{av}	Simulated Q_{av} 2006-2010
Mekong	Chiang Saen	TH	1961-1993	2682.72	1436.01
Mekong	Luang Prabang	LA	1960-1993	3892.07	2384.70
Mekong	Chiang Khan	TH	1968-1992	4136.41	2817.63
Mekong	Nong Khai	TH	1970-1993	4417.56	3249.48
Mekong	Nakhon Phanom	TH	1962-1993	6826.04	4811.68
Mekong	Mukdahan	TH	1925-1993	7939.28	5409.22
Mekong	Pakse	LA	1960-1993	9633.12	8092.93
Mekong	Stung Treng	KH	1960-1994	13151.65	10442.68
Mekong	Kratie	KH	1960-1969	13479.44	10808.92

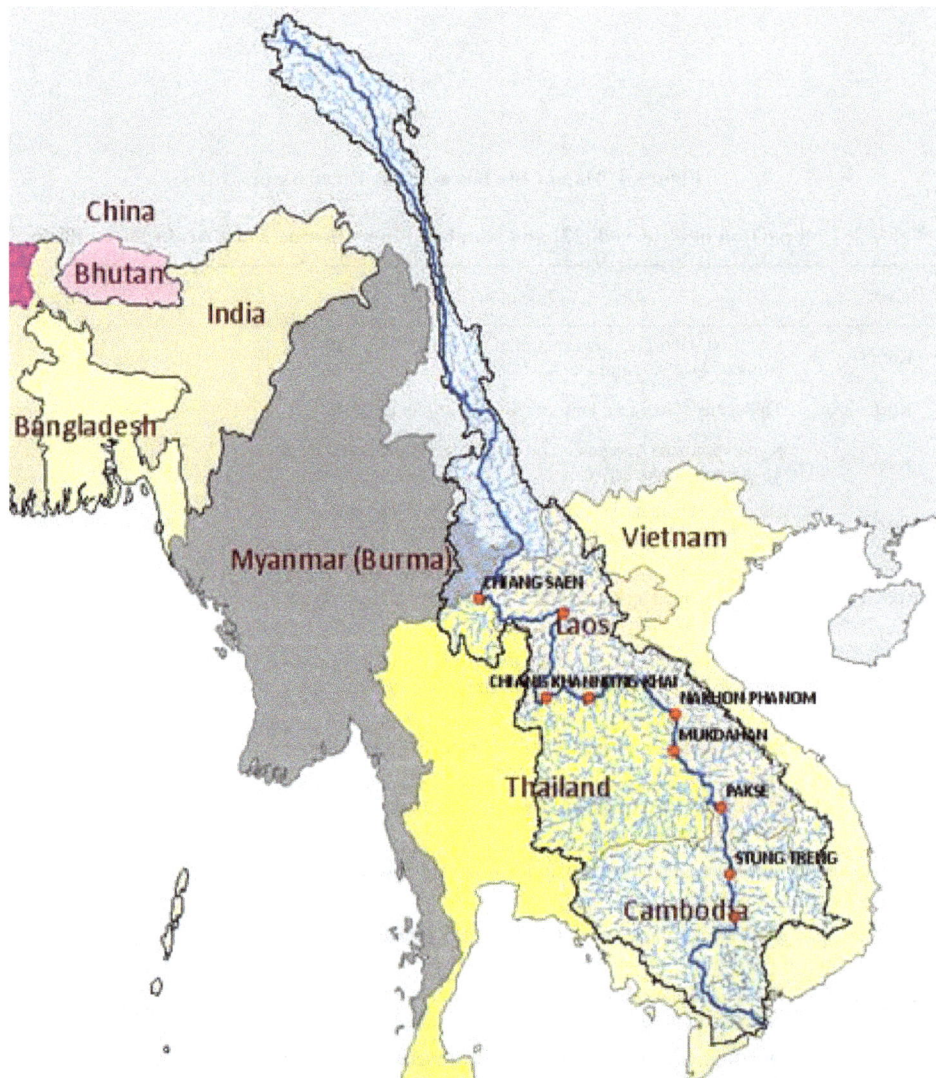

Figure 5. Map of the Mekong River Basin.

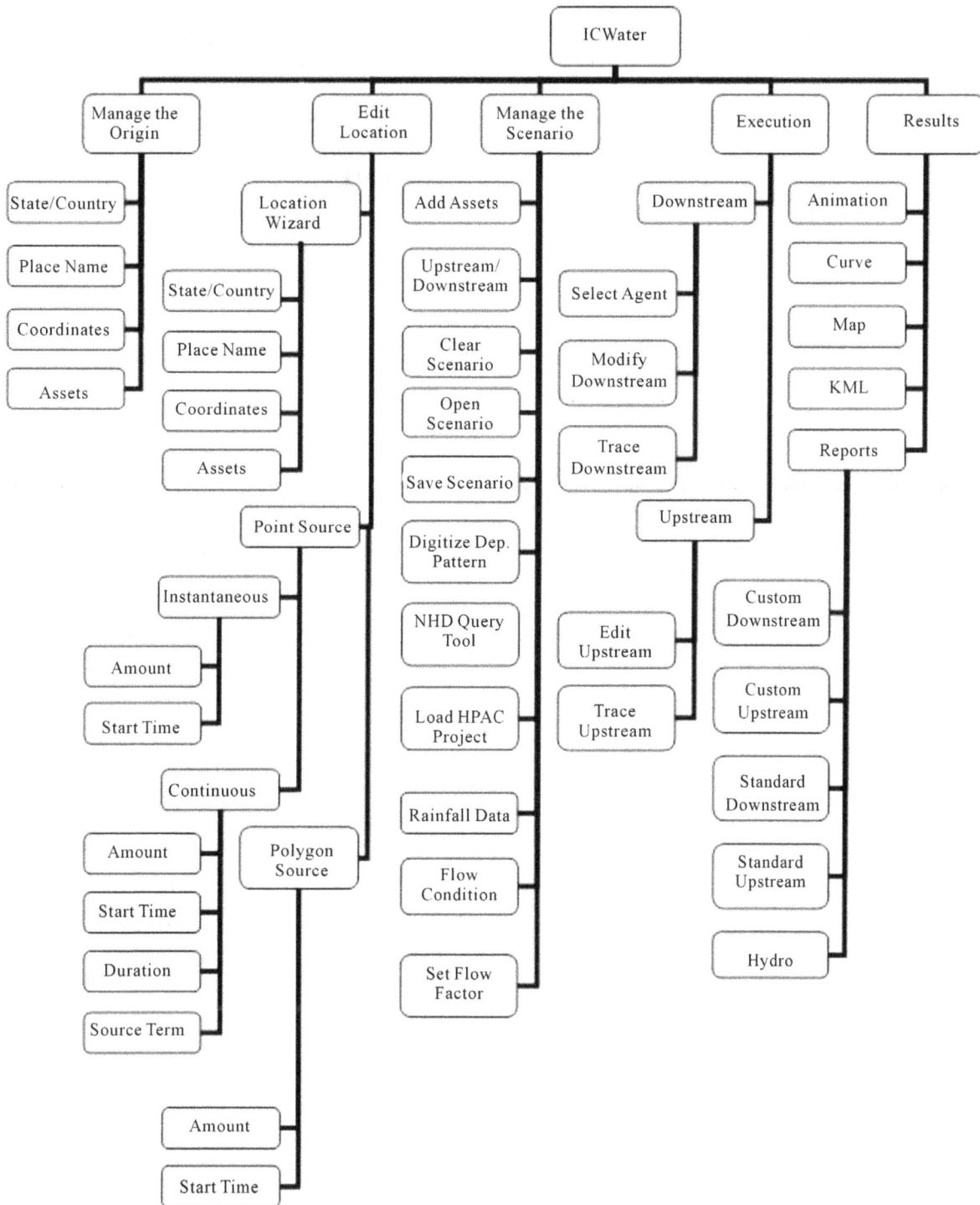

Figure 6. ICWater functionality for riverine contaminant transport modeling.

pheric deposition of contaminants is modeled as non-point source pollution. For non-point source pollution, CMORPH rainfall data are used to calculate runoff of contamination from the land surface to the receiving stream. The dispersion equation used in ICWater characterizes one-dimensional turbulent diffusion in constant density flow. The concentration is considered to be a function only of time and distance along the longitudinal axis. Reach velocities, estimated from real-time measurements reported from stream gauging stations, are applied over the uniform cross sections along a reach (defined from confluence to confluence). Substance decay is modeled as a first order exponential process.

A case study for a toxic chemical spill in the Danube River basin is shown in **Figure 7**. On January 30th 2000, the dam containing toxic waste material from the Baia Mare Aurul gold mine in North Western Romania burst and released 100,000 cubic meters of waste water, heavily contaminated with cyanide, into the Lapus and some tributaries of the river Tisza, one of the biggest in Hungary [25-27]. The cyanide concentration at the accident site was 7800 mg/l. ICWater was run with this source term and the results are shown in **Figure 8**. On February 4, the cyanide concentration at Szeged, Hungary was reported to be 1.1 mg/l. The model predicted the concentration to be 1.25 mg/l. After sixteen days, the concentration in the Danube River as measured at 0.06 mg/l (the model prediction was 0.1 mg/l).

5. Summary and Conclusions

This approach for watersheds and rivers uses the Geo-Spatial Stream Flow Model to generate river networks, catchments, flows and velocities for input to the Incident

Figure 7. Map of the Danube River Basin.

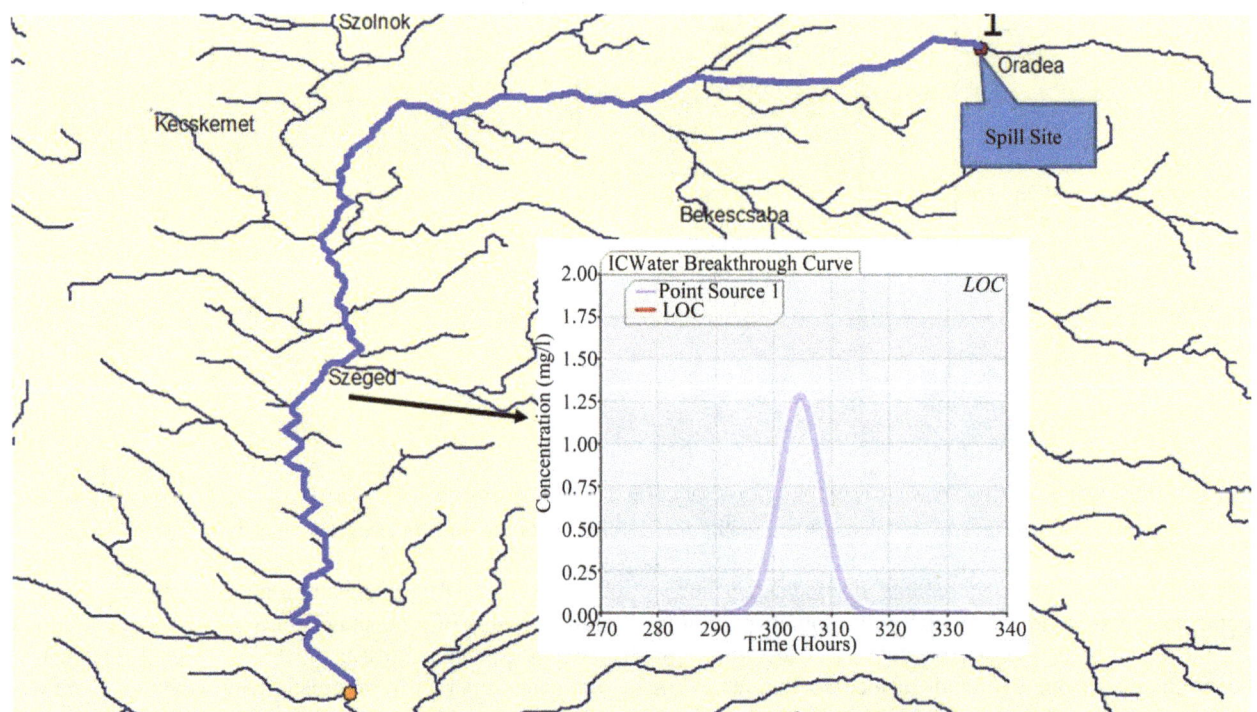

Figure 8. ICWater simulation (downstream trace) of cyanide spill in the Danube River Basin.

Command Tool for Drinking Water Protection toxic spill model. The process followed is:

- Prepare the input data for GeoSFM from a set of global databases (terrain, land use, soils, rainfall, and evapotranspiration).
- Organize and integrate the input databases so that the specification of an area of interest triggers the extraction of data to run the model.
- Automate Hydrology Data Extraction based on Polygon of Interest.
- Run GeoSFM
- Calibrate and validate the flow/velocities with observed data from the Global Runoff Data Center database or any other set of available observations.
- For the contaminant transport specify an area of interest; extract the river network, catchments, flows, velocities and assets and import to ICWater.
- Run the ICWater model to predict downstream time of travel and concentration of toxic spills.

There are requirements for quantified river flow to support the prediction and analyses of contaminant transport and dispersion worldwide. Procedures have been developed for use throughout the US which rely on the existing network of real-time gages. In many parts of the world there are few or no real-time river gages or the gage observations may not be accessible. A capability for determining global river flows has been developed by integrating HydroSHEDS with GeoSFM to calculate river discharge in regions with few or no stream gauges or other databases describing river networks and catchments. Atmospheric forcing is provided by satellite derived global forecasts of rainfall and evapotranspiration. A contaminant transport application has been validated against a chemical spill in the Danube River Basin.

The architectural framework in ICWater relies on the Environmental Systems Research Institute, Inc.'s Geographic Information System and various interface modules have been created to enable the seamless and transparent communication of the software components with the common map background and with the databases. ICWater operates as an extension to ArcGIS or as a stand-alone code using the ArcGIS Engine runtime libraries.

6. Acknowledgements

This project was supported by a contract to Science Applications International Corporation from the US Department of Defense.

REFERENCES

[1] C. R. Horn, S. A. Hanson and L. D. McKay, "History of the US EPA's River Reach File: A National Hydrographic Database Available for ARC/INFO Applications," 1994. www.epa.gov/waters/doc/historyrf.pdf

[2] C. R. Horn and W. M. Grayman, "Water Quality Modeling with the EPA Reach File System," *Journal Water Resources Planning and Management*, Vol. 119, No. 2, 1993, pp. 262-274.

[3] D. R. Maidment, "Arc Hydro GIS for Water Resources," ESRI Press, Redlands, 2002.

[4] B. DeVantier and A. Feldman,"Review of GIS Applications in Hydrologic Modeling," *Journal Water Resources Planning Management*, Vol. 119, No. 2, 1993, pp. 246-261.

[5] W. B. Samuels, and D. Ryan, "ICWater: Incident Command Tool for Protecting Drinking Water," *Proceedings ESRI International User Conference*, San Diego, 25-29 July 2005.

[6] GTN-R, "Global Terrestrial Network for River Discharge (GTN-R)," 2007. http://www.bafg.de/GRDC/EN/04_spcldtbss/44_GTNR/gtnr_node.html

[7] K. O. Asante, G. A. Artan, S. Pervez, C. Bandaragoda and J. P. Verdin, "Technical Manual for the Geospatial Stream Flow Model (GeoSFM): US Geological Survey Open-File Report 2007-1441," US Geological Survey, Reston, 2008, p. 65

[8] G. Artan, J. Verdin and K. Asante, "A Wide-Area Flood Risk Monitoring Model," *Proceedings of the Fifth International Workshop on Application of Remote Sensing in Hydrology*, Montpellier, 2-5 October 2001.

[9] J. P. Pickus, J. Johnson, M. Chehata, R. Bahadur, D. E. Amstutz and W. B. Samuels, "Development of a Global River Transport and Observation Network," *Proceedings AWRA GIS Specialty Conference*, San Mateo, 17-19 March 2008.

[10] K. O. Asante, R. D. Macuacua, G. A. Artan, R. W. Lietzow and J. P. Verdin, "Developing a Flood Monitoring System from Remotely Sensed Data for the Limpopo Basin," *IEEE Transactions on Geoscience and Remote Sensing*, Vol. 45, No. 6, 2007, pp. 1709-1714.

[11] USGS, "HYDRO1K Elevation Derivative Database," 2006. http://eros.usgs.gov/products/elevation/gtopo30/hydro/index.html

[12] W. B. Samuels and D. E. Amstutz, "Using HydroSHEDS and GeoSFM to Calculate River Discharge," *Proceedings International Perspective on Environmental and Water Resources*, Bangkok, 5-7 January 2009.

[13] WWF, "HydroSHEDS Overview," 2007. http://www.worldwildlife.org/hydrosheds

[14] FAO, "Digital Soil Map of the World and Derived Properties on CD-ROM," 1997. http://www.fao.org/AG/agl/agll/dsmw.htm

[15] USGS, "Global Land Cover Characterization (GLCC)," 2006. http://eros.usgs.gov/products/landcover/glcc.html

[16] R. J. Joyce, J. E. Janowiak, P. A. Arkin and P. Xie, "CMORPH: A Method that Produces Global Precipitation Estimates from Passive Microwave and Infrared Data at High Spatial and Temporal Resolution," *Journal Hydro-*

meteorology, Vol. 5, 2007, pp. 487-503.

[17] W. J. Shuttleworth, "Evaporation," In: D. Maidment, Ed., *Handbook of Hydrology*, McGraw-Hill, Inc. New York, 1992.

[18] USGS, "Global Potential Evapotranspiration (PET)," 2007.
http://earlywarning.usgs.gov/adds/global/web/imgbrowsc2.php?extent=glpt

[19] WMI, "Major watersheds of the world," 2003.
http://www.wri.org/publication/watersheds-of-the-world

[20] WRI, "Major watersheds of the world," 2003.
http://pdf.wri.org/watersheds_2003/gm1.pdf

[21] OSCE/UNECE, "Transboundary Diagnostic Study for the Dniester River Basin," Environment, Housing and Land Management Division, United Nations Economic Commission for Europe, November 2005.

[22] ECONOMIC COMMISSION FOR EUROPE, "Convention on the Protection and Use of Transboundary Watercourses and International Lakes. Our Waters: Joining Hands Across Borders," First Assessment of Transboundary Rivers, Lakes and Groundwaters, 2007.
http://www.unece.org/fileadmin/DAM/env/water/blanks/assessment/assessment_full.pdf

[23] Mekong River Commission, "Overview of the Hydrology of the Mekong Basin," *Mekong River Commission*, Vientiane, November 2005, p. 73.

[24] W. B. Samuels, D. E. Amstutz, R. Bahadur and J. M. Pickus, "RiverSpill: A National Application for Drinking Water Protection," *Journal of Hydraulic Engineering.* Vol. 132, No. 4, 2006, pp. 393-403.

[25] Independent of UK, "Cyanide Leak Heads towards Danube Killing Every Living Thing in Its Path," 2000.
http://www.commondreams.org/headlines/021400-02.htm

[26] BBC, "Cyanide spill reaches Danube," 2000.
http://news.bbc.co.uk/2/hi/europe/641566.stm

[27] New York Times, "Cyanide Spill Kills Danube Fish," 2000.
http://www.nytimes.com/2000/02/14/world/cyanide-spill-kills-danube-fish.html

7

Climate Change and Variability in Southeast Zimbabwe: Scenarios and Societal Opportunities

David Chikodzi, Talent Murwendo, Farai Malvern Simba
Department of Physics, Geography and Environmental Science, Great Zimbabwe University, Masvingo, Zimbabwe

ABSTRACT

A lot of researches have been done on the negative impacts and challenges caused by extreme weather conditions due to climate change and variability. Not many researches have been focused on the positive side in form of opportunities presented due to climate change. The study aimed to show the climate change scenarios and explore possible opportunities that could be derived from such scenarios in the southeastern region of Zimbabwe. The research used climate data records from three Zimbabwe Meteorological Services Department run weather stations in the region. The time series data were analyzed to show trends of rainfall and temperature over time. A questionnaire survey was also carried out to enquire from the farmers if they perceived climate change to have any opportunities. The rainfall trend analysis showed that rainfall amounts have declined at two of the three stations used. Rainfall total was also shown to be variable from year to year at all the stations. Ambient temperatures at all the stations were shown to have increased for both winter and summer. Opportunities that could be derived from climate change in the region were identified as the hydrological, agricultural and industrial. The research concludes that taking advantages of opportunities offered by climate change and variability provides the quickest way of embracing climate change adaptation.

Keywords: Opportunities; Climate Change; Climate Variability; Adaptation; Rainfall; Temperature; Southeast Zimbabwe and Masvingo Region

1. Introduction

The occurrence of climate change in Zimbabwe and all over the world is no longer debatable [1,2]. Recent reports produced by the Intergovernmental Panel on Climate Change (IPCC) conclude not only that green-house gas emissions are already beginning to change the global climate, but also that Africa will experience increased water stress, decreased yields from rain-fed agriculture, increased food insecurity and malnutrition, sea level rise, and an increase in arid and semi-arid land as a result of this process [3]. Extreme weather events, notably floods, droughts and tropical storms are also expected to increase in frequency and intensity across the continent. These projections are consistent with recent climatic trends in southern Africa, including Zimbabwe. The effects of this exposure to changes in climate are exacerbated by the high levels of sensitivity of the social and ecological systems in the region, and the limited capacity of civil society, private sector and government actors to respond appropriately to these threats [2].

Zimbabwe is particularly vulnerable to climate change due to its heavy dependence on rain-fed agriculture and climate sensitive resources [4]. Agriculture's sensitivity to climate induced water stress is likely to intensify the existing problems of declining agricultural outputs, declining economic productivity, poverty and food insecurity, with smallholder farmers particularly affected. Extreme weather events, notably drought, flood and tropical storms, are also likely to threaten development gains across a variety of sectors and intensify existing natural hazard burdens for the vulnerable populations in both rural and urban areas [5]. Consequently, climate change presents risks to lives and livelihoods at the individual level and to the economy and infrastructure at the regional and national levels and impacts on the ability to achieve sustainable development [6]. Climate change adaptation is therefore a principal development challenge in Zimbabwe.

The potential societal consequences of climate change over time in Zimbabwe are clearly game-changing. Adapting to a changing climate ought to be viewed as a business and development opportunity. While the threat posed by climate change through droughts, cyclones, heat waves and other severe weather events seem overwhelming, it presents unique opportunities for innovation, development and employment creation. To reduce the economic and social impact of climate change in Zimbabwe the focus has to be on adaptation to the changes that are necessary to thrive in changing climatic conditions. Adaptation to climate resilient smart initiatives such as reforestation, renewable energy use, recycled water for agriculture, rainwater harvesting, water use efficiency, and improved environmental planning and enforcement requires funding and implementation and should be promoted to provide opportunities to market innovations, generate employment and build climate resilient communities [4, 6].

Investment in alternative environmental practices that are resilient to climate change, will not only address the challenges presented by the climate change crisis, will but also help the economy, which is currently beset by high rates of unemployment and a growing debt to GDP ratio [6].

It is against this background that there is now a realisation that climate change opportunities to affected communities can be exploited to help adaptation [7]. Research has shifted to concentrating on the opportunities and potentialities that could be derived from extreme weather events [8]. What is now important is to focus on how people can sustain their livelihoods in the face of climate change and variability [9].

The major objective of this study is therefore to focus on the positives that can be exploited by farmers as a result of climate change. Specifically, the study analyses the climate change scenarios for south eastern Zimbabwe. Secondly the opportunities offered by these climate change scenarios to local farmers will also be explored.

1.1. Study Area

The area of study comprises the whole southeastern parts of Zimbabwe were mainly Masvingo province is located. The region is found in the south-eastern parts of Zimbabwe comprising of seven districts Bikita, Chiredzi, Chivi, Gutu, Masvingo, Mwenezi and Zaka. The region is 56,566 km^2 in area with a total population of 1,318,705 of which 616,243 are male and 702,462 are female [10-12]. The main economic activities in the region include farming of mainly maize, groundnuts, roundnuts and small grains, commercial sugar plantation agriculture at Hippo Valley and Triangle Estates, cattle ranching and animal, tourism and a bit of mining [11]. The drainage of the region is dominated by the Save, Runde, Mwenezi, Mutirikwi, Tokwe and Limpopo Rivers [12]. The region has a number of dammed rivers which pass through it as it straddles the Save-Limpopo Catchment. The countryside is dotted by kopjes, hills and mountain ranges. The area is

dominated mostly by sandy soils and the rich basaltic soils of the South Eastern Lowveld. Miombo woodlands dominate the wetter parts while Mopani trees, which are drought tolerant and sturdy, are found throughout the region. **Figure 1** shows the study area of the research.

1.2. Climate of Southeastern Zimbabwe

The region is located in the south-east low veld of the country where rainfall is minimal and uncertain. The southern parts of the region are drought prone and occur in agro-ecological region 5 of Zimbabwe's climatic regions. The area receives an average of 620 mm of rainfall per annum and average potential evapo-transpiration between 600 - 1000mm which by far exceeds the available water supply. The area has got an aridity index of between 0.2 - 0.5 and means that the region is semi-arid. **Figure 2** shows the average temperature of the region [11,13].

The average temperature of southeastern Zimbabwe is 19.4 C. The range of average monthly temperatures is 9.5 C. The warmest average max temperature is 34 C in October. The coolest average min temperature is 5 C in July. Temperatures in parts of southeastern Zimbabwe, especially Chiredzi have risen by 0.6 C since 1966 [9, 14]. Both the maximum and the minimum temperatures

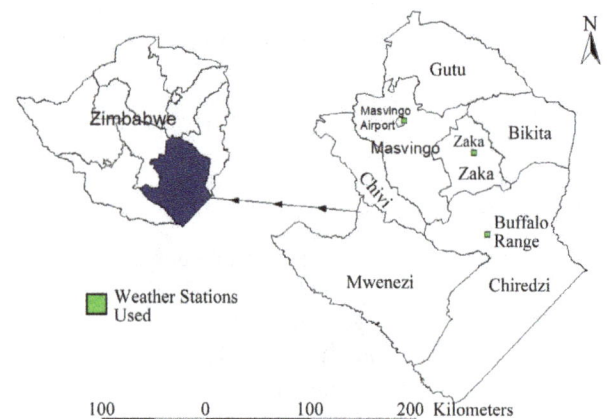

Figure 1. Southeastern Zimbabwe showing the surface water characteristics within the districts of the region.

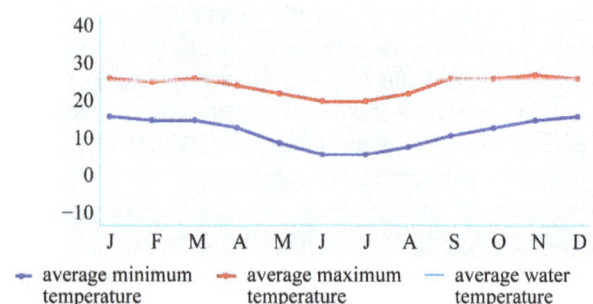

Figure 2. Average temperatures of southeastern Zimbabwe.

have increased. In some instances the temperature increase has surpassed the average global temperature increase of 0.4 C [6,15]. Tentatively while the summer temperatures have become warmer the winters have also become warmer. **Table 1** shows the relative humidity, hours of sunshine and precipitation of southeastern Zimbabwe.

Southeastern Zimbabwe receives on average 622 mm of precipitation annually or 52 mm each month. On average there are 75 days annually on which greater than 0.1 mm of precipitation occurs or 6.3 days on an average month. The month with the driest weather is July when on balance 1 mm of rain, drizzle and hail falls across 2 days. The month with the wettest weather is December when on balance 153 mm of rain, sleet and hail falls across 13 days [11].

Rainfall variability has become markedly wide in the country. The onset of the rains has been changing with the rains coming in late most of the times [2]. The dry spells are also increasing and this is having profound impact on the crop production rhythm of the area [10]. The cessation of rainfall is also unpredictable with rainfall periods extending into June and July and sometimes ending much earlier in March. The frequency and magnitude of flooding has also increased perhaps due to increasing frequency of land falling tropical cyclones within the Save-Limpopo Basin [11,14]. Even though these floods might appear to be far apart water logging conditions have tended to reduce the harvest in the area [14,15]. Despite, the seasonal variability in rainfall in southeastern Zimbabwe, it was realised that the average rainfall amount for the area has not changed significantly [15].

Mean relative humidity for an average year is recorded as 41.7% and on a monthly basis it ranges from 27% in September to 55% in February. Hours of sunshine range between 6.7 hours per day in January and 9.1 hours per day in September. On balance there are 2991 sunshine hours annually and approximately 8.2 sunlight hours for each day [11-12].

2. Research Materials and Methods

2.1. Materials Used

1) Daily climatic data for Masvingo Airport; Zaka and Buffalo Range stations.
2) Questionnaires.

3) GPS Handsets.
4) Statistical Package for Social Scientists (SPSS).
5) Arcview GIS.

2.2. Methods

This study is a result of a combination of data collection procedures and analysis. Data on the climate change scenarios in southeastern Zimbabwe was mainly obtained from daily climatic records of the Zimbabwe Meteorological Services Department (ZMSD). Daily rainfall and temperature data were obtained for three principal weather stations in southeastern Zimbabwe which are: Buffalo Range, Masvingo Airport and Zaka.

The weather stations provided weather data between 1900 to 2011. The data needed to be extended for a period of more than 30 years in order to be usable. This was meant to enable the determination of climatic variability and change in terms of rainfall and temperature. Similarly, climate change data need to cover a long period of time so that weather extremes can be determined for a number of stations. The climatic scenarios for each individual station were then derived and analysed as time series trend analysis.

A questionnaire survey was carried out in southeastern Zimbabwe to capture the views and perceptions of regions households on issues of climate change and variability. Households selected to take part in the survey were randomly chosen using the Arcview GIS random points generator points. A total of 200 points were generated and questionnaires administered. Global Positioning Systems (GPS) receivers were then used to navigate to the selected points.

Purposive sampling targeting the elderly of about 60 years and above was also used. The elderly were targeted because they would be in a position to make significantly reliable comparisons with past climate swings [16,17]. The questionnaire was targeting the household head. In the absence such persons the second influential person in the household was considered. However, in some cases informed males and females of twenty years and above in age were also considered suitable. The questionnaire dealt with the views and perceptions of households on environmental and climate change issues. Further, respondents were probed whether climate change has led to changes in livelihood systems for better or for worse.

Focus group discussions (FGD) were held to obtain

Table 1. Shows the relative humidity, hours of sunshine and precipitation of southeastern Zimbabwe.

	J	F	M	A	M	J	J	A	S	O	N	D
Relative humidity	75	87	83	76	71	69	68	56	49	50	63	74
Rainfall sum	143	136	78	23	8	3	3	3	8	28	101	160
Rainfall days	14	22	16	6	5	3	4	2	4	5	13	16

more information on the views of men and women on climate change and variability and possible adaptation measures. The first FGD was held at Chief Nemamwa's Homestead in Masvingo Rural District. The focus group consisted of eight men and eight women. The second FGD was held in Matibi 2 Area close to Gonarezhou National Park. In Matibi 2 Area the focus group consisted of the male chief, village headman, eight men and seven women. The age groups of the focus group participants were varied to capture the interests of everyone in the communities. The FDGs were used to identify the opportunities offered by climate change in southeastern Zimbabwe. In addition the exercise presented an opportunity to identify adaptation measures related to opportunities in the region.

Key informants at institutions with specialised knowledge on climate change were used in the study. These included chief researchers at Research Stations in, Agricultural Research and Extension (AREX) officers, community development officers from non-governmental organisations working on climate change issues in southeastern Zimbabwe were also interviewed. ZMSD officers also had an input into the study. Key informants helped in the provision of informed assessment of the role that could be played by exploitation of climate change opportunities and food availability in the region.

The questionnaires were then checked for completeness and other data quality related issues before being coded and entered in SPSS for further statistics exploration and analysis.

3. Results

3.1. Climate Scenarios for Southeastern Zimbabwe

The long-term climate scenarios for southeastern Zimbabwe show declining rainfall amounts and increasing temperatures. The rainfall amounts received in the region over time as compared to the national average rainfalls received over time are shown on **Figure 3**. **Figure 3** shows that there is decline in both the National and southeastern Zimbabwe rainfall totals over time, however the annual rate of rainfall decline for region is lower than the national decline figure.

Figure 3 shows that starting from the 1960s to the 2010s there were marked variations in the precipitation pattern southeastern Zimbabwe. The years with extreme above normal rainfall were 1973/74, 1977/78 and 1999/2000 and this were replicated country wide. Years with extreme below normal rainfall were numerous and dominant in 1963/64, 1967/68, 1972/73, 1982/83, 1986/87 and 1991/92. There is a noticeable increase in departure from normal between 1960 and 1990 and the 1991/92 was the season of extreme drought for southeastern Zimbabwe and the southern African region.

Figure 4(a) shows the time series of rainfall at Zaka station in southeastern Zimbabwe. Rainfall over Zaka shows a fluctuating trend about the mean from year to year, however there is a marked increase in the magnitude of years of below average rainfall since 1980. Year of above normal rainfall have also declined but occa-

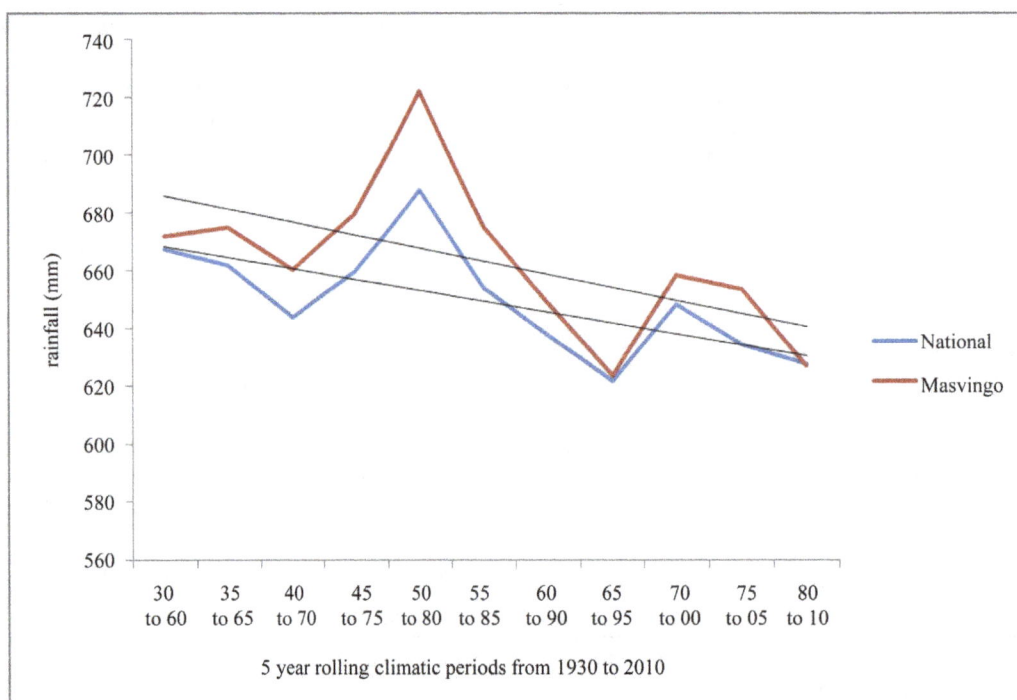

Figure 3. The national and southeastern Zimbabwe average rainfall for 30 years climatic periods from 1930-2010.

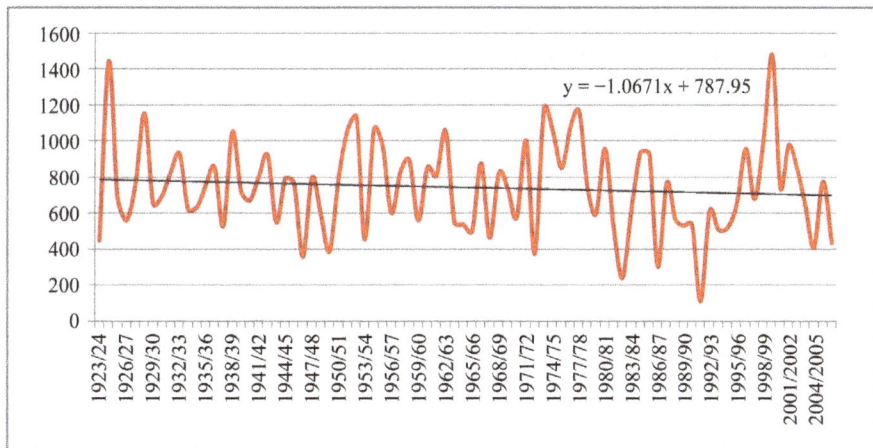

(a)

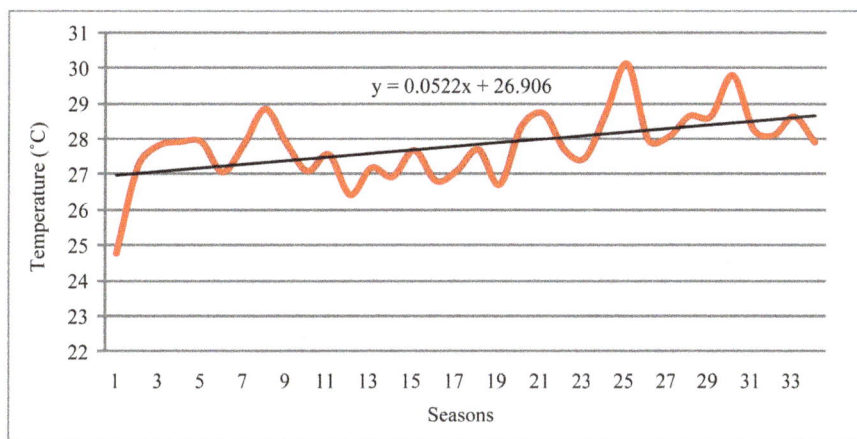

(b)

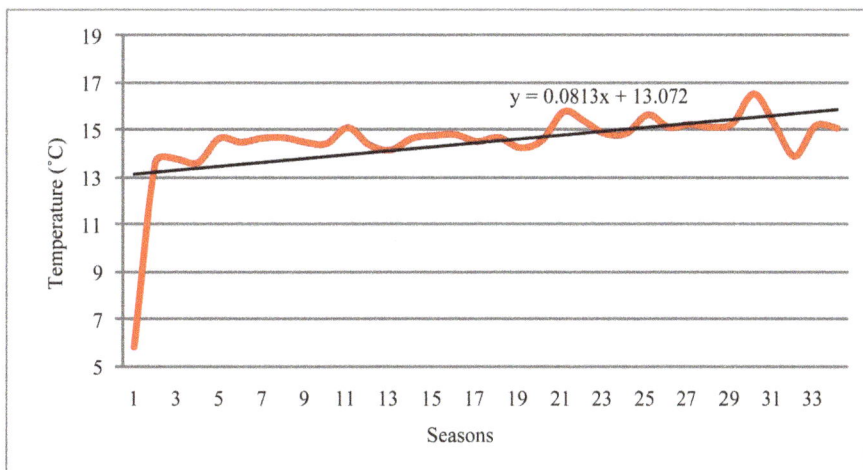

(c)

Figure 4. (a) Longterm rainfall trend scenerio for Zaka station; (b) Longterm maximum temperatures for Zaka; (c) Longterm minimum temperatures for Zaka.

sional occurrence of extreme rainfall events above 1400 mm is also shown on the rainfall trend (**Figure 4(a)**). On average rainfall shows a declining trend at Zaka station as shown by the negative trend line equation on **Figure 4(a)**.

Figures 4(b) and **(c)** show that both mean summer and mean winter temperatures have increased over the past 33 seasons at Zaka stations. It is mainly the summer temperatures that have increased by an average of almost 2 C. Winter temperatures have increased steadily by almost 1.5 C over the past 33 years.

Figure 5(a) shows the rainfall trend of Masvingo Airport station from 1950-1980. The rainfall pattern shows a fluctuating pattern from year to year but with more inci-

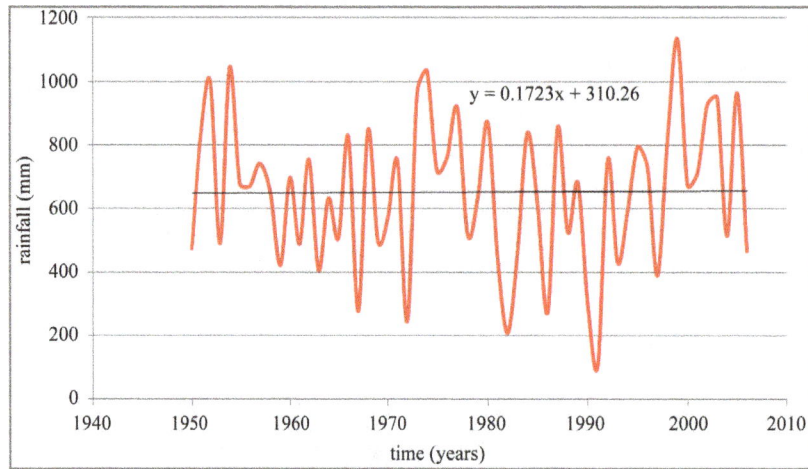

(a)

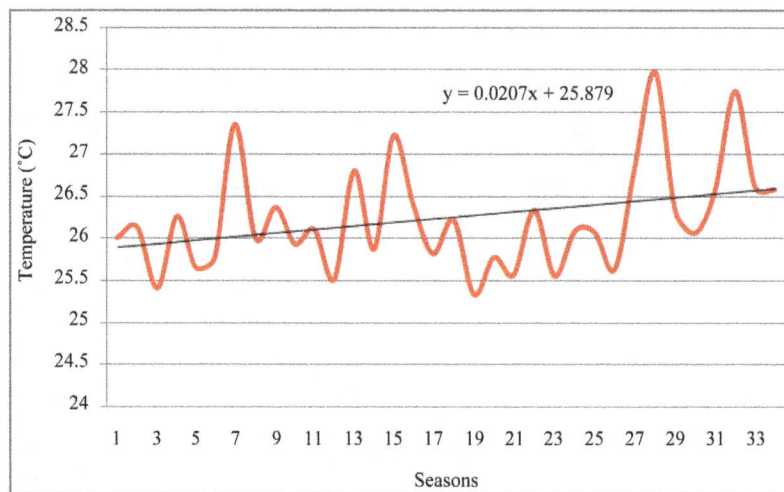

(b)

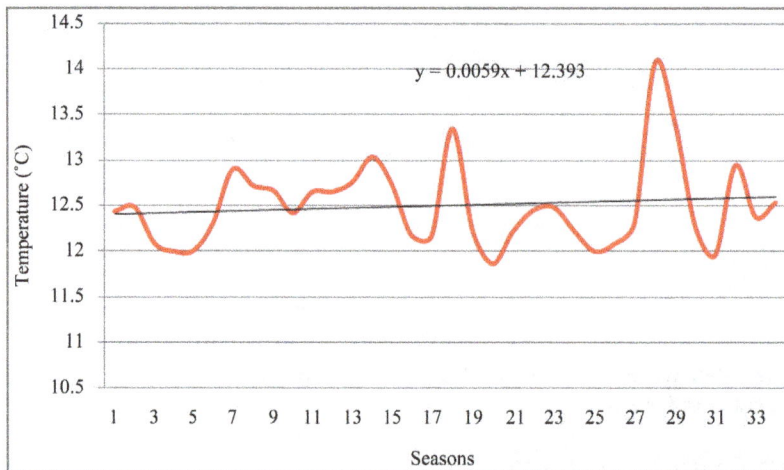

(c)

Figure 5. (a) Longterm rainfall trend scenerio for Masvingo station; (b) Longterm maximum temperatures for Masvingo; (c) Longterm minimum temperatures for Masvingo.

dents of below normal rainfall than above normal. The trend also shows that seasons of extreme sudden peak of above normal rainfall occur occasionally. On average, the rainfall pattern of Masvingo Airport station has remained more or less constant with just a marginal increase as shown by the positive trend line on **Figure 5(a)**.

Figures 5(b) and **(c)** show that both summer and winter temperatures have been on the increase especially the summer temperatures that have gone up by an average of 0.5 C over the past 33 seasons. Winter temperatures have only marginally increased by less than 0.01 C.

Figure 6(a) shows the rainfall trend for Buffalo Range

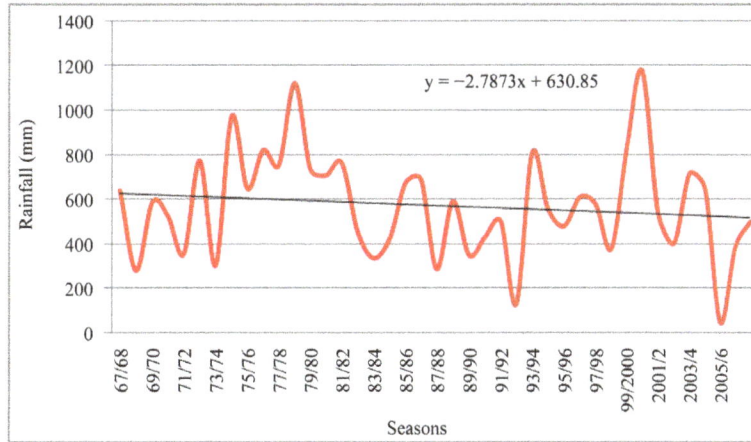

(a)

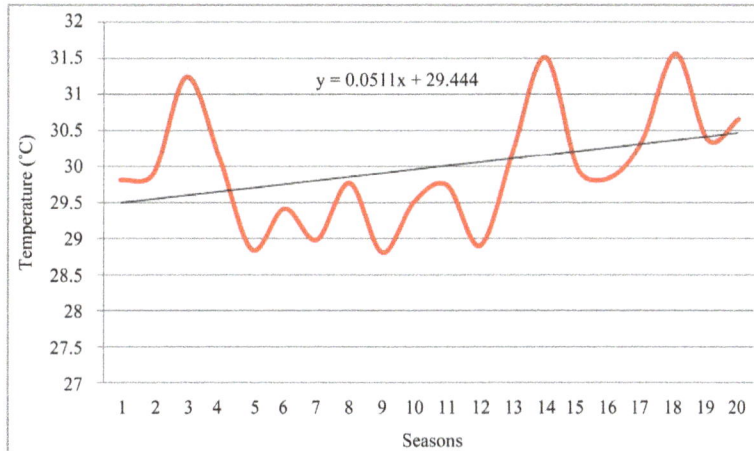

(b)

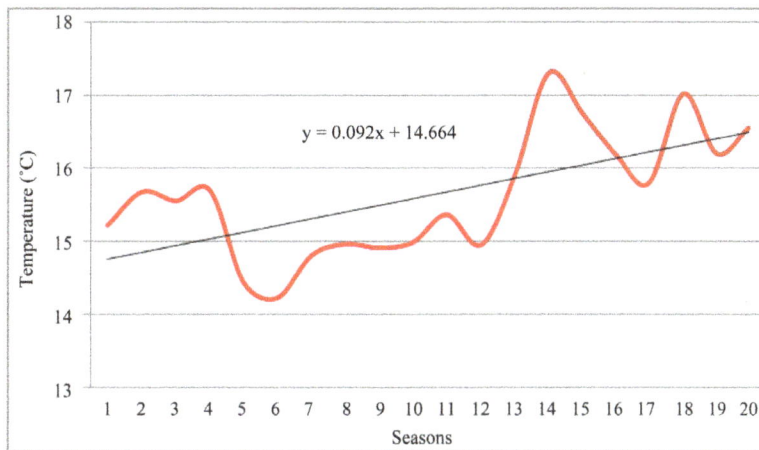

(c)

Figure 6. (a) Longterm rainfall trend scenerio for Bufalo Range station; (b) Longterm maximum temperatures Buffalo Range; (c) Longterm minimum temperatures for Buffalo Range.

Airport in Chiredzi District. The trend shows declining rainfall totals for the station over time as shown by the negetive trend line and a tendency to have occassional high peak rainfall events. The station trend aslo shows that the years of below normal rainfall are now frequently occuring more than those of above normal.

Figures 6(b) and **(c)** show the long-term trends of minimum and maximum temperature for Buffalo Range station. The figure shows that both summer and winter temperatures are on the increase. Maximun temperatures have gone up by an average of 1 C from an average of about 29.5 C to an average of 30.5 C. **Figure 6(b)** also shows that winter temperatures have incresed by about 1.5 C from an average close to 15 C to about 16.5 C.

Table 2 shows aspects of seasonal rainfall change and variability as perceived by farmers in southeastern Zimbabwe.

The major area of concern for southeastern Zimbabwe is the occurrence of drought at seasonal level. About 32% of the respondents were worried about the end of season drought which would normally wipe away their produce when they would be expectant of a better yield at the same time the season would have looked promising. About 26% of the responses were worried about the mid season drought which would normally occur at the middle of the growing season. Twenty (20%) of the respondents singled out beginning of season drought as having devastating effects on their yields as some crops and farming activities have strict schedules. About 16% of the respondents singled out terminal drought as a threat to their activities and also affects the hydrological set up of an area. Nearly 6% of the respondents singled out a whole range of other seasonal rainfall anomalies within the rainy season especially inadequate rainfall during rainy periods.

3.2. Climate Change Opportunities

Table 3 shows the climate change opportunities that can be derived from climate change as perceived by the local farmers.

Table 2. Aspects of seasonal variations used in the study.

Aspect	Number of respondents	Percentage of respondents
End of season drought	64	32
Mid season drought	51	25.5
Beginning of season drought	40	20
Terminal drought	33	16.5
Other	12	6
Total	200	100

Table 3. Opportunities presented by climate change.

Activity and opportunity	Number of respondents	Percentage of respondents
Irrigation and related activities	58	29
Short/medium term growing crops	41	20.5
Introduction of new agricultural practices	38	19
Energy production	23	11.5
Mineral exploitation	16	8
Agribusiness	13	6.5
Other	11	5.5
Total	200	100

About 29% of the respondents saw climate change potential and opportunities for southeastern Zimbabwe emanating from development of irrigation practices. Adoption of short-medium season crop varieties was singled out by 20.5% of the respondents. Because of the seasonal variability experienced in the region farmers saw it prudent to the grow small grains and short to medium term growing crops which are early maturing and drought resistant. About 19% of the respondents indicated that the opportunities can arise from the introduction of new agricultural practices. In as much as crop production is vulnerable to climate change farmers saw it prudent practice agricultural activities which are more favorable to the prevailing dry conditions.

About 11.5% indicated that the existence of large water bodies in the area generates large interest in energy production. The major inland lake of Mutirikwi has potential to generate hydro-electricity power. About 6.5% indicated that the region has potential to develop agribusiness ventures. These can use agricultural products as raw materials, sugar cane has already acted as a pointer towards this direction. About 8% of the respondents indicated that mineral exploitation in southeastern Zimbabwe can absorb people who are being pushed off their main activity of crop and livestock production. Other opportunities singled out by about 5% of the respondents include the introduction of new farming techniques that hinge other activities like agro-forestry, intercropping, indigenous tree plantations among others.

Opportunities that can be derived from the increasing temperatures as indicated by the people of southeastern Zimbabwe are shown on **Table 4**. The respondents say that if exploited well these can be used as the best alternatives that do not involve lots of capital and is adaptive to local communities in southeastern Zimbabwe.

Close to 31.5% of the respondents saw rising temperatures as increasing the potential for using organic material as a soil conditioner to improve its fertility and productiveness. Nearly 26.5% of the respondents indi-

Table 4. Opportunities derived from high temperatures.

Activity and opportunity	Number of respondents	Percentage of respondents
Use of organic manure	63	31.5
Extended cropping season	53	26.5
Crop diversification	33	16.5
Greenhouse farming	23	11.5
Energy exploitation	15	7.5
Other	13	6.5
Total	200	100

cated that they apply the spontaneous response technique of extending the cropping season. About 16.5% of the respondents indicated that crop diversification and intensification is of great potential in the face of climate change in the area. There is a possibility of maintaining the same crops as in the past as well as diversifying in terms of types of crops to be grown since adaptive crops can be cultivated.

About 11.5% of the respondents singled out greenhouse farming as a chance of creating real opportunities and investment in southeastern Zimbabwe especially in peri-urban environments. This will attract new farmers and increase land under intensive farming. Close to 7.5% of the respondents indicated that energy exploitation alternatives will also be beneficial to the populace. About 6.5% of the respondents were singling out other measures that would be used to take advantage of the vagaries of climate change.

4. Discussion

Rainfall at all stations in southeastern Zimbabwe has been declining except for Masvingo Airport while as temperatures have been on the increase. Temperature increases for both the mean annual maximum and minimum temperatures has shortened the winter period and extended the growing season for crops [2,10,14], hence temperature upswings have advantages which can be exploited for the development of the Region.

Southeastern Zimbabwe has the largest Dam capacity in Zimbabwe. This gives the region the best opportunity to harness these waters, develop low cost and water efficient irrigation systems that can practice crop farming throughout the year. Priority should therefore be allocated to water harnessing, harvesting and consumption [11,16,17]. The water from rivers should be captured, stored and used before leaving the catchment areas of the region.

During periods of floods and flooding that can occur in the region as periodic intense rainfall events, there is addition of fertility to the soil which can be used to enhance crop productivity on the croplands. Flooding is also known for exposing new minerals to the surface [18]. Flooding also increases the levels of both surface and ground water reservoirs which can be used for a number of purposes that include irrigation, consumption by humans and animals, biodiversity support and energy generation [19].

Climate change as shown in the results section provides an opportunity for diversification from dryland agriculture in the region. Diversification includes cattle and wildlife ranching, fish and crocodile farming [20-23]. Researchers have indicated that crop production in the region will diminish and wild stock production will increase [22,23]. However, crop production in southeastern Zimbabwe remains a possibility as rainfall amount hovers around 600 mm mark annually. Small grains thrive in regions of low rainfall and high temperatures. Other crops with much resilience include cassava, yams, and sweet potatoes among others can be adopted at a large scale.

The presence of large and numerous water bodies in southeastern Zimbabwe like Lake Mutirikwi, Bangala, Siya, MacDougall, Mushandike, Ruti and the giant Tokwe-Mukorsi Dam can make the region potentially self sufficient in energy production from small scale hydroelectric plants and bio-fuels. The irrigated sugar cane plantations of Hippo Valley, Mkwasine, Mwenezi and Triangle have potential to generate electricity from baggase and produce ethanol which can be used to blend fuel used by motor vehicles. All these energy sources could be used to drive the regions general development [23].

Apart from hydro-electricity, high temperatures resulting from incident solar radiation from the sun to the atmosphere and on the surface of the earth would result in increased solar energy which can be harnessed and help the region to produce affordable and sustainable clean energy [24,25]. Wind energy can also be exploited. High temperatures have an effect on increasing high wind speeds and an investment in wind energy can be worthwhile and sustainable in the near future [24].

Industrial development in the region is possible when irrigation related potential is exploited. Diversified irrigation farming will include cotton production as is at Chisumbanje and Mkwasine Estates as well as for other crops which will act as raw materials for these industries. Livestock and wilds tock can be used to create industries related to these products like tourism development, tanneries, abattoirs, meat processing among other factories. These factories will generate employment for the local community and also provide market for the products they produce.

Increasing temperatures observed at all the weather stations in the region can increase the rate of organic matter decay rates and provide a conducive environment for micro-organism activity in the breakdown of the matter. This provides an opportunity use of compost manure in conservation agriculture first at small scale, then transformed to large scale and latter mechanized to improve the practicality of the processes. In rural southeastern Zimbabwe, the use of organic manure is more practical in the sense it increases the fertility of the soil for a longer period than artificial fertilizers [10, 21].

5. Conclusions

From the study, it can be realized that not all impacts of climate change are bad. Climate change can be used in southeastern Zimbabwe as catalyst to reorient the development goals and targets of the region to those that are climate change resistant. In the process of trying to overcome the new threats, the region also creates opportunities for new development [10].

Adaptation in its many forms can be enhanced by many opportunities that are derived in climate change and variability. There is a need for capacity and intellect for full realisation of the advantages and merits that extreme weather events avail. Most communal farmers do not either have one or both, a situation that compromises their adaptive capacity. Stake holders and government intervention in tapping into the resources availed by these conditions can go a long way in changing livelihoods in the region for the better.

REFERENCES

[1]   P. Z. Yanda, "Climate change impacts, vulnerability and adaptation in Southern Africa," *Sarua Leadership Dialogue Series*, Vol. 2, No. 4, 2010, pp. 11-30.

[2]   L. Unganai, A. Murwira, B. Nherera, J. Troni and D. Mukarakate, "Historic and Future Climate Change Implications for Crop and Livestock Production in Chiredzi District," National Climate Change Symposium, Rainbow Towers, Harare, 2010.

[3]   IPCC, "Climate Change 2007: the Physical Science Basis, Contribution of Working Group I to the Fourth Assessment Report of the Intergovernmental Panel on Climate Change".

[4]   T. Chagutah, "Climate Change Vulnerability and Preparedness in Southern Africa: Zimbabwe Country Report," Heinrich Boell Stiftung, Cape Town, 2010.

[5]   C. Mutasa, "Evidence of Climate Change in Zimbabwe. Paper Presented at the Climate Change Awareness and Dialogue Workshop for Mashonaland Central and Mashonaland West Regions," Kariba, 29-30 September, 2008.

[6]   M. E. Hellmuth, A. Moorhead, M. C. Thomson and J. Williams, "Climate Risk Management in Africa: Learning from Practice," International Research Institute for Climate and Society (IRI), Columbia University, New York, 2007.

[7]   B. Smit, D. McNabb and J. Smithers, "Agricultural Adaptation to Climatic Variation," *Climate Change*, Vol. 33,

1996, pp. 7-29.

[8]   E. Sundbland, "People's Knowledge about Climate Change: Uncertainty as a Guide to Future Comments," University of Gothenburg, Gothenburg, 2008.

[9]   P. Kachere, "Zimbabwe Meteorological Services Department Blasted for Inaccurate Weather Reports," *The Sunday Mail*, 2011, p. 11.

[10] F. Simba, T. Murwendo, D. Chikodzi, B. Mapurisa, A. Munthali and L. Seyitini, "Environmental Changes and Farm Productivity: An Assessment of Masvingo Region of the Zimbabwe," *SACHA Journal of Environmental Studies*, Vol. 2, No. 1, 2012, pp. 114-129.

[11] J. Malherbe, F. A. Engelbrecht, W. A. Landman and C. J. Engelbrecht, "Tropical Systems from the Southwest Indian Ocean Making Landfall over the Limpopo River Basin, Southern Africa: A Historical Perspective," *International Journal of Climatology*, Vol. 32, No. 7, pp. 1018-1032.

[12] Central Statistics Office (CSO), "Zimbabwe Population Profile," 2002.
http://www.zimstat.co.zw/dmdocuments/Census/Census.pdf

[13] D. Chikodzi and G. Mutowo, "Agro-Ecological Zonation of Masvingo Region: Land Suitability Classification Factoring In Climate Change, Variability Swings and New Technology," 2012.

[14] J. M. Makadho, "Potential Effects of Climate on Corn Production in Zimbabwe," *Climate Research*, Vol. 16, 1996, pp. 146-151.

[15] D. Mazvimavi, "Investigating possible changes of extreme annual rainfall in Zimbabwe Hydrol," *Hydrology and Earth System Sciences*, Vol. 5, 2008, pp. 1765-1785.
www.hydrol-earth-syst-sci-discuss.net/5/1765/2008/

[16] F. T. Mugabe, M. G. Hodnet, A. Senzanje and T. Gonah, "Spatio-Temporal Rainfall and Runoff Variability of the Runde Catchment, Zimbabwe, and Implications on Surface Water Resources," *African Water Resources Journal*, Vol. 1, No. 1, 2010, pp. 66-79.

[17] United Nations Convention to Combat Desertification, "Global Dry Lands," 2012.
http://www.unccd.int/Lists/SiteDocumentLibrary/Publications/Desertification-EN.pdf

[18] USGS, "Minerals and the Environment," 2013.
http://minerals.usgs.gov/granted.html

[19] A. E. De Jonge, "Farmers' perception on adaptation to climate change: A case study of irrigators in the Riverland, South Australia," Master Thesis, Wageningen University, 2010.

[20] B. L. Dhaka, K. Chayal and M. K. Poonia, "Analysis of Farmers' Perception and Adaptation Strategies to Climate Change," *Libyan Agriculture Research Center Journal International*, Vol. 1, No. 6, 2010, 388-390.

[21] FAO/WFP, "Crop and Food Security Assessment Mission to Zimbabwe," Special Report, FAO Global Information and Early Warning System on Food and Agriculture, 2010.

[22] M. Parry, C. Rosenzweig, A. Iglesias, G. Fischer and M. Livermore, "Climate Change and World Food Security. A

New Assessment," *Global Environmental Change*, Vol. 9, Suppl. 1, 1999, pp. 51-67.

[23] FAO, "Building Adaptive Capacity to Climate Change: Policies to Sustain Livelihoods and Fisheries. New Directions in Fisheries: A Series of Policy Briefs on Development Issues," No. 08, Rome, 2007.

[24] K. Voss, J. Herkel, G. Pfafferott, C. Lohnert and A. Wagner "Energy Efficient Office Buildings with Passive Cooling. Results and Experiences from a Research and Demonstration Programme," *Solar Energy*, Vol. 81, No. 3, 2007, pp. 424-434.

[25] WEC, "New Energy Resources World Energy Resources," WEC, London, 1994.

8

Assessing the Presence or Absence of Climate Change Signatures in the Odzi Sub-Catchment of Zimbabwe

Kosamu Nyoni[1], Evans Kaseke[2], Munashe Shoko[1]

[1]Faculty of Agriculture and Natural Sciences, Great Zimbabwe University, Masvingo, Zimbabwe
[2]Shared Watercourses Support Project between Zimbabwe and South Africa, Masvingo, Zimbabwe

ABSTRACT

Climate change and potential adverse impacts on water availability for the purposes of sustaining competing demand uses are causes of concern among water resources managers. This study focused on assessing rainfall and runoff data of a micro catchment in Save's Odzi sub-catchment to determine if any trends existed and how far the results indicated climate change. The study had four rainfall stations (Rusape, Nyanga, Mukandi and Odzi Police Rail) and five runoff stations (E32, E72, E73, E127 and E129). Mann Kendall's test was applied for determining trends in the two variables. The results obtained do not point to climate change. This study recommended that issues of current land use patterns and water abstractions be thoroughly understood for the area under study. It also recommended that techniques which promote terrestrial carbon sequestration should be introduced in the micro catchment.

Keywords: Climate Change; Trend Analysis; Sustainable Development; Human Activities; Mann Kendall; Terrestrial Carbon Sequestration

1. Introduction

Climate change according to IPCC usage refers to any change in climate over time, whether due to natural variability or as a result of human activity [1,2]. IPCC [3] contended that climate change is a serious and urgent issue since the earth's climate is changing, and the scientific consensus is not only that unsustainable human activities have contributed to it significantly, but that the change is far more rapid and dangerous than thought earlier. Human existence is becoming more and more threatened. [4] argues that sustainable agricultural production will decrease as a result of climate change. [5] argue that climate change induced reductions in rainfall amount and raised temperature will lead to reduced runoff and increased water stress. This will disrupt water dependent activities including those on which livelihoods and food security are based. Climate change tends to significantly affect sustainable development. [3] point out that justified pleas have been made to address climate change from a sustainable development viewpoint.

Over the past three decades, there have been several studies addressing the issue of climate change both globally and regionally [6-19]. These studies attributed extreme changes to increased global mean temperatures (caused by increased green-house gases), population increase and changing land use, thus causing increased frequencies and severity of droughts and floods. They also dealt with effects of changes in temperature and precipitation on mean monthly, seasonal or annual runoff [20]. This was mainly done through trend analysis. In some instances, significant trends pointing to existence of climate change were reported [16,21-23]. These studies dealt with trend analysis of temperature, rainfall and runoff at regional, national or basin level.

This current study assessed the presence or absence of climate change signatures through trend analysis of rainfall and runoff data for a micro catchment of the Odzi sub catchment. Micro catchment study tends to reduce the size of area of focus. Impacts of climate change vary depending on the geographical location of the study area [3]. This study is not going to deal with trend detection in temperature as studies by [18,22], proved that variations in flow from year to year are much more strongly related to precipitation changes than to temperature changes. Studies by [24-26] also predicted that changes in rainfall have significantly greater impacts on stream flow than predicted changes in temperature.

2. Overview of Some Previous Trend Detection Studies

[11] asserted that detection of changes in long time series of hydrological data is an important scientific issue: It is necessary if we are to establish the true effect of climate change on our hydrological systems, and it is fundamental for planning of future water resources and flood protection. This will answer the question of sustainability in water resources planning and development. The process of river flow has been directly influenced by changes caused by man (e.g. land-use changes: urbanization, deforestation, changes in agricultural practices, and engineering works: drainage systems, dam construction and river regulation).

Many studies have shown a trend that runoff tends to increase with increase in precipitation and decrease with decrease in precipitation, depending on availability of data [2]. [27] analyzed time series of discharge in four rivers in Germany. After having smoothed the year-to-year oscillation of annual peak discharge, he found a marked recent increase in the amplitude of floods. He also compared floods of different recurrence intervals for two consecutive sub periods. The 100-year flood determined from the older data in the first sub-period corresponds to much lower return periods (between 5 and 30-year-flood) for the more recent data. Large flows are therefore becoming more frequent. However, no space-covering study placing these results in a truly regional perspective has been available yet.

[28] analyzed the flood trend in Austria. They considered different periods of observation (40 year interval: 1952-1991 and parts thereof). Only in a portion of cases, a significant trend was detected. The quantitative results depended on the sub-period and the characteristics studied (whether annual maxima, or number of floods per year, or partial duration series). The portion of cases for which a significant trend was detected ranged from 4.3% to 31.5%. Among those cases where a significant trend was detected; there were more examples of positive trend (64.3%) than of negative trend (35.7%). Analysis of the full 40 year period resulted in detecting a positive trend in 66.3% of the cases with significant trend. [23] investigated the impact of climate change on the temporal and spatial distribution of precipitation, temperature, evapotranspiration and surface runoff in the Volta Basin (400,000 km^2) of West Africa. Trend analysis showed clearly positive trends with high level of significance for temperature time series. Precipitation time series showed both positive and negative trends, although most significant trends were negative. In the case of river discharge,

a small number of (mostly positive) significant trends for the wet season were observed.

The study also by [16] concluded that significant changes in the river flow regimes in Southern Africa were identified in the 1970s through early 1980s. These changes led to decrease in river flows in western Zambia, Namibia and northeast South Africa considerably affecting the flows during the high flow months in which 34% - 80% of annual flow volumes are observed. [29] described the development and application of a procedure that identifies trends in the hydrological variables. The non-parametric Mann-Kendall (MK) statistic test to detect trends was applied to assess the significance of the trends in the time series. Different parts of the hydrologic cycle were studied through 15 hydrologic variables, which were analysed for a network of Upper Mazowe catchment. The distribution of the significant trends indicated that monthly flows significantly decreased with the exception of the month of September for the less than 30 years series. The field significance of trends was evaluated by the bootstrap test at the significance level of 0.05 and none of the two flow regimes expressed field significant changes.

[21] ran three transient climate change experiments which showed a significant warming in all seasons that is most pronounced in September-November. Precipitation was predicted to decrease during June-November on the continental scale while simulated changes in December-May were found to be small. The experiment ran scenarios of Pungwe basin rainfall for the 1991-2020, and 2021-2050 periods compared to the 1961-1990 period and these indicated that approximately 10% reduction of annual runoff, with no significant variability between sub-basins was expected. The results also showed that a slight decrease of rainfall for 2021-2050, compared to the 1991-2020 was also expected.

The current study intended to add to the already existing knowledge on climate change and adaptation in Zimbabwe. Such studies are important for Zimbabwe as it was reported that the worst impacts of climate change would fall on developing countries (of which Zimbabwe is one of them), in part because of their geographical location, in part because of weak coping capacities, and in part because of more vulnerable social, institutional and physical infrastructures [3].

3. Study Area

The Odzi sub catchment is shown in **Figure 1**. **Figure 2** shows the micro catchment of the Odzi sub catchment. Area description in terms of climate, hydrology, land use, population, physiography and possible impacts of climate change in study area shall be presented.

The Odzi sub catchment studied is found in the eastern mountainous areas of Zimbabwe (**Figure 1**). Odzi River

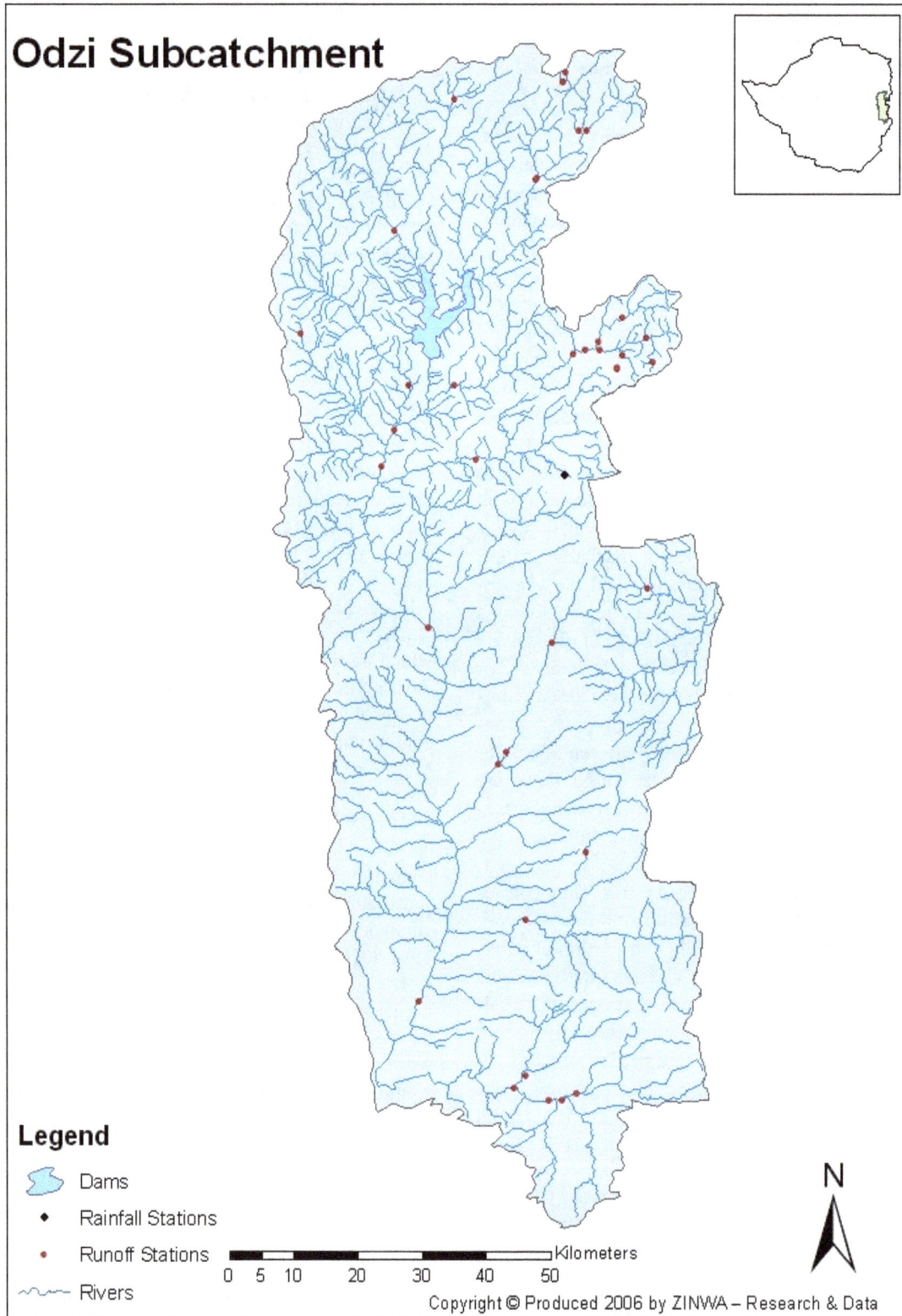

Figure 1. Map showing Odzi Sub catchment.

is the main river in this sub-catchment. It is a perennial river that rises in the Eastern Highlands and flows to the south before feeding the Save River. Total catchment area is 2486 km^2; altitude difference is large, 950 to 2160 m a.s.l [30]. However this study shall focus on the upper sections of the sub catchment, as shown in **Figure**

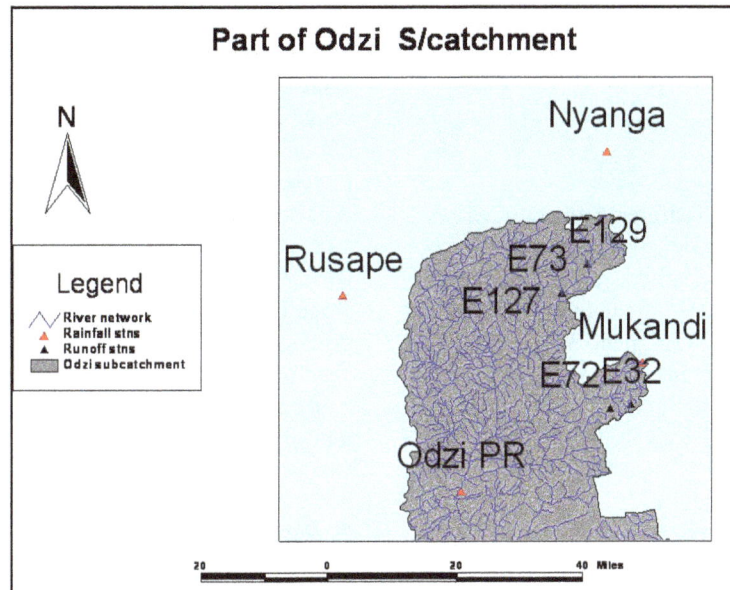

Figure 2. Part of Odzi sub catchment showing runoff and rainfall station locations.

2 hereafter termed micro catchment.

3.1. Demography and Land Use

Population density of whole catchment is above 60 persons per km^2 in large parts of the catchment; land use is a mixture of communal lands, commercial farms and forests [30]. On the commercial areas, large farms are found, which run extensive farming. The main crop in this part of Zimbabwe is tobacco during the rainy season and wheat during the dry season; and this is made possible by irrigation. Most of the farms also have cattle [31].

In the communal areas, farming is performed on a much more intensive basis. Small fields are mainly used for maize cultivation; in the drier areas; however, rapocko (a maize-relative) and sorghum (a relative to sugarcane) are cultivated [31]. Cattle and goats are the most common livestock. In the mountainous parts of the catchment, land is mainly covered with forests and in some areas companies carry on forestry. Big plantations of pines and gum trees, which are clear cut when it is time for harvesting, are common. Also in the drier areas of the communal lands, there are irrigation systems such as canals extracting water from the rivers.

The shortage of land and sometimes the unwillingness to carry water has forced people to cultivate on the riverbanks in many places in the catchment area. This is a serious problem since the soil is bare when the first rains come and it is very easily eroded, which causes land degradation and river siltation. Another problem is the removal of vegetation cover due to cattle and goat activities close to the riverbanks. However, most people seem to be aware of the problem but in times of increasing population and scarce resources it is not easy to find a

solution [31].

3.2. Physiographic

The geology of the area is dominated by granitic bedrock and *in-situ* weathered sandy soils [30]. The landscape is hilly, specially in the upper parts where the mountains reach heights up to 2000 m but also the western spur of Chimanimani mountains where some tributaries rise [31].

Granite forms the foundation of the area. The bedrock is very old; around 3.8 billion years. In fact, it is among the oldest geological formations that can be found on earth since the area is situated in the middle of the continent and therefore not recently exposed to tectonic movements. This also means that the bedrock is strongly weathered and bare rock is rarely seen except for the mountain peaks [31]. The soils within the river basin are mostly of igneous origin and rather thick. The parent material is the underlying bedrock since most soils have been formed *in situ* [32]. In some areas, red late rite clay is found. One important factor could be that compared to a clayey soil the vegetation is sparser on a sandy soil and thereby the soil is more easily eroded.

3.3. Hydrology

The climate is seasonal with a rainy season from November to March. The upper catchment area receives up to 1500 mm rainfall per year while the area west of the river is much drier and receives about 450 mm/year and average runoff is 150 to 400 mm/yr with a coefficient of variation of 174%. The upper mountainous parts receive the highest amounts [31]. The irregular seasonal availability of water has led to the establishment of irrigation systems [30]. The average potential evapo-transpiration is around 4 mm/day in the catchment area.

4. Research Methods

4.1. Rainfall Data

Monthly rainfall data for the aforementioned area were collected from the meteorological department for four stations namely, Mukandi (Latitude 18°41", Longitude 32°49"), Nyanga (Latitude 18°13", Longitude 32°44"), Odzi Police rail (Latitude 18°58", Longitude 32°24") and Rusape (Latitude 18°32", Longitude 32°08") (see **Figure 2**). Data were from July 1959 to June 2006 for all the stations.

4.2. Runoff Data

Data for five stations (E32, E72, E73, E127 and E129) were collected from ZINWA which pertain to the upper part of the Odzi Sub catchment (see **Figure 2**). Station data E32 (Latitude 18°47", Longitude 32°37") are from 1957 to 2005. Station E72 (Latitude 18°47", Longitude 32°45") data are from 1961 to 2005. Station data E73 (Latitude 18°32", Longitude 32°38") are from 1961 to 2005. Station data E127 (Latitude 18°32", Longitude 32°38") are from 1969 to 2005. Station data E129 (Latitude 18°28", Longitude 32°41") are from 1970 to 2005.

4.3. Data Analysis

The statistical methodology used in this report follows that of [11].

4.3.1. Hypothesis
For this study, to test for significant changes in the rainfall and runoff time series, the null hypotheses (H_0) are that there are no changes. The alternative hypotheses (H_1) are that there are changes (increasing or decreasing over time).

4.3.2. Rainfall and Runoff Analysis
The Mann-Kendall test was used to test for trends in the rainfall and runoff data. It is a distribution-free method which is frequently applied to detect trends. This testing approach was selected because it allows the investigation to have minimum assumptions (constancy of distribution and independence) about the data. It is possible to avoid

assumptions about the form of the distribution that the data derive from. For example, there is no need to assume data are normally distributed [33]. This was done in conjunction with tests for independence and equal distribution. To do this, Kendall's Turning point test was used to test for independence and equal distribution of the time series data. Monthly data was used.

If a series is not independent, it means there is auto correlation (serial correlation). Suggestions have been made to remove the serial correlation from the data set prior to applying a trend test. Approaches common have been to resample the data set. This is done by generating many random time series with distribution identical to that of the original time series [11]. The significance of changes was computed using standard formulae [34].

5. Results

Test for independence was also carried out on the data using the Kendall's turning point method and the data were found to be dependent and thus had auto correlation. Hence the data had to be re-sampled using block permutation or block bootstrapping after which trend analysis was carried out for both rainfall and runoff data.

5.1. Rainfall Data

The following results were obtained from the Mann Kendall's test done on the rainfall stations' data (**Table 1**).

Total annual rainfall data (mm/annum) and trend lines were also plotted against time (years) for the specified time periods provided for each station in order to augment findings. These are shown in **Figures 3-6**.

Rusape Rainfall Station

Figure 3 shows a time series graph with a trend line whose gradient is −0.18. This shows that over time, rainfall is decreasing though at a very small rate. However, the Mann Kendall test for the station (**Table 1**) showed that there was no significant trend at 95% confidence level.

Mukandi Rainfall Station

Figure 4 shows a time series graph with a trend line whose gradient is −10.8. This shows that over time, rainfall is decreasing significantly. The Mann Kendall test for the station (**Table 1**) also showed that there was a significant trend at 95% confidence level. There is therefore, a significant drop in rainfall on this station over time.

Table 1. Mann Kendall test results and interpretation for rainfall stations Rusape, Mukandi, Odzi Police rail and Nyanga.

Rainfall Station	Test Statistic (S)	P	H_0 Decision	Rainfall Trend Description
Rusape	−0.72	0.32	Retained	No Significant Trend
Mukandi	−2.56	0.0027	Rejected	Significant Trend and Decreasing
Odzi Police Rail	−0.36	0.60	Retained	No Significant Trend
Nyanga	+2.46	0.0016	Rejected	Significant Trend and Increasing

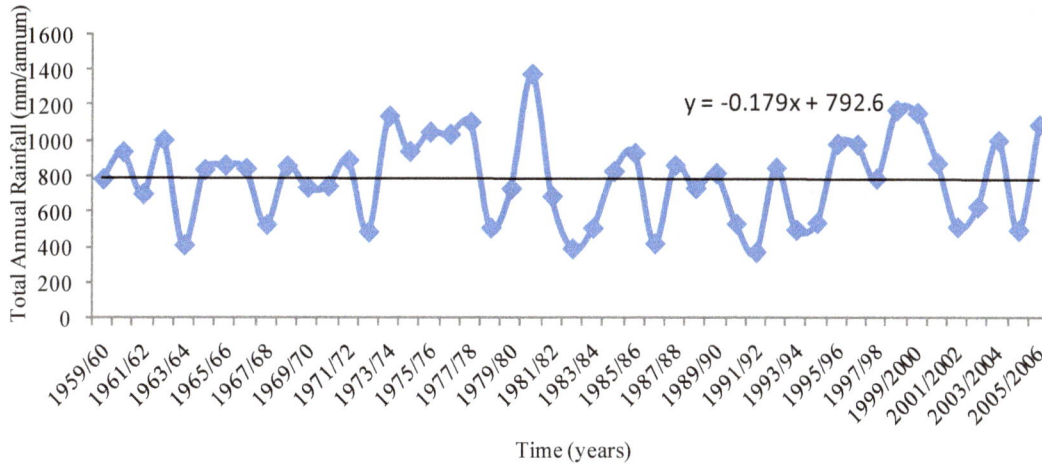

Rusape Rainfall Station graph.

Figure 3. Rusape Rainfall Station time series and trend line.

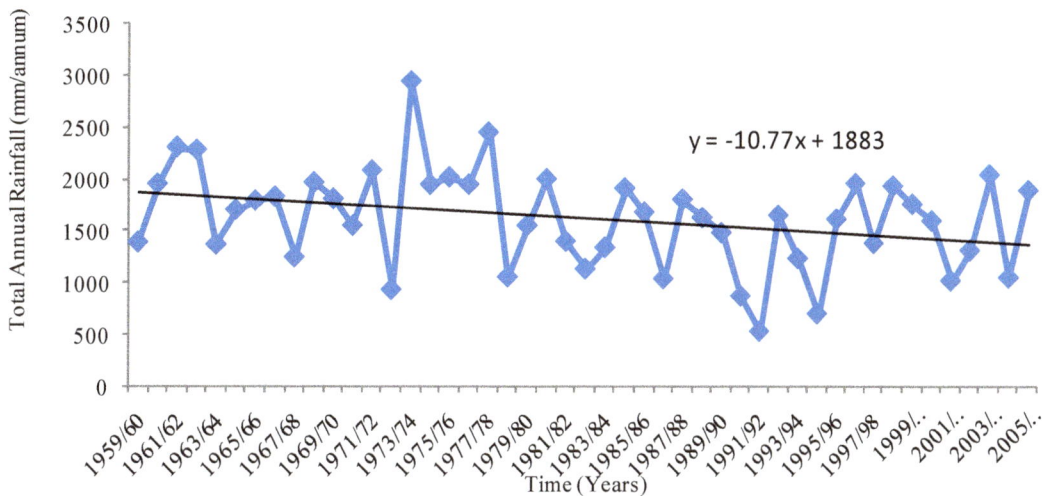

Mukandi Rainfall Station graph.

Figure 4. Mukandi Rainfall Station time series and trend line.

Odzi Police Rail Rainfall Station

Figure 5 shows a time series graph with a trend line whose gradient is +1.2. This shows that over time, rainfall is increasing but at a small rate. However, the Mann Kendall test for the station (**Table 1**) showed that there was no significant trend at 95% confidence level.

Nyanga Rainfall Station

Figure 6 shows a time series graph with a trend line whose gradient is +8.4. This shows that over time, rainfall is increasing significantly. The Mann Kendall test for the station (**Table 1**) also showed that there was a significant trend at 95% confidence level. There is therefore, a rise in rainfall over Nyanga Rainfall station for the period under consideration.

5.2. Runoff Data

The following results were obtained from the Mann Kend-

all's test done on the runoff stations' data (**Table 2**).

Total Annual runoff data (10^3*m^3/annum) and trend lines were also plotted against time (years) for the specified time period provided for each station to augment findings. These are shown in **Figures 7-11**.

Station E32

Figure 7 shows a time series graph with a trend line whose gradient is −691. This shows that over time, runoff passing through station E32 is decreasing very significantly. The Mann Kendall test for the station (**Table 2**) also showed that there was a significant trend at 95% confidence level with a negative S value. There is therefore, a significant drop in runoff passing through this station over time.

Station E72

Figure 8 shows a time series graph with a trend line whose gradient is +13.4. This shows that over time,

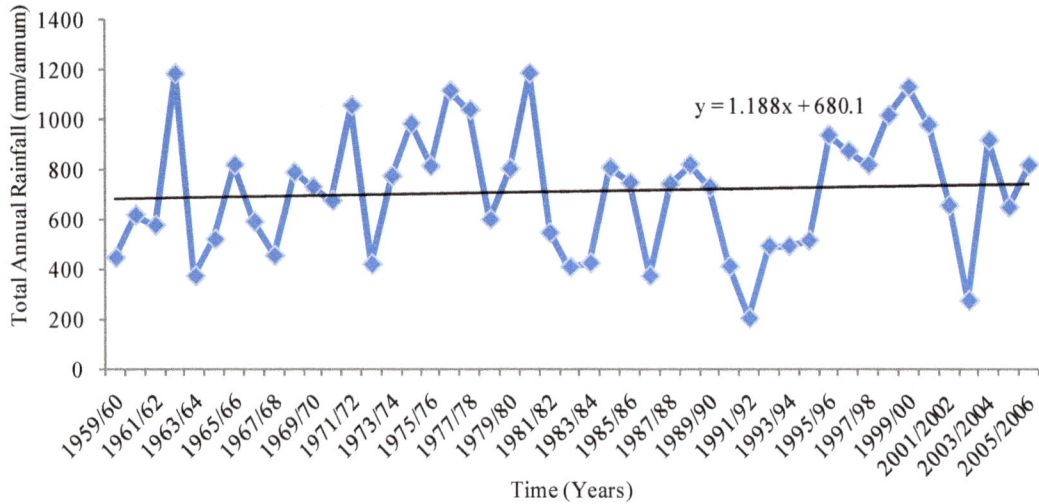

Odzi Police Rail Rainfall Station graph.

Figure 5. Station Odzi Police Rail time series and trend line.

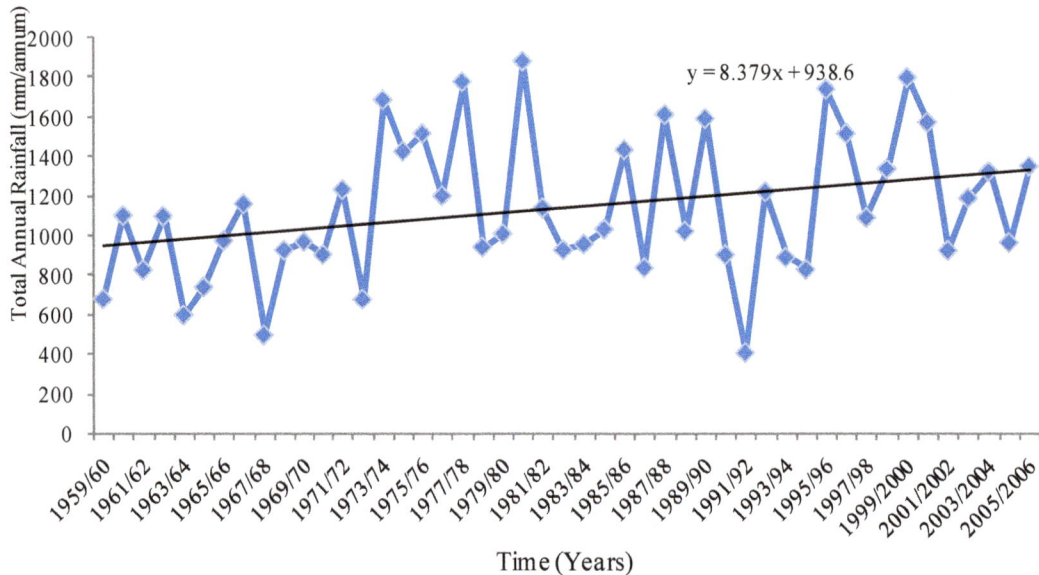

Nyanga Rainfall Station graph.

Figure 6. Nyanga Rainfall Station time series and trend line.

Table 2. Mann Kendall test results and interpretation for runoff stations E32, E72, E73, E127 and E129.

Runoff Station	Test Statistic (S)	P	H_0 Decision	Runoff Trend Description
E32	−4.68	0.04	Rejected	Significant Trend and Decreasing
E72	−1.59	0.43	Retained	No Significant Trend
E73	−3.63	0.07	Retained	No significant Trend
E127	−4.34	0.07	Retained	No Significant Trend
E129	−2.85	0.14	Retained	No Significant Trend

flows passing through station E72 are slightly increasing. The Mann Kendall test for the station (**Table 2**), however, showed that there was no significant trend at 95% confidence level. There is therefore, no significant change in runoff passing through this station over time.

Station E73

Figure 9 shows a time series graph with a trend line whose gradient is −76. This shows that over time, runoff passing through station E73 is decreasing. The Mann Kendall test for the station (**Table 2**), however, showed

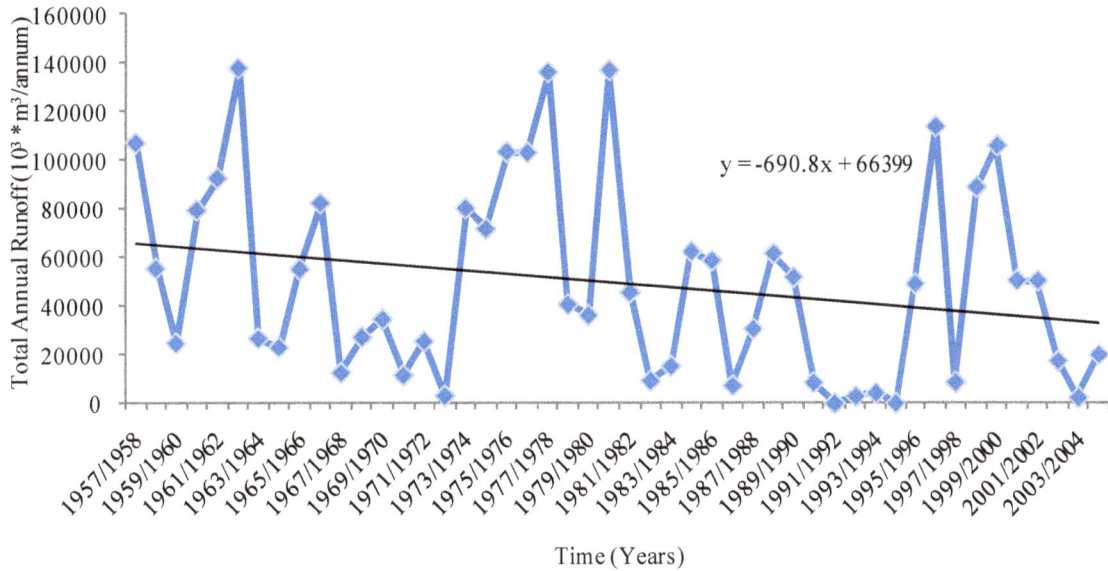

Station E32 graph.

Figure 7. Station E32 time series and trend line.

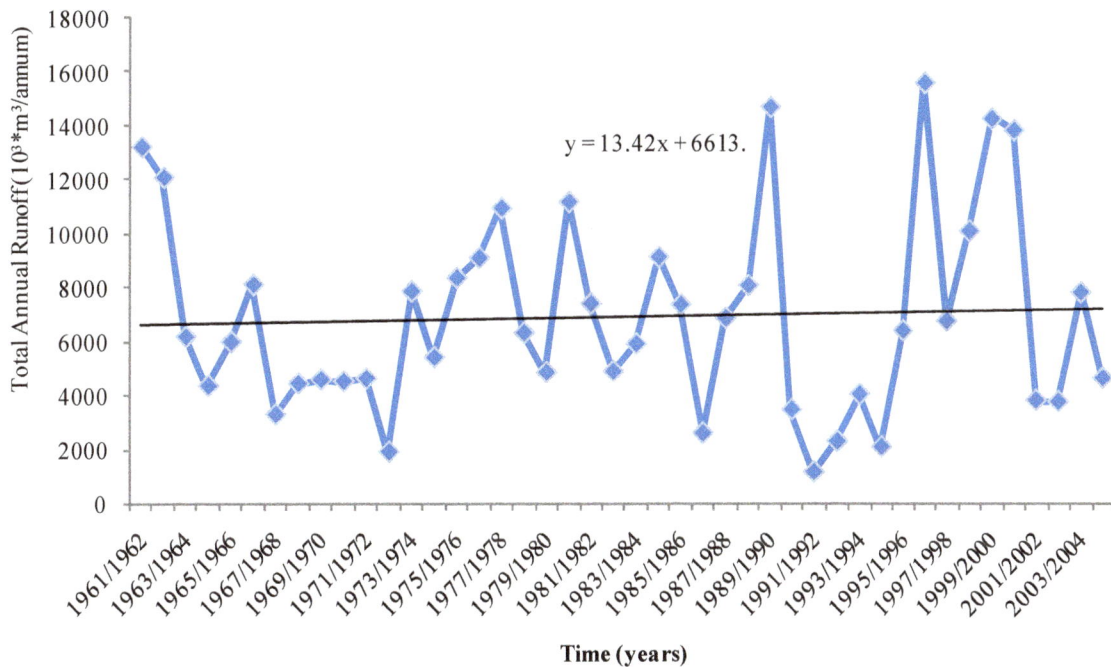

Station E72 graph.

Figure 8. Station E72 time series and trend line.

that there was no significant trend at 95% confidence level. There is therefore, no significant drop in runoff passing through this station over time.

Station E127

Figure 10 shows a time series graph with a trend line whose gradient is −156. This shows that over time, runoff passing through station E127 is decreasing. The Mann Kendall test for the station (**Table 2**), however, showed that there was no significant trend at 95% confidence

level. There is therefore, no significant drop in runoff passing through this station over time.

Station E129

Figure 11 shows a time series graph with a trend line whose gradient is −184. This shows that over time, runoff passing through station E129 is decreasing. The Mann Kendall test for the station (**Table 2**), however, showed that there was no significant trend at 95% confidence level. There is therefore, no significant drop in runoff

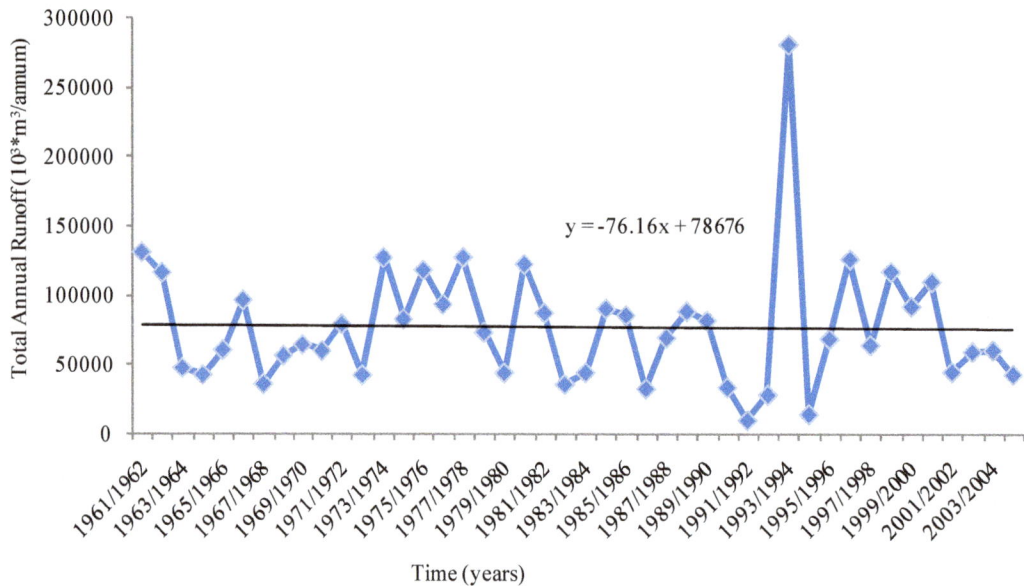

Station E73 graph.

Figure 9. Station E73 time series and trend line.

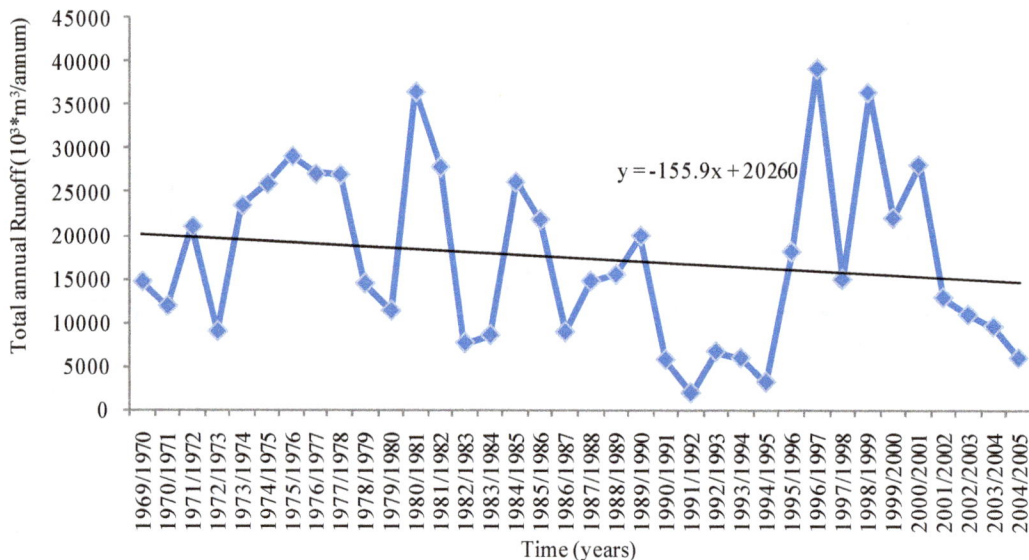

Station E127 graph.

Figure 10. Station E127 time series and trend line.

passing through this station over time.

6. Discussion of Results

6.1. Rainfall Results

Rusape and Odzi Police rail stations showed no significant change in rainfall over the 1959-2006 period. This agreed with results obtained by [21] who ran scenarios of Pungwe basin rainfall for the 1991-2020, and 2021-2050 periods compared to the 1961-1990 period. Results obtained showed that a slight decrease of rainfall for 2021-2050, compared to the 1991-2020 was also expected.

Mukandi showed a negative trend or drop in rainfall over the same period of 1959-2006. However, Nyanga station showed a rising trend in rainfall over the same period [35]. Such contrasting results are possible and this is supported by [23] whose study showed both positive and negative trends in precipitation time series, although most significant trends were negative. However, a further study is necessary to establish the main causes of these two contrasting outcomes [35].

Further analyses using the total annual rainfall data plotted over time for all the four stations (**Figures 3-6**) augmented the results obtained using the Mann Kendall

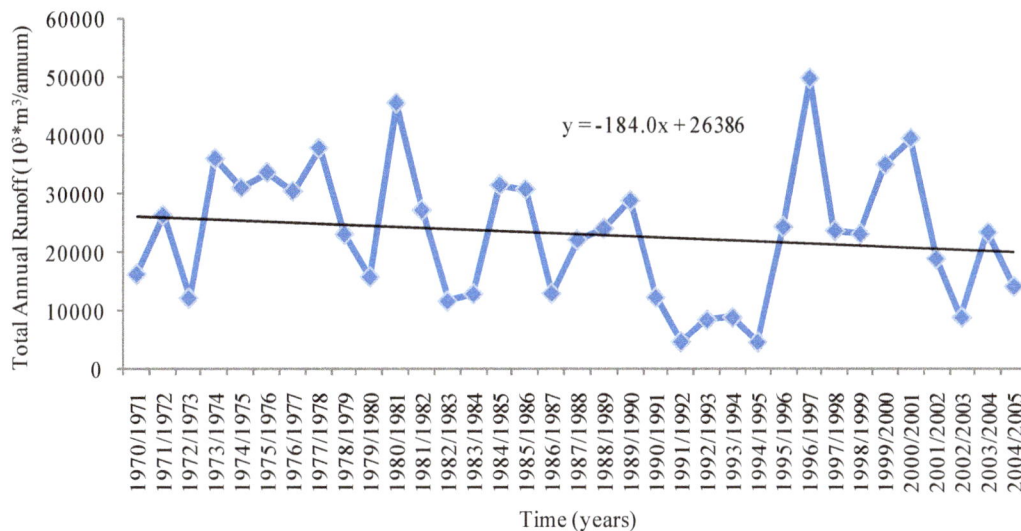

Station E129 graph.

Figure 11. Station E129 time series and trend line.

test. Slight discrepancies were, however shown on Rusape and Old Police Rail stations where the gradients of the trend lines for the time series graphs for the stations were −0.18 and +1.2 respectively, yet the Mann Kendall test showed no significant change in rainfall over the two stations. Further investigations are necessary to establish the reason(s) behind the discrepancies.

6.2. Runoff Results

Of the five stations, only station E32 showed a significant change in runoff over time [35]. This agrees with other studies done on trend detection in runoff [16,29]. The rest of the stations had no significant downward trends at 95% confidence level. Caution must be taken, however, when interpreting the physical importance (or lack thereof) of non-statistically significant trends in annual runoff. A trend in runoff may not be statistically significant, but could have important effects on water resources [20]. Further analyses using the total annual runoff data plotted over time for all the five stations (**Figures 7-11**) augmented the results obtained using the Mann Kendall test. Discrepancies were, however, shown for all but one station (E32). The gradients of the trend lines for the time series graphs for the stations (E72, E73, E127 and E129) were +13.4, −76, −156 and −184 respectively, yet the Mann Kendall test showed no significant change in runoff over the four stations. Further investigations are necessary to establish the reason(s) behind the discrepancies.

7. Conclusions and Recommendations

Rainfall results have shown that rainfall is both decreas-

ing (Mukandi station) and increasing (Nyanga station) in the micro catchment. The remaining stations showed no positive or negative trends. Going by these results, it could be concluded that the climate change signature was not very clear. Further study needs to be done to ascertain whether or not climate change exists with in the study area. Runoff analysis results showed that in princeple climate change could not be justified hence the study's Null hypothesis of no change was retained.

This study recommended that issues of current land use patterns and water abstractions be thoroughly understood within the study area. It is also known that land use activities have an impact on the quantity and quality of catchment water, therefore, sustainable land use cannot be separated from proper catchment management. Thus, it is strongly recommended that water resource management be integrated with land use management. Finally, though results did not point to existence of climate change signatures, the people dwelling in the study area (micro catchment of the Odzi sub catchment) must be taught of the importance of using their natural resources, especially water, in a sustainable manner. This will go a long way in promoting social and economic progression. Rampant cutting down of trees should be avoided and instead reforestation should be promoted. Forests and woodlands are very good carbon sinks. This move tends to reduce the amount of atmospheric carbon dioxide which has been known to promote global warming.

8. Acknowledgements

All blessings, and honour, and glory, and power be unto the Ancient of Days from this time hence forth and forever, amen and amen. I am also grateful to WATERNET for sponsoring the research which was part of my MSc studies in Integrated Water Resources Management. This paper is a portion of my full MSc thesis. I also would like to appreciate WATERNET staff for their professionalism and commitment to their work. On this note, special thanks go to Mrs. Sadazi for her tireless work and support she offered to the students. I also express my gratitude to my supervisors, Eng. E. Kaseke and Mr. E Madamombe for all their support during the period of research. Post humously, I would like to express my thanks to the late Mr. Merka for the support he also gave me at the start of this project. I also would like to express my gratitudes to Mr. Tererai (ZINWA) as well as Dr. L. Katiyo (Chinhoyi University of Technology) for their great and unspeakable support as resource persons. Many thanks also go to WATERNET-DCE lecturers also for their constructive criticisms throughout the course of this project. This goes especially to Mr. A. Mhizha and the current Coordinator Mr. Siwadi. I would like to express my special gratitudes to the Inaugural Dean of the Faculty of Agricultural Sciences, Great Zimbabwe University, Professor Munashe Shoko for his support and encouragements. Last but not least, I would like to thank my wife Dorcas and our children, Joshua, Caleb and Jephthah for their continual support.

REFERENCES

[1] IPCC, "Summary for Policymakers Report," Prepared by the Intergovernmental Panel on Climate Change Working Group I. World Meteorological Organization, United Nations Environmental Program, Geneva, 1995.

[2] IPCC, "Synthesis Report, Third Assessment, Working Group III. Summary for Policymakers," IPCC, Paris, 2001.

[3] T. Banuri and H. Opschoor, "Climate Change and Sustainable Development," Department of Economic and Social Affairs Working Paper No. 56, 2007.

[4] J. Chigwada, "Adverse Impact of Climate Change and Development Challenges: Integrating Adaptation in Policy and Development in Malawi," IIED, London, 2004.

[5] S. Eriksen and L. O. Naes, "Pro-Poor Climate Adaptation: Norwegian Development Corporation and Climate Change Adaptation. An Assessment of Issues, Strategies and Potential Entry Points," CICERO, Oslo, 2003.

[6] F. A. Bernham, "Development and Application of a GIS-Based Regional Hydrological Variability and Environmental Assessment System (RHVEAS) for Southern Africa Region," Ph.D. Thesis, University of Dar Es Salaam, Dar Es Salaam, 1999.

[7] F. A. Bernham, B. T. Zaake and R. K. Kachroo, "A Study of Variability of Annual River Flow of the Southern African Region," Hydrological Science Journal, Vol. 46, No. 4, 2001, pp. 513-524.

[8] M. Hulme and N. Sheard, "Climate Change Scenarios for Zimbabwe," Climate Research Unit, University of East Anglia, Norwich, 1999.

[9] G. Bergkamp, B. Orlando and I. Burton, "Change. Adaptation of Water Management to Climate Change," IUCN, Gland and Cambridge, 2003.

[10] B. Moyo, "Impact and Adaptation of Climate Variability on Water Supply Reservoirs Yields for the City Of Bulawayo-Mzingwane Catchment," M.Sc. Thesis, University of Zimbabwe, Mount Pleasant Harare, 2005.

[11] Z. W. Kundzewicz and A. Robson, "Change Detection in River Flow Records—Review of Methodology," Hydrological Sciences Journal, Vol. 49, No. 2, 2004, pp. 7-19.

[12] T. J. Osborn, M. Hulme, P. D. Jones and T. A. Basnet, "Observed Trends in the Daily Intensity of United Kingdom Precipitation," International Journal of Climatology, Vol. 20, No. 4, 2000, pp. 347-364.

[13] W. N. Ward, "Diagnosis and Short-Lead Time Prediction of Summer Rainfall in Tropical North Africa at Interannual and Multidecadal Timescales," Journal on Climate, Vol. 11, No. 12, 1998, pp. 3167-3191.

[14] S. E. Nicholson and I. M. Palao, "A Re-Evaluation of Rainfall Variability in the Sahel, Part I. Characteristics of Rainfall Fluctuations," International Journal of Climatology, Vol. 13, No. 4, 1993, pp. 371-389.

[15] J. Albergel, "Genese Et Predertermination Des Crues Au Burkina Faso, Du M^2 Au Km^2, Etude Des Parametres Hydrologiques Et De Leur Evolution," Etudes et Thèses, Edition de L'ORSTOM, 1988.

[16] P. Valimba, "Rainfall Variability in Southern Africa, Its Influences on Stream-Flow Variations and Its Relationships with Climatic Variations," Ph.D. Thesis, Rhodes University, Grahamstown, 2004.

[17] D. Stone, A. J. Weaver and F. W. Zwiers, "Trends in Canadian Precipitation Intensity," Atmosphere-Ocean, Vol. 38, No. 2, 2000, pp. 321-347.

[18] J. S. Risby and D. Entekhabi, "Observed Sacremento Basin Stream Flow Response to Precipitation and Temperature Changes and Its Relevance to Climate Impact Studies," Journal of Hydrology, Vol. 184, No. 3-4, 1996, pp. 209-223.

[19] K. R. Pankaj and M. Asis, "Regional Hydrological Impacts of Climate Change—Impact Assessment and Decision Making," Proceedings of Symposium S6 Held during the 7th Iahs Scientific Assembly, Foz Do Iguacu, April 2005.

[20] G. J. McCabe and D. M. Wolock, "Climate Change and the Detection of Trends in Annual Runoff," Climate Research, Vol. 8, No. 2, 1997, pp. 129-134.

[21] K. Losjö, L. Andersson and P. Samuelsson, "Report on Climate Change Impact on Water Resources in the Pungwe Dainage Basin," Final Report, Swedish Meteorological and Hydrological Institute, No: 2006-45. http://www.bgs.ac.uk/sadcreports/mozambique2006andersonclimatechangepungwebasin.pdf

[22] I. Krasovskaia, "Quantification of the Stability of River Flow Regimes," *Hydrological Sciences Journal*, Vol. 40, No. 5, 1995, pp. 587-598.

[23] H. Kunstmann and G. Jung, "Investigations of Feed-Back Mechanisms between Soil Moisture, Land Use and Precipitation in West Africa," In: S. Franks, G. Blöschl, M. Kumagai, K. Musiake and D. Rosbjerg, Eds., *Water Resources Systems-Water Availability and Global Change*, IAHS Press, Wallingford, 2003, pp. 149-159.

[24] K. D. Frederick and D. C. Major, "Climate Change and Water Resources," *Climatic Change* Vol. 37, No. 1, 1997, pp. 7-23.

[25] P. H. Gleick and E. L. Chalecki, "The Impacts of Climatic Changes for Water Resources of the Colorado And Sacramento-San Joaquin River Basins," *Journal of the American Water Resources Association*, Vol. 35, No. 6, 1999, pp. 1429-1441.

[26] D. M. Wolock and G. J. McCabe, "Estimates of Runoff Using Water-Balance and Atmospheric General Circulation Models," *Journal of the American Water Resources Association*, Vol. 35, No. 6, 1999, pp. 1341-1350.

[27] H. J. Caspary, "Increased Risk of River Flooding in Southwest Germany Caused by Changes of the Atmospheric Circulation across Europe," In A. Bronstert, Ch. Bismuth and L. Menzel, Eds., *Proceedings of the European Conference on Advances in Flood Research*, PIK Report No. 65, 2000, pp. 212-223.

[28] F. Nobilis and P. Lorenz, "Flood Trends in Austria," In: G. H. Leavesley, H. F. Lins, F. Nobilis, R. S. Parker, V. R. Schneider and F. H. M. van de Ven, Eds., *Destructive Water: Water-Caused Natural Disasters, Their Abatement and Control*, IAHS Publishing, Wallingford, 1997.

[29] W. Chingombe, J. E. Gutierrez, E. Pedzisai and E. Siziba, "A Study of Hydrological Trends and Variability of Upper Mazowe Catchment, Zimbabwe," 2013. http//www.jsd-africa.com/Jsda/Spring2005/article.htm

[30] R. Lidén, J. Harlin, M. Karlsson and M. Rahmberg, "Hydrological Modeling of Fine Sediments in the Odzi River, Zimbabwe," *Water SA*, Vol. 27, No. 3, 2001, pp. 303-314.

[31] C. Magnus and R. Magnus, "Assessment of Suspended Sediment Variability in the Odzi River, Zimbabwe," Working Group for Tropical Ecology at Uppsala University, Uppsala city, 1999.

[32] J. G. Thompson and W. D. Purves, "A Guide to the Soils of Zimbabwe," *Zimbabwe Agricultural Journal*, 1978, p. 64.

[33] WMO Report No. 1239, "Detection of Change in World—Wide Hydrological Time Series of Maximum Annual Flow," WMO, Geneva, 2004.

[34] M. Radzeijewski and Z. Kundzewicz, "Development, Use and Application of the Hydrospect Data Analysis System for the Detection of Change in Hydrological Time Series for Use in WPC-Water and National Hydrological Services-WCASP-65," WMO/TD-No 1240, Poznan, 2004.

[35] K. Nyoni, "An Assessment of Climate Change and Possible Impact on Available Water Resources on the Odzi Sub-Catchment in Zimbabwe," M.Sc. Thesis, University of Zimbabwe, Harare, 2007.

9

The Composition and Structure of Reef Community at Tho Chu Island (South China Sea) after Ketsana Typhoon

Yuri Ya. Latypov

A.V. Zhirmunsky Institute of Marine Biology, Far East Branch of Russian Academy of Sciences (FEB RAS), Vladivostok, Russia

ABSTRACT

Fringing the reefs of the island Tho Chu explored a quarter century later now reexamined in the Gulf of Thailand of the South China Sea. It was found that most of the reefs of the island were lost after the passage of typhoon "Ketsana" accompanied by heavy rains and strong sea waves higher than 2 meters in October 2009. Because of these natural phenomena, physical destruction of many of the coral communities occurred. Removal of terrigenous sediments from the islands in the water has led to increased sedimentation and loss of almost all the scleractinian family Acroporidae, the main building corals on the reefs of the Indo-Pacific, at present, in places where once there was full-reef community succession restoring the reef due to the survivors of the typhoon colonies of coral genera *Pocillopora* and *Acropora*.

Keywords: Typhoon Ketsana; Destruction; Reef Community; Change; Succession; Vietnam

1. Introduction

Coral reefs are under threat worldwide. An estimated 58% of reefs are classified as threatened [1], and 11% of the original extent of coral reefs have already been lost [2]. The composition of remaining coral reefs is also changing rapidly. For example, coral cover on reefs across the Caribbean has decreased by 80% in the past three decades [3], and some formerly abundant coral species have almost disappeared from the region [4]. The causes of coral decline are thought to include a combination of direct anthropogenic factors, such as overfishing, pollution, and sedimentation [5]. Hurricanes and tropical storms are perhaps the most obvious and frequent natural disturbances affecting reef communities. They have long been recognized as being important determinants of both the structure and function [6-8] of reef ecosystems. A number of studies have documented the severe immediate consequences of hurricane impacts at single sites in terms of reduced coral cover [9,10], highlighting the effects as being impressive in magnitude, speed, and patchiness.

Vietnamese island of the Gulf of Thailand for the most part is open all year co wind waves in all directions. Their slopes are formed by underwater boulders and boulder heaps of moving from deep stony deposits, and on to the platform of sand and silt. Openness and vulnerability of islands, shallow water of the coastal zone make them prone to intense excitement even during strong winds. In times of typhoons, such excitement leads to severe physical damage and destroying fragile coral colonies falling asleep and their bottom sediments. This peculiarity of the coastal geomorphology and hydrology of the region affects the formation of his few reefs, which are also often subject to strong typhoons. So in October 2009, the Vietnam landfall of typhoon "Ketsana", which reached gusts of wind up to 165 km/h corresponded to category 2 on a scale Saffir-Simpson Hurricane. Typhoon was accompanied by heavy rainfall (about 200 mm of rain), and the excitement of the sea more than two meters tall. After comparable in strength typhoon corals are destroyed at depths greater than 12 m—60% to 80%—between 12 m and 30 m and 100%—beyond 35 m, where as earlier living coral cover age ranged from 60 to 75% in these zones [11]. As is well known, flows of storm (flash flood) of wastewater and desalination are characterized by strong turbidity due to the large number of different particulate matter, which have a detrimental effect on the existence of coral communities, leading to their partial and sometimes total loss [12,13]. Most of the reefs of the island Tho Chu were virtually destroyed. When the coastal zone at depth of 5.3 m, 40% of its corals were destroyed; at depth of 5 - 8 m in the settlement of Acropora—100%; at depth of 8 - 12 m —60%; 5.3 m, 40% of its corals were destroyed; at depth of

5 - 8 m in the settlement of Acropora—100%; at depth of 8 - 12 m—60%; at depth of 12 m—10%. In this regard, it is appropriate to go back briefly to the information about the state of these reefs of quarter century ago.

First reefs in this region have been studied in detail in the joint Soviet-Vietnamese expeditions in 1986 and 2007, and then World Wildlife Fund (WWF) in 1992-1993 [14-17]. There are clarified species composition, population density of dominant species of macrobenthos and the degree of coverage of the substrate corals and macrophytes. Through identifying these reefs based on the data points of the coral community, it shows that the reefs are quite satisfactory and similar with those of North and South Vietnam and the Indo-Pacific [18-21].

Tho Chu Island (9°18'N, 103°28'E), together with surrounding rocks and small islets is about 11 km in diameter (**Figure 1**). The relief of the island represents a high plateau bordered by steep abrasion denudation slopes. It is a remnant of a large sub platform structure destroyed in the Pleistocene because of a tectonic immersion of the bottom in the Gulf of Siam. The island consists of coarse layered sub horizontal sandstones and conglomerates of coastal-marine genesis, which is evidenced by well-rolled gravels, shoestring distribution of pebbles, and the pattern of rhythmic and textured sediments that is characteristic of coastal shallows. The development of the ruined-rocky submarine relief is, to a significant degree, due to these geomorphologic peculiarities. The submarine slopes represent boulder-block bottoms transforming in deeper depths into stony and gravel deposits, which, in deeper depths, replaced by sandy-corallogenic deposits with great amounts of organic detritus. The northern and northeastern slopes have ratios of 0.05 - 0.08 going down to a platform at a depth of 35 - 40 m. In the south and southeast, the slope ratio is 0.032 - 0.044 and the slope transforms into the platform at a depth of 23 - 25 m. An ingressive sandy bay with a slope ratio of 0.026 and a wide zone of wave accumulation in the innermost part is set into the western coast.

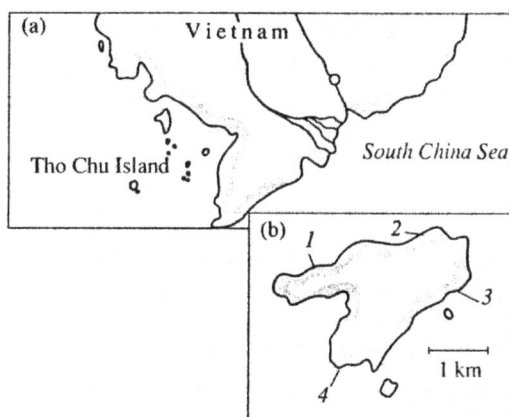

Figure 1. Schematized map showing the location of Tho Chu Island (a) and the position of transects 1-4 (b).

The re-examination reefs in 2010 and 2013 on the same cuts as in the 80 s of the last century have been found significant changes in the composition and structure of reef communities. Therefore, these results of the studied reefs in the region, in our view, are interesting, and as an independent study, they can also be used as material for comparison with the changes taking place on the reefs not only in the central and southern Vietnam, but also throughout the entire Pacific.

2. Materials and Methods

Using SCUBA equipment, we investigated the composition, distribution of scleractinian corals and mass species of macrobenthos, structure of communities in reef zones at four sections in sandy and stony inlets and near rocky coasts (**Figure 1**). Investigations were carried out in accordance with the standard hydrobiological technique using quadrats and transects [22,23]. Abundance of mass species of mollusks and echinoderms, branched, massive, encrusted and funnel-shaped scleractinian colonies, as well as the degree of substrate cover by corals were estimated along a 150 m transect frame divided into 100 squares with the area of 10 cm^2 each. Photographing of reef landscapes, and their flora and fauna was conducted. More than 750 photos by Olympus and Lumix digital cameras were made for later analysis of species composition [24, 25] and structure of community of coral reef survey methods [26-28]. Coefficients of species diversity corals were calculated by the formula:

$H = -\sum \left[(ni/N) x (\ln ni/N) \right]$, where H—Shannon Diversity Index, ni—number of individuals belonging to i species, N—total number of individuals [29].

3. Results and Discussion

In 1986, the diversity of species on different reefs varied from 335 to 387 species, of which 275 species in total were scleractinian. Ubiquitous and often dominated by 37 species: sponge *Petrozia testudinaria*; corals *Sarcophyton trocheliophorum, Sinularia dura, Lobophytum pauciflorum, Junceella fragilis, Seriatopora hystrix, Pocillopora damicornis, Acropora nobilis, A. cytherea, Pavona decussata, Pachyseris rugosa, Diaseris distorta, Polyphyllia talpina, Galaxea fascicularis, Lobophyllia hemprichii, L. hattai, Goniopora stokesi, Platygyr daedalea, Diploastrea helioporau Millepora platyphylla*; mollusks *Cyprea tigris, Beguina semiorbiculata и Malleus malleus*; echinoderms *Holothuria atra, H. edulis, Stichopus variegatus, Bohadshia graeffei, Diadema setosum, Echinotrix diadema, Culcita novaeguineae и Linckia laevigata*; polychaetes *Spirobranchus giganteus*; and algae *Turbinaria ornata, Padina australis, Laurencia papilosa, Caulerpa racemosaи Sargassum duplicatum*.

Distribution of certain dominant species or groups of

the same species in different zone of the same reefs re-vealed several macrobenthos of communities. The com-position and characteristics of formation of which are discussed below.

In the coastal zone of the studied island, except the southern sandy bay, a coral formed polyspecific commu-nity arises extending through 15 - 35 m from the shore-line to a depth of 2 m. Usually, in this area, there are separate spots of settlements with an area of up to a few tens of square meters. In terms of frequency of occurrence (75% - 100%) predominated scleractinian *P. daedalea*, *P. verrucosa*, *Acropora millepora*, *A. robusta* and hydroid *M. platyphylla*, were also distributed to individual colo-nies and settlements of soft corals *S. trocheliophorum* (biomass of 3050 g/m^2) and *L. pauciflorum* (2680 g/m^2). The largest coral cover substrate does not exceed 40%. Constant components of the coral community were algae *L. papilosa* and *T. ornata*, with a predominance of the first of them (up to 376 spec./m^2 at biomass 3516 g/m^2), clam *C. arabica*, echinoderms *D. setosum* and *N. atra*. A polychetes *S. giganteus* (up to 179 spec./m^2) is always pre-sent in the branches of the colonies hydroid *M. platyphylla*.

The coastal polyspecific coral-algal community is re-placed by a **community of *Acropora* + *Diploastrea***, which develops in the area from the reef front zone down to the lower part of reef slope, extending for 50 - 100 m along the slope, in depths of 2 to 15 m deep. No absolute domination of any *Acropora* species is observed there. This community represents extended spots of solid populations, with 100% substrate coverage by one of the following species: *A. nobilis*, *A. cytherea*, *A. rnicroph-thalma*, *A. divaricata*, or *A. florida*. The subdominants are *D. heliopora* (large massive colonies up to 2 - 3 m in diameter) and encrusting-lamellate colonies of *Euphyllia orphensis* (up to 2 m in diameter); they encountered in 70% - 100% of cases. This community is characterized by rich taxonomic diversity of scleractinian (more than 60% of the total species diversity). Each of the genera *Mon-tipora*, *Pontes*, *Fungia*, *Symphyllia*, *Favia*, *Favites*, *Pavona*, *Echinophyllia*, *Montastrea*, and *Lobophyllia* is often represented there by several species. The degree of sub-strate coverage by corals totals, as a rule, more than 60% and reaches 100% in the spots of monosettlements. In the windward reefs of the northern side of the island, in the community of *Acropora* + *Diploastrea*, on a submerged terrace, a facies of *A. nobilis* + *A. microphthalma* has developed with almost solid coverage by substrate by the corals of these two species.

A **community** of ***Junceella fragilis* + *Diaseris distorta*** develops in gravel-sandy bottoms with numerous small coral fragments, at a depth of 15 - 18 m. The basis of this community is constituted by gorgonian *J. fragilis* and single mushroom-shaped corals *D. distorta* that dominate all other macrobenthos species by population density (18

and 20 spec./m^2 respectively). Besides the dominating species, constant components of this community are other species:corals *Verrucella umbraculum*, *S. dura*, *Cyclos-eris costulata*, *Fungia fungites*, and *Polyphyllia talpina*, sponge *P. testudinaria*, echinoderms *B. graeffei*, *H. edulis* and *Toxopneustes pileolus*. Among the associated macro-benthos, the corals *Turbinaria peltata*, *S. hystrix*, *Go-niopora stokesi*, *Leptastrea pruinosa*, and *Favia speciosa*; holothurians *S. variegates* and *Halodeima edulis*, seastar *Culcita novaeguineae* and gastropods *Cyprea tigris* and *Cassis cornuta* are most often encountered.

A **community** of the **reef slope** is characterized by a great degree of substrate coverage (75% - 100%) and the greatest species diversity of macrobenthos (about 70% of the registered species diversity), which is characteristic of the reef zone in the Indo-Pacific [14,20,30-35]. It is formed in the depth of 3 - 12 m and extends for 120 - 150 m along the reef slope. The upper part of the slope, as in the reef flat, is dominated by *A. cytherea*, *A. hyacinthus*, and *A. nasuta*. Large lamellate colonies (1.2 - 2 m in di-ameter) of *M. aequituberculata* and *M. hispida* distrib-uted amongst the colonies of these three species. The area of coverage by monospecific settlements of *P. rus* and large colonies of *D. heliopora* increases up to 30% - 40% in the middle part of the reef slope. The diversity of other scleractinian species (faviids, fungiids, and encrust-ing-lamellate colonies of the genera *Euphyllia*, *Echini-pora*, *Pachyseris*, *Micedium*, and *Merulina*) increases and reaches 147 species. The dominance of any coral species is not observed in the lower part of the slope. A signifi-cant role-played in this part of the reef by funnel form and encrusting species and colonies with rather large corallites is represented by different species of the genera *Turbinaria*, *Pavona*, *Euphyllia*, *Physogyra*, *Galaxea*, *Lobo-phyllia*, and *Symphyllia*. The degree of substrate cover-age by corals falls down to 20% - 30%. In the massive colonies of the corals *Porites*, *Platygyra*, and *Astreopora*, there are settlements of mollusks *B. semiorbiculata*, *A. ventricosa*, and *P. pinguin* with the mean density of 2.8 - 4.2 spec./m^2 and the domination of *A. ventricosa* (up to 10.9 spec./m^2, with the biomass of 455.6 g/m^2). The oys-ter *L cristagalli* (0.5 spec./m^2) and the mollusk *T. squamosa* (1.5 - 2.0 spec./m^2) are continuously encountered. The sea urchins *D. setosum* and *E. diadema* and holothurians *S. variegatus* and *B. graejfei* are common in the commu-nity of the reef slope in this part of Tho Chu Island, as well as in the communities described above.

Typhoon caused significant damage of richness scler-actinian species declined by almost a third (from 275 to 95 species). There are decreased by 2 - 3 times of sub-strate coral cover as well as an index of species diver-sity (**Figures 2** and **3**). The number of related corallo-bionts: clams, gastropods, sea cucumbers and sea stars, sea urchins, with the exception of *D. setosum*, *E.*

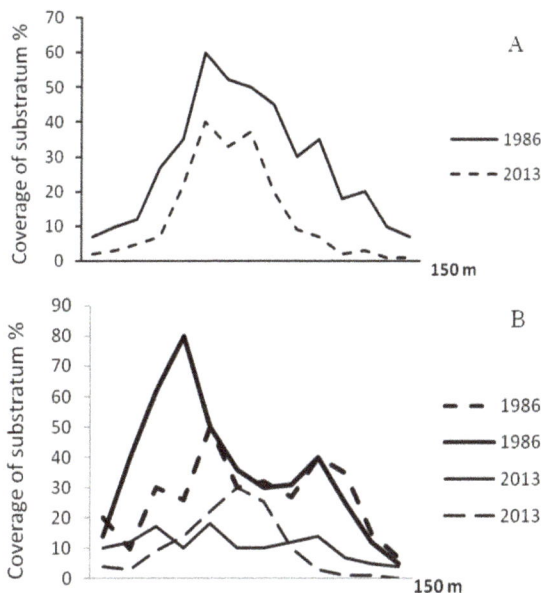

Figure 2. Coverage of the substrate corals: A. Common, B. Branched (solid line) and massive (dashed line) colonies.

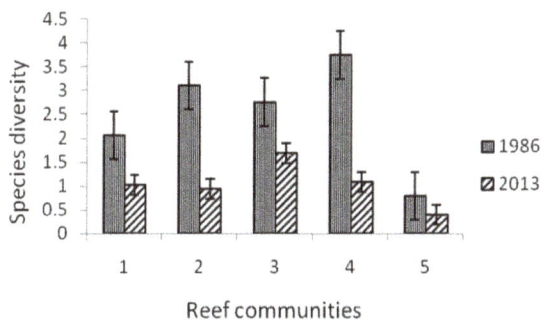

Figure 3. Variations of a specific diversity on the investigated reefs. 1. coastal polyspecific community; 2. fascia *A. nobilis* + *A. microphthalma*; 3. community *Acropora* + *Diploastrea*; 4. community of reef slope; 5. community *Junceella fragilis* + *Diaseris distorta*.

diadema and *T. pileolus* is also reduced. Although this reduction is probably not due to the influence of the typhoon, and the omnipresence of the hose diving equipment, with which listed aquatic actively fished by all Vietnamese reefs. Recovery of structure reefs zone, composition and structure of coral communities occurred mainly in the south-eastern side of the island (**transect 3**, 9°18'N., 108°29'E). This typhoon outlives specimen of corals with large corallites: *Diploastrea heliopora*, *Turbinaria peltata*, *Astreopora microphthalma*, up to 2 - 3 species from Euphyllidae (*Euphyllia*, *Plerogyra*) and Mussidae (*Lobophyllia*, *Symphyllia*), large specimen of *Fungia*, and large massive colonies of *Porites*. The abundance of poritids accounted for by their ability to secrete a firm mucous covering and starter production 1 - 2 months earlier than other coral species. These peculiarities favor their better adaptation to water overheating,

and desalination under other stressful conditions of silted shallow water [36-39]. Scleractinian and several species of *Pocillopora* come off outlive remnant animated due to its farness from shorefront to depth 12 - 15 meters. This is known that *Pocillopora* even damaged 25% preserve to two thirds of productive capacity and settling [40].

The community *Junceella fragilis* + *Diaseris distorta* of for reef platform (depth of 16 m, 125 m from the shoreline) was transformed into settlements that are only one of the gorgonian with a density of 4.5 to 12.0 spec./m² and individual colonies *Sinularia*, *Astreopora*, *Pocillopora*, *Porites*, *Favia*, *Favites* and single young *Fungia*, still attached to the substrate. Here there are uncommon sponge *P. testudinaria* and sea urchins *E. diadema* and *T. pileolus*, single starfish *C. novaeguineae*. Dominated in this community fungiids *D. distorta*, as well as the formerly common species of 4 - 5 other single mushroom scleractinian corals were not met.

At a depth of 6 - 8 meters and a distance of 50 - 40 meters from the shoreline spreads the zone of dead *Acropora* (**Figure 4**) spreads on the site of pre-existing fasces *A. nobilis* + *A. microphthalma*. Currently, there is formed polyspecific settlement of corals survived the typhoon and re-inhabiting species of scleractinian and alcyonarian. The most common here are Faviidae (*D. heliopora*, *Favia*—5 species, *Favites*—3 species), Poritidae—7 species, Fungiidae—6 species for 3 species of genus *Lobophyllia*, *Symphyllia* and *Pavona*. The most diverse are coral genus *Acropora*—12 species, but they met only unitary colonies, with the exception of *A. cytherea* and *A. gemmifera*, 4 and 3 colonies are observed in the meter area transect. Here, there are settlements of soft corals *S. trocheliophorum*, covering the surface of the substrate to a rock 40% - 70%. On clearing free of dead *Acropora* formed settlements of various *Fungia* (14 - 16 specimen/m² and 4 - 6 species) of different age and size characteristics.

Figure 4. Completely destroyed by the typhoon *Acropora* from the former fasces *A. nobilis* + *A. microphthalma*, depth 4 m.

There are also visible remains of the dead fungiids (**Figure 5**). In general, there is another stage of succession reef community—its rebirth. Ubiquitous young colonies (size 3 - 5 cm) *Acropora*, *Favia*, *Favites*, *Goniastrea*, and *Fungia* settle on a free dead *Acropora* substrate. So, newly settled *Acropora* successfully grow (**Figure 6**).

With 100 meters of the coast at a depth of 12 meters is now formed **community *Pocillopora* + *Junceella***, which is based on the survival of the typhoon and re-settling colonies of these cnidarians (**Figure 7**). This community extends over a distance of 40 - 60 meters to a depth of 8 meters with a population density of gorgonians to 25 spec./m^2 and coating the surface of the substrate of *Pocillopora* to 70%. Besides the three species of *Pocillopora* (*damicornis*, *meandrina* and *verrucosa*) and large colonies *Diploasrea heliopora* in this community are common, individual colonies include *Porites australiensis*, *Lobophyllia robusta*, *L. hemprichii*, *L. flabelliformis*, *Galaxea fascicularis*, *Euphyllia divisa*, *Fungia fungites*, *Pavona explanulata* and

Figure 5. Settlement of *Fungia*, seen dead corals, depth 5 m.

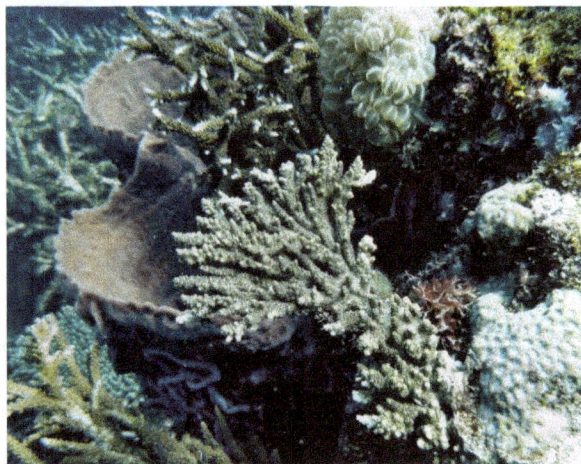

Figure 6. Scleractinian settlement on the free hard substrate in the former zone of *Acropora* + *Diploastrea*, depth 3 m.

Figure 7. Recovery settlements of *Pocillopora* (existed in 1986), coated surface of the substrate of *Pocillopora* to 70%, depth 11.5 m.

single *Acropora abrolhosensis*, *A. gemmifera*, and *Montipora grisea*. Concomitant spread of macrobenthos sponge *P. testudinaria* is up to 1.2 meters high and 0.6 meters in diameter, sea urchins *D. setosum* to 14 spec./m^2, *E. diadema* to 2 spec./m^2, *T. pileolus*—0.2 spec./m^2, as well as individual *Tridacna crocea*, *Trochus niloticus*, *Pteria penguin*, *Mauritia arabica* and *Holothuria atra*.

Rocky coastal area was the most taxonomically rich. It was most likely to be preserved corals with massive and encrusting forms colonies between large rocky boulders several feet across. This area is still formed polyspecific community scleractinian and alcyonarian consisting of 40 - 50 species, of which the most common are *Acropora millepora*, *Diploastrea heliopora*, *Goniastrea aspera*, *Echinopora lammelosa*, *Euphyllia divisa*, *Lobophyllia robusta*, *Montipora grisea*, *Pavona decussata*, *Plerogysa sinuosa*, *Symphyllia recta*, and by 2 - 3 species of *Favia* and *Porites*. There are also common colony *S. troheliophorum*, *Lobophytum pauciflorum* and *Zoanthus* ssp. near the water's edge marked by some bushes macrophytes *T. ornatum* and *Laurencia papilosa*. In the colonies of *Porites* constantly meet bivalve's *B. semiorbiculata* (from two to seven individuals per colony) and the polychetes *S. giganteus* and 12 individuals per colony. Distributed earlier hydroid *M. platyphylla* was not met in any area of the transect, or in the immediate vicinity.

The reefs on the Tho Chu now meet only a few members of the family Acroporidae, which are the basic element of all living reefs and usually form the basis of their (reef) species diversity and a high degree of coverage of live coral substrate [35,39-43]. Species diversity of corals, their coverage of the substrate and slow of their growth with increasing amounts of sediment have been noted in many publications [16,44-48]. Further, the strong deposit may hold coral larvae from settling and cause high mortality after attachment to the substrate by mechanical abrasion. All this leads to a fundamental change in the community scleractinian and possible power macrophytes and other competitors of corals [49].

In general, clearly visible irregularities and changes are observed in the zonal structure of the reef, so the composition and structure of reef communities of the island Tho Chu are changed. Nevertheless, it is equally obvious that reef communities are gradually recovering. In addition, it is hoped that after some time it will return to its previous state and species diversity, which consisted of more than five hundred species of coral and abundant species associated macrobenthos.

Especially because there are, opportunities to restore reefs through the replenishment of coral, larvae and aquatic organisms associated with other reef community. To the south of the island (in the distance of one mile 9°16'N, 103°21'E) there is a bank with depths ranging from 4 to 13 meters from the optimally developed noticeably damaged by typhoon coral community. There's the usual form for the settlement of various coral reefs more than 100 species, many common, at least 20 species *Acropora*, (**Figure 8**) and related corallobionts typical Vietnamese optimal reef community: *T. crocea, T. squamosa, L. cristagalli, A. ventricosa Lambis lambis, L. hiragra, C. tigris, M. arabica, Acantaster planci, C. novaeguineae, L. laevigata, H. atra, S. variegatus. D. setosum, E. diadema* and so on. In 2 - 3 years after calcareous substrate formation the dominance of reef building coral colonies becomes appreciable. The climax of the secondary succession, where Acropora community is formed, seems to be reach 4 - 6 years after the start of the succession, as long as the coral reefs remain in good health and are not subjected to large disturbance such as coral bleaching, sedimentation, typhoons or other bad factors [50,51].

4. Acknowledgements

The author is grateful to I. Budin, A. G. Goloseev, A. A. Gutnick, N. I. Selin, Tran Dinh Nam, Dao Tan Ho for his help in the study communities, the definition of flora and fauna.

Figure 8. Reef communities in optimal conditions at the bank, depth 7 m.

REFERENCES

[1] D. L. Bryant, J. W. Burke, Mc. Manus and M. Spalding, "Reefs at Risk. A Map Based Indicator of Threats to the World's Coral Reefs," World Resources Institute, Washington DC, 1998.

[2] C. R. Wilkinson, "Status of Coral Reefs of the World: 2000," Australian Institute of Marine Science, Townsville, 2000.

[3] T. A. Gardner, I. M. Côté, J. A. Gill, A. Grant and A. R. Watkinson, "Long-Term Region-Wide Declines in Caribbean Corals," *Science*, Vol. 301, No. 5635, 2003, pp. 958-960.

[4] W. F. Precht, A. W. Bruckner, R. B. Aronson and R. J. Bruckner, Endangered Acroporid Corals of the Caribbean," *Coral Reefs*, Vol. 21, No. 1, 2002, pp. 41-42.

[5] C. S. Rogers and J. Beets, "Degradation of Marine Ecosystems and Decline of Fishery Resources in Marine Protected Areas in the US Virgin Islands," *Environmental Conservation*, Vol. 24, No. 4, 2001, pp. 312-322.

[6] J. Geister, "The Influence of Wave Exposure on the Ecological Zonation of Caribbean Reefs," *Proceedings of the 3rd International Coral Reef Symposium*, Miami, 1977, Vol. 1, pp. 23-39.

[7] P. Blanchon, "Architectural Variation in Submerged Shelf-Edge Reefs: The Hurricane-Control Hypothesis," *Proceedings of the 8th International Coral Reef Symposium*, Okinawa, 1997, Vol. 1, pp. 547-554.

[8] J. H. Connell, "Diversity in Tropical Rainforests and Coral Reefs," *Science*, Vol. 199, No. 4335, 1978, pp. 1302-1309.

[9] M. Harmelin-Vivien, "The Effects of Storms and Cyclones on Coral Reefs: A Review," *Journal of Coastal Research*, Vol. 12, 1994, pp. 211-231.

[10] M. Harmelin-Vivien and P. Laboute, "Catastrophic Impact of Hurricanes on Atoll Outer Reef Slopes in Tuamotu (French Polynesia)," *Coral Reefs*, Vol., 5, No. 2, 1986, pp. 55-62.

[11] Yu. Ya. Latypov and N. I. Selin, "Changes of Reef Community near Ku Lao Cham Islands (South China Sea) after Sangshen Typhoon," *American Journal of Climate Change*, Vol. 1, No. 1, 2012, pp. 41-47.

[12] L. M. Harmelin-Vivien and P. Laboute, "Catastrophic Impact of Hurricanes on Atoll Outer Reef Slopes in the Tuamotu (French Polynesia)," *Coral Reefs*, Vol. 5, No. 2, 1986, pp. 55-62.

[13] D. Yellowleess, "Land Use Patterns and Nutrient Loading of the Great Barrier Reef Region: Proceedings of the Workshop Held at the James Cook University of North Queensland," Sir George Fisher Centre for Tropical Marines Studies and James Cook University of North Queensland, Queensland, 1991.

[14] P. R. F. Bell, "Eutrophication and Coral Reefs—Some Examples in the Great Barrier Reef Lagoon," *Water Research*, Vol. 26, No. 5, 1992, pp. 553-568.

[15] Y. Y. Latypov, "Coral Communities of the Namsu Islands (Gulf of Siam, South China Sea)," *Marine Ecology Progress Series*, Vol. 29, 1986, pp. 261-270.

[16] Yu. Ya. Latypov, "Macrobenthos Communities of Coral-Reefs of the An Thoi Archipelago, South China Sea," *Russian Journal of Marine Biology*, Vol. 26, No. 1, 2000, pp. 22-30.

[17] Yu. Ya. Latypov, "Communities of Coral Reefs of Central Vietnam," *Russian Journal of Marine Biology*, Vol. 27, No. 4, 2001, pp. 197-200.

[18] WWF Vietnam Marine Conservation Southern Survey Team, "Survey Report on the Biodiversity Resource Utilization and Conservation Potential of PhuQuoc Islands, Kien Giangprovince, Gulf of Thailand," Gland, Switzerland, 1994.

[19] Yu. Ya. Latypov, "Community Structure of Scleractinian Reefs in the Baitylong Archipelago (South China Sea)," *Asian Marine Biology*, Vol. 1-2, 1995, pp. 27-37.

[20] Yu. Ya. Latypov, "Coral Reefs of the Gulf of Tonkin," *Vestnik DVO RAN*, No. 2, 1997, pp. 92-98.

[21] Yu. Ya. Latypov, "Benthic Communities of Coral Reefs of Tho Chu Island, Bay of Thailand, South China Sea," *Biologia Morya*, Vol. 25, No. 3, 1999, pp. 201-208.

[22] Yu. Ya. Latypov, "Reef-Building Corals and Reefs of Vietnam. 1. The Gulf of Siam," *Biologia Morya*, Vol. 29, No. 3, 2003, pp. 155-165.

[23] C. G. J. Petersen, "The Animal Association of the Sea Bottom in the North Atlantic," *Kobenhavn Berlin Biology Stantion*, Vol. 22, 1914, pp. 89-98.

[24] Y. Loya and L. B. Slobodkin, "The Coral Reefs of Elate (Gulf of Elate. Red Sea)," *Journal of Sampling Zoology-Society London*, Vol. 28, 1971, pp. 117-140.

[25] J. E. N. Veron and M. G. S. Smith, "Coral ID: An Electronic Key to the Zooxanthellate Scleractinian Corals of the World," Australian Institute of Marine Science, Town-Swill, 2004.

[26] Yu. Ya. Latypov, "The Common Coral of Vietnam: Field Handbook," Far Eastern National University Press, Vladivostok, 2006.

[27] S. Weinberg, "A Comparison of Coral Reef Survey Methods," *Bijdragen Tot De Dierkunde*, Vol. 51, No. 2, 1981, pp. 199-218.

[28] W. Leujak and R. F. G. Ormond, "Comparative Accuracy and Efficiency of Six Coral Community Survey Methods," *Journal of Experimental Marine Biology Ecology*, Vol. 351, No. 1-2, 2007, pp. 168-187.

[29] S. Mandaville, "Benthic Macroinvertebrates in Freshwater—Taxa Tolerance Values, Metrics, and Protocols, Project H—1. (Nova Scotia. Soil & Water Conservation Society of Metro Halifax)," 2002.

[30] Y. Loya, "The Red Sea Coral Stylophora Pistillata Is an Rstrategist," *Nature*, Vol. 259, No. 5543, 1976, pp. 478-480.

[31] M. Pichon, "Dynamic Species of Coral Reef Benthic Structures and Zonation," *Proceeding 4th International Coral Reef Symposium*, Manila, 1981, Vol. 1, pp. 581-594.

[32] C. R. C. Sheppard, "Coral Population on Reef Slopes and Their Major Controls," *Marine Ecology Progress Series*, Vol. 7, 1982, pp. 83-115.

[33] Yu. Ya. Latypov, "Encrusting Protected Reef Hon Nai in Cam Ranh Bay in the South China Sea," *Natural Science*, Vol. 4, No. 1, 2012, pp. 14-21.

[34] Yu. I. Sorokin, "Ekosistemykorallovykhrifov," (Ecosystems of Coral Reefs), Nauka, Moscow, 1990.

[35] Yu. Ya. Latypov, "Benthic Communities of Coral Reefs of Con Dao Islands, South China Sea," *Biologia Morya*, Vol. 5-6, 1993, pp. 40-53.

[36] Yu. Ya. Latypov, "Reef-Building Corals of Vietnam as a Part of the Indo-Pacific Reef Ecosystem," *Russian Journal of Marine Biology*, Vol. 31, 2005, pp. S34-S40.

[37] H. Ditlev, "Zonation of Corals (Scleractinia Coelenterata) on Intertidal Reef Flats at Ko Phuket, Eastern Indian Ocean," *Marine Biology*, Vol. 47, No. 1, 1978, pp. 29-39.

[38] H. W. Ducklow and R. Mitchell, "Bacteria in Mucus Layers on Living Corals," *Limnology and Oceanography*, Vol. 24, No. 4, 1979, pp. 715-725.

[39] P. F. Holthus, J. E. Maragos and C. W. Evans, "Coral-Reef Recovery Subsequent to the Freshwater Kill of 1965 in Kaneohe Bay, Hawaii," *Pacific Science*, Vol. 43, No. 2, 1989, pp. 122-134.

[40] K. O. Amar, N. E. Chadwick and B. Rinkevich, "Coral Planulae as Dispersion Vehicles: Biological Properties of Larvae Released Early and Late in the Season," *Marine Ecology Progress Series*, Vol. 350, 2007, pp. 71-78.

[41] Yu. Ya. Latypov, "Scleractinian Corals of South Vietnam," *Biologia Morya*, No. 5, 1987, pp. 111-119.

[42] M. B. Best, B. W. Hoeksema, W. Moka, H. Moli and I. N. Sutarna, "Recent Scleractinian Corals Species Collected during the Snellius-II Expedition in Eastern Indonesia," *Netherlands Journal of Sea Research*, Vol. 23, No. 2, 1989, pp. 7-115.

[43] J. E. N. Veron and G. Hodgson, "Annotated Checklist of the Hermatypic Corals of the Philippines," *Pacific Science*, Vol. 43, No. 3, 1989, pp. 234-287.

[44] J. E. N. Veron, "Corals in Space and Time: The Biogeography and Evolution of the Scleractinia," Cornell University Press, New York, 1995.

[45] S. Vo and G. Hodgson, "Coral Reefs of Vietnam: Recruitment Limitation and Physical Forcing," *Proceeding of the 8th Internet Coral Reef Symposium*, Okinawa, 1997, Vol. 1, pp. 477-482.

[46] K. Sakai and M. Nishihira, "Immediate Effect of Terrestrial Runoff on a Coral Community near a River Mouth in Okinawa," *Galaxea*, Vol. 10, 1991, pp. 125-134.

[47] K. P. Sebens, "Biodiversity of Coral Reefs: What Are We Losing and Why?" *American Zoologist*, Vol. 34, 1994, pp. 115-133.

[48] N. G. Andres and J. D. Witman, "Trends in Community Structure on a Jamaican Reef," *Marine Ecology Progress Series*, Vol. 118, 1985, pp. 305-310.

[49] B. Salvat, "Dredging in Coral Reefs," In: B. Salvat, Ed., *Human Impact on Coral Reefs*: *Facts and Recommendations*, Museum National D'histoire Naturelle et École-Pratique des Hautesétudes, Antenne de Tahiti & Centre de l'Environnement, California, 1987, pp. 165-184.

[50] D. R. Choi, "Ecological Succession of Reef Cavity-Dwellers (Coelobites) in Coral Rubble," *Bulletin Marine of Science*, Vol. 35, No. 1, 1984, pp. 72-79.

[51] H. Ohba, K. Hashimoto, K. Shimoike, T. Shibuno and Y. Fujioka, "Secondary Succession of Coral Reef Communities at Urasoko Bay, Ishigaki Island, the Ryukyus (Southern Japan)," *Proceeding of the* 11*th Internet Coral Reef Symposium Ft Lauderdale*, Florida, 7-11 July 2008, pp. 319-327.

10

Phytoremediation: A Green Technology to Remove Environmental Pollutants

Annie Melinda Paz-Alberto[1], Gilbert C. Sigua[2*]

[1]Institute for Climate Change and Environmental Management, Central Luzon State University,
Science City of Muñoz, Philippines

[2]Coastal Plains Soil, Water & Plant Research Center, Agricultural Research Service,
United States Department of Agriculture, Florence, USA

ABSTRACT

Land, surface waters, and ground water worldwide, are increasingly affected by contaminations from industrial, research experiments, military, and agricultural activities either due to ignorance, lack of vision, carelessness, or high cost of waste disposal and treatment. The rapid build-up of toxic pollutants (metals, radionuclide, and organic contaminants in soil, surface water, and ground water) not only affects natural resources, but also causes major strains on ecosystems. Interest in phytoremediation as a method to solve environmental contamination has been growing rapidly in recent years. This green technology that involved "tolerant plants" has been utilized to clean up soil and ground water from heavy metals and other toxic organic compounds. Phytoremediation involves growing plants in a contaminated matrix to remove environmental contaminants by facilitating sequestration and/or degradation (detoxification) of the pollutants. Plants are unique organisms equipped with remarkable metabolic and absorption capabilities, as well as transport systems that can take up nutrients or contaminants selectively from the growth matrix, soil or water. As extensive as these benefits are, the costs of using plants along with other concerns like climatic restrictions that may limit growing of plants and slow speed in comparison with conventional methods (*i.e.*, physical and chemical treatment) for bioremediation must be considered carefully. While the benefits of using phytoremediation to restore balance to a stressed environment seem to far outweigh the cost, the largest barrier to the advancement of phytoremediation could be the public opposition. The long-term implication of green plant technology in removing or sequestering environmental contaminations must be addressed thoroughly. As with all new technology, it is important to proceed with caution.

Keywords: Phytoremediation; Green Technology; Pollutants; Contaminants; Toxic Metals

1. Green Technology

The success of green technology in phytoremediation, in general, is dependent upon several factors. First, plants must produce sufficient biomass while accumulating high concentrations of metal. In some cases, an increased biomass will lower the total concentration of the metal in the plant tissue, but allows for a larger amount of metal to be accumulated overall. Second, the metal-accumulating plants need to be responsive to agricultural practices that allow repeated planting and harvesting of the metal-rich tissues. Thus, it is preferable to have the metal accumulated in the shoots as opposed to the roots, for metal in the shoot can be cut from the plant and removed. This is manageable on a small scale, but impractical on a large scale. If the metals are concentrated in the roots, the entire plant needs to be removed. Yet, the necessity of full plant removal not only increases the costs of phytoremediation, due to the need for additional labor and plantings, but also increases the time it takes for the new plants to establish themselves in the environment and begin accumulation of metals. **Table 1** lists some of the common pollutant accumulating plants found by phytoremediation researchers.

The availability of metals in the soil for plant uptake is another limitation for successful phytoremediation. For example, lead (Pb^{2+}), an important environmental pollutant, is highly immobile in soils. Lead is known to be "molecularly sticky" since it readily forms a precipitate within the soil matrix. It has low aqueous solubility, and, in many cases, is not readily bioavailable. In most soils capable of supporting plant growth, the soluble Pb^{2+} levels are relatively low and will not promote substantial uptake by the plant even if it has the genetic capacity to accumulate the metal. In addition, many plants retain Pb^{2+} in their roots via absorption and precipitation with only minimal transport to the aboveground harvestable

*Corresponding author.

Table 1. Selected pollutant accumulating plants.

SCIENTIFIC NAME	COMMON NAME
Armeria maririma	Seapink thrift
Ambrosia artemisiifolia	Ragweed
Brassica juncea	Indian mustard
Brassica napus	Rape, Rutabaga, Turnip
Brassica oleracea	Flowering/ornamental kale and cabbage, Broccoli
Festuca ovina	Blue/sheep fescue
Helianthus annuus	Sunflower
Thalspi rotundifolium	Pennycress
Triticum aestivum	Wheat (scout)
Zea mays	Corn

plant portions. Therefore, it is important to find ways to enhance the bioavailability of Pb^{2+} or to find specific plants that can better translocate the Pb^{2+} into harvestable portions [1].

Although there are some challenges associated with the phytoremediation, it remains a very promising strategy and feasible alternative. However, in many situations, soil contamination may have unique factors that require special evaluation. Some plants may only accumulate these essential elements and prevent all others from entering. For plants termed as "hyperaccumulators" can extract and store extremely high concentrations (in excess of 100 times greater than non-accumulator species) of metallic elements [2]. It is believed that these plants initially develop the ability to hyperaccumulate non-essential metallic compounds as a means of protecting themselves from herbivorous predators that would experience serious toxic side effects from ingestion of the hyperaccumulator's foliage [3].

1.1. Plants as Phytoremediators

The principal application of phytoremediation is for lightly contaminated soils and waters where the material to be treated is at a shallow or medium depth and the area to be treated is large. This will make agronomic techniques economical and applicable for both planting and harvesting. In addition, the site owner must be prepared to accept a longer remediation period. Plants that are able to decontaminate soils does one or more of the following: 1) plant uptake of contaminant from soil particles or soil liquid into their roots; 2) bind the contaminant into their root tissue, physically or chemically; and 3) transport the contaminant from their roots into growing shoots and prevent or inhibit the contaminant from leaching out of the soil.

Moreover, the plants should not only accumulate, degrade or volatilize the contaminants, but should also grow quickly in a range of different conditions and lend themselves to easy harvesting. If the plants are left to die *in situ*, the contaminants will return to the soil. So, for complete removal of contaminants from an area, the plants must be cut and disposed of elsewhere in a non-polluting way. Some examples of plants used in phyoremediation practices are the following: water hyacinths (*Eichornia crassipes*); poplar trees (*Populus* spp.); forage kochia (*Kochia spp*); alfalfa (*Medicago sativa*); Kentucky bluegrass (*Poa pratensis*); Scirpus spp, coontail (*Ceratophyllum demersum* L.); American pondweed (*Potamogeton nodosus*); and the emergent common arrowhead (*Sagittaria latifolia*) amongst others [4].

Four heavy metal concentrations in soils (Cu, Cr, As, and Pb) were examined to see if removal through the process of phytoremediation was possible. Tomato and mustard plants were able to extract different concentrations of each heavy metal from the soils. The length of time that the soils were exposed to the contaminants affected the levels of heavy metals accumulation. Today, many institutions and companies are funding scientific efforts to test different plants' effectiveness in removing wide ranges of contaminants. Scientists favor *Brassica juncea* and *Brassica olearacea*, two members of the mustard family, for phytoremediation because these plants appeared to remove large quantities of Cr, Pb, Cu, and Ni from the soil [5].

1.2. Grasses as Potential Phytoremediators

1.2.1. Vetiver Grass (*Vetiveria zizanioides* L.)
Vetiver (*Vetiveria zizanioides* L.) belongs to the same grass family as maize, sorghum, sugarcane, and lemon grass. It has several unique characteristic as reported by

the National Research Council [6]. Vetiver grass is a perennial grass growing two meters high, and three meters deep in the ground. It has a strong dense and vertical root system. It grows both in hydrophilic and xerophytic conditions. The leaves sprout from the bottom of the clumps and each blade is narrow, long and coarse. The leaf is 45 - 100 cm long and 6 - 12 cm wide.

Vetiver grass is highly suitable for phytoremedial application due to its extraordinary features. These include a massive and deep root system, tolerance to extreme climatic variations such as prolonged drought, flood, submergence, fire, frost, and heat waves. It is also tolerant to a wide range of soil acidity, alkalinity, salinity, sodicity, elevated levels of Al, Mn, and heavy metals such as As, Cr, Ni, Pb, Zn, Hg, Se, and Cu in soils [7]. The roots of vetiver are the most useful and important part. Its root system does not expand horizontally, but penetrates vertically deep into the soil, whether it is the main, secondary or fibrous roots. The horizontal expansion of the vetiver grass root system is limited up to only 50 cm. The root vertical penetration expends up to 5 meters. Normally, yield levels of the leaves is 15 - 30 tons·ha^{-1} (15,000 - 30,000 kg·ha^{-1}) while vetiver grass roots can produce a dry matter yield of about 1428.6 to 2142.9 kg·ha^{-1} [8].

Various uses of vetiver grass are known worldwide. In South Africa, it was used effectively to stabilize waste and slime dams from Pt and Au mines [9]. In Australia, vetiver grass was used to stabilize landfill and industrial waste sites contaminated with heavy metals such as As, Cd, Cr, Ni, Cu, Pb, and Hg [7]. In China, vetiver grass was planted in large scale for pollution control and mine tail stabilization [10]. In Thailand, vetiver grass is found widely distributed naturally in all parts of the country. It has been used for erosion control and slope stabilization. Vetiver hedges had an important role in the process of captivity and decontamination of pesticides, preventing them from contaminating and accumulating in crops [11]. When compared with other plants, vetiver grass is more efficient in absorbing certain heavy metals and chemicals due to the capacity of its root system to reach greater depths and widths [7]. As confirmed by Roongtanakiat and Chairoj [12], vetiver grass was found to be highly tolerant to an extremely adverse condition. Therefore, vetiver grass can be used for rehabilitation of mine tailings, garbage landfills, and industrial waste dumps which are often extremely acidic or alkaline, high in heavy metals, and low in plant nutrients.

1.2.2. Cogon Grass (*Imperata cylindrica* L.)

Cogon grass, generally occurs on light textured acid soils with clay subsoil, and can tolerate a wide range of soil pH ranging from strongly acidic to slightly alkaline [13]. It is hardy species, tolerant of shade, high salinity, and drought. It can be found in virtually any ecosystem, especially those experiencing disturbances [8]. It is a perennial grass up to 120 cm high with narrow and rigid leaf-blades.

The roots can penetrate to a soil depth of about 58 cm in alluvial soil. More than 80 percent of shoots can originate from rhizomes less than 15 cm below the soil surface. The average number of shoots of cogon grass was about 4.5 million per hectare, producing 18,500 kg·ha^{-1} of leaves and rhizomes (11,500 kg of leaves and 7000 kg of rhizomes) [13].

1.2.3. Carabao Grass (*Paspalum conjugatum* L.)

Carabao grass is a vigorous, creeping perennial grass with long stolons and rooting at nodes. Its culms can ascend to about 40 to 100 cm tall, branching, solid, and slightly compressed where new shoots can develop at every rooted node. Under a coconut plantation, a yield of about 19,000 kg·ha^{-1} of green materials was obtained. It grows from near sea-level up to 1700 m altitude in open to moderately shaded places. It is adapted to humid climates and found growing gregariously under plantation crops and also along stream banks, roadsides, and in disturbed areas. This grass can adapt easily to a wide range of soils [14].

2. Phytoremediation as a Cleansing Tool: An Overview

Phytoremediation is described as a natural process carried out by plants and trees in the cleaning up and stabilization of contaminated soils and ground water. It is actually a generic term for several ways in which plants can be used for these purposes. It is characterized by the use of vegetative species for *in situ* treatment of land areas polluted by a variety of hazardous substances [15].

Garbisu [16] defined phytoremediation as an emerging cost effective, non-intrusive, aesthetically pleasing, and low cost technology using the remarkable ability of plants to metabolize various elements and compounds from the environment in their tissues. Phytoremediation technology is applicable to a broad range of contaminants, including metals and radionuclides, as well as organic compounds like chlorinated solvents, polychloribiphenyls, polycyclic aromatic hydrocarbons, pesticides/insecticides, explosives and surfactants. According to Macek [17], phytoremediation is the direct use of green plants to degrade, contain, or render harmless various environmental contaminants, including recalcitrant organic compounds or heavy metals. Plants are especially useful in the process of bioremediation because they prevent erosion and leaching that can spread the toxic substances to surrounding areas [18].

Several types of phytoremediation are being used today. One is phytoextraction, which relies on a plant's natural ability to take up certain substances (such as heavy metals) from the environment and sequester them in their cells until the plant can be harvested. Another is phytodegredation in which plants convert organic pollutants into a non-toxic form. Next is phytostabilization, which makes plants release certain chemicals that bind with the contaminant to make it less bioavailable and less mobile in the surrounding environment. Last is phytovolitization, a process through which plants extract pollutants from the soil and then convert them into a gas that can be safely released into the atmosphere [19]. Rhizofiltration is a similar concept to phytoextraction, but mainly use with the remediation of contaminated groundwater rather than the remediation of polluted soils. The contaminants are either absorbed onto the root surface or are absorb by the plant roots. Plants used for rhizofiltration are not planted directly *in situ*, but are acclimated with the pollutant first. Until a large root system has developed, plants are hydroponically grown in clean water rather than in soil. Once a large root system is in place, the water supply is substituted for polluted water supply to acclimate the plant. After the plants become acclimatized, they are planted in the polluted area. As the roots become saturated, they are harvested and disposed of safely.

Phytoremediation is a naturally occurring process recognized and documented by humans more than 300 years ago [2]. Since then, humans have exploited certain plant abilities to survive in contaminated areas and to assist in the removal of contaminants from the soil. However, scientific study and development of these plants' unique qualities were not conducted until the early 1980's [2]. At this time, it was recognized that certain species of plants could accumulate high levels of heavy metals from the soil while continuing to grow and proliferate normally [2]. Research has been slow and tedious due to scientists' incomplete understanding of the generalized cellular mechanisms of plants. However, the advent of new genetic technology has allowed scientists to determine the genetic basis for high rates of accumulation of toxic substances in plants [20]. Using genetic engineering, scientists may soon be able to exploit plants' characteristics that can provide faster and more efficient means of removing contaminants from the soil. Genetic engineering will also be crucial for the creation of transgenic plants that will be able to combine the natural agronomic benefits associated with plants (ease of harvest and rapid, expansive growth) with the remediation capabilities of bacteria-a traditional organism used in bioremediation [21].

Phytoremediation of heavy metals from the environment serves as an excellent example of plant-facilitated bioremediation process and its role in removing environmental stress. Traditionally, when an area becomes contaminated with heavy metals, the area must be excavated and the soil should be removed and put to a landfill site [2]. This process is extremely expensive and, therefore not entirely appealing despite recent discoveries regarding phytoremediation [2]. Analysts have estimated the cost of cleaning one hectacre of highly contaminated land at a depth of one meter. The estimated cost would range from $600,000 to $3,000,000 depending on the extent of the pollution and the toxicity of the pollutants [21]. The cost of phytoremediation could be as much as 20 times less expensive, making this practice far less prohibitive than conventional methods [2]. The ideal type of phytoremediator is a species that creates a large biomass, grows quickly, extensive root system, and can be easily cultivated and harvested [20]. The only problem is that natural phytoremediators often lack most of the qualities described above. Therefore, scientists have been forced to become very creative in developing effective transgenic phytoremediators.

Many human diseases result from the buildup of toxic metals in soil, making remediation crucial in protecting human health. Lead is one of the most difficult contaminants to be removed from the soil and one of the most dangerous. According to Lasat [2], the presence of Pb in the environment can have devastating effects on plant growth and can result in serious side effects-including seizures and mental retardation if ingested by humans or animals. Much of the global Pb contamination has occurred as a result of mining and iron smelting activities [22]. Phytoremediation of Pb contaminated soil involves two of the aforementioned strategies-phytostabilization and phytoextraction. It is believed that plants' ability to phytoextract certain metal is a result of its dependence upon the absorption of many metals such as Zn, Mn, Ni, and Cu [2].

2.1. Phytoremediation of Water Pollutants

In 2005, Cortez [23] conducted a study to assess pollution and survey the potential plants that can be used as phytoremediators of heavy metals in Nueva Ecija, Philippines. Water and plant samples were taken near the dumpsites, which is about 500 m away from the creek. Results of the water analysis showed that the dumpsite and Panlasian Creek were slightly polluted with considerable amount of phosphate. Results of the plant chemical analysis showed that kangkong (*Ipomea aquatic*) and Hydracharitaceae (*Ottelia alismoides* L.) were both efficient in phytoremediating Pb. Analysis of the plants further suggests that the concentrations of Pb in morning glory (*Ipomea violacea* L.) and hydracharitaceae (*Ottelia alismoides* L.) was about 210% more than the concentration of Pb in the water [23].

Xia and Ma [24] in 2005 investigated the potential of

water hyacinth (*Eichhornia crassipes*) in removing a phosphorus pesticide ethion. The disappearance rate constants of ethion in culture solutions were -0.01059, -0.00930, -0.00294 and -0.00201 for the non-sterile planted, sterile planted, non-sterile unplanted and sterile unplanted treatment, respectively. The accumulated ethion in live water hyacinth plant decreased by 55% - 91% in shoots and 74% - 81% in roots after the plant growing 1 week in ethion free culture solutions, suggesting that the plant uptake and phytodegradation might be the dominant process for ethion removal by the plant. Given the promising result of the study, water hyacinth could be utilized as an efficient, economical and ecological alternative to accelerate the removal and degradation of agro-industrial wastewater polluted with ethion.

Letachowicz *et al.* [25] conducted a study on the phytoremediation capacity on heavy metals accumulation in different organs of *Typhia latifolia* L. The concentrations of Cd, Pb, Cu, Ni, Mn, Zn, and Fe were determined in different organs of *Typhia latifolia* from seven water bodies in the Nysa region in Poland. The *Typhia latifolia* species that can absorb heavy metals can be used as bio-indicator of pollutants is a macrohydrophyte and is widely present in the entire lowland and lower mountain sites. It is linked with nutritious water and organic or inorganic mineral bottom sediments. *Typhia latifolia* is a strongly expansive species because it can control water space due to intensive growth of rhizomes and often creates almost mono-species group, though it can also be found in various groups of rushes.

2.2. Phytoremediation Species in Coastal Water

The Philippines is blessed to have relatively high mangrove diversity having 35 species [26] including five major families, namely: Avicenniaceae; Arecaceae; Combretaceae; Lythraceae; and Rhizophoraceae [27]. Though Philippines has high mangrove diversity, it was reported that there was a drastic decline of mangrove resources from 450,000 hectares in 1918 to 120,000 hectares in 1995. The decrease of the mangrove forests was due to human activities, such as fish pond conversion, human settlement, and salt production [28]. However, with the alarming rate of mangrove forest degradation, Philippines strived to continue greater conservation of mangroves and reforestation of the coastal areas [29].

Mangroves are higher plants, which are found mostly in the intertidal areas of tropical and subtropical shorelines and show remarkable tolerance to high amounts of salt and oxygen poor soil. The mechanisms of mangrove to keep the salt away from the cytoplasm of the cell were through the excretion of salt in their salt glands found in the leaves and roots and through storage of salts in the mature leaves, bark and wood [26]. Mangroves developed unique body features in order to cope up with harsh environment. There are different types of roots, such as prop roots in *Rhizophora*, pencil-like pneumatophores in *Avicennia*, and cone-like pneumatophores in *Sonneratia* that have large lenticels to permit gas exchange. The leaves of mangroves have characteristics to survive from dessication and conserve water like the presence of thick epidermis, waxy cuticle, and presence of hypostomata [26]. Mangrove ecosystem is exposed to different pollutants such as heavy metal, sewage wastes, pesticides and petroleum products. Heavy metal accumulation in the mangrove sediment can result in biological and ecological effects. Even though, mangrove trees may have the immunity against the toxic effects of the heavy metals, but the animals thriving in the ecosystem are vulnerable to the negative effects of heavy metals [30].

Few studies were conducted about phytoremediation potential of mangroves and other wetland plant species. However, those researchers paved the way to explore more species of mangroves particularly the native species present in the area, for their feasibility to accumulate heavy metals. Zheng *et al.* [31] studied the different metal concentrations of Cu, Ni, Cr, Zn, Pb, Cd, and Mn in *Rhizophora stylosa* at Yingluo Bay, China. The study showed less pollution due to relatively low concentration of metals especially Pb, Mn, Zn and Cd.

MacFarlane and Burchett [32] examined the cellular distribution of Cu, Pb and Zn in grey mangrove, *Avicennia marina* (Forsk.) using scanning electron microscope X-ray microanalysis and atomic absorption spectroscopy. They reported that metals mostly accumulate in plants' cell walls. Their study showed that certain parts of mangroves have the ability to control the entrance of heavy metals in other parts of the plants. The laboratory research of MacFarlane and Burchett [33] contributed information on the accumulation, growth effect, and toxicity of Cu, Pb and Zn in grey mangrove, *Avicennia marina* (Forsk.). Accumulation of the different metals occurred at varying concentrations in the roots and leaf tissue. In the roots, Pb accumulated lesser than the other metals while high concentration of Zn was found in the leaf tissue. The effects of excessive Cu and Zn on young mangrove were reductions in seedling height, leaf number, total biomass, and root growth. The germination of mangrove was inhibited at 800 $\mu g \cdot g^{-1}$ Cu and 1000 $\mu g \cdot g^{-1}$ Zn. The Pb showed only little negative effects in the growth of the plant due to low absorption of this metal.

Cheng [34] cited heavy metals can be absorbed by plants using their roots, or via stems and leaves, and stored the metals into different plant parts. Moreover, the distribution and accumulation of heavy metals in the plants depend on plant species and chemical factors. The *Avicennia marina*, a salt-excretive mangrove, and *Rhizophora stylosa*, a salt-exclusion mangrove, have different accumulation potential of different heavy metals. In terms of Pb absorption, *A. marina* was able to accumulate more concentrations of heavy metals than in *R. stylosa*. However, the purification processes of plants were affected by different factors such as heavy metal concentration, plant species, and exposure duration.

Sari *et al.* [35] conducted an *in-situ* experiment on the bioaccumulation of Pb in two mangrove species, *Avicennia alba* and *Rhizophora apiculata* using hydroponics culture. The mangroves were grown in 0%, 15% and 30% salinity and 0.03, 0.3, and 3 mg·L^{-1} of Pb concentration. They observed that both mangroves had significantly lower Pb accumulation in leaves than in roots. They claimed that the mobility of Pb in the aerial part of the plant can be related to its mechanism associated with the accumulation of sodium in the salt glands found in the leaves. Saenger and McConchie [36] evaluated the accumulation trend of Pb in the tissues, barks and woods, old and young leaves and fruits of different mangrove species. They discovered that Pb concentrated more in the bark than in other tissues of mangroves because of atmospheric Pb due to vehicle exhausts from nearby major roads.

Shete *et al.* [37] revealed in their study entitled, "Bioaccumulation of Zn and Pb in *Avicennia marina* (Forsk.) from urban areas of Mumbai (Bombay), India," that the mangrove species can bioaccumulate and survive despite heavy metal contamination. Results showed that mangroves have greater uptake of heavy metals. Variations on the concentrations of Zn were found from the different plant parts while high accumulation of Pb was focused in the roots. They found out that Pb concentrations were present in the leaves and roots. Kamaruzzaman *et al.* [38] studied the cumulative partitioning of Pb and Cu in the *Rhizophora apiculata* in the Setiu mangrove forest, Terengganu. Results showed increasing concentration of Cu and Pb from the leaf, bark, root, and sediments. The study by Pahalawattaarachchi *et al.* [39] reported the absorption, accumulation, and partitioning of eight different metals specifically Cu, Cd, Cr, Fe, Mg, Ni, Pb and Zn by mangrove species, *Rhizophora mucronata* (Lam.) at Alibag, Maharashtra, India. They revealed that Cu, Mn and Fe showed limited mobility due to their accumulation in the roots while other metals (Cd, Zn, Ni and Pb) were concentrated in the aerial part of the plant. They concluded that *Rhizophora mucronata* (Lam.) was more capable of phytostabilization rather than phytoextraction because of low uptake capacity of different metals.

Nazli and Hashim [40] revealed that *Sonneratia caseolaris* was a potential phytoremediation species for selected heavy metals in Malaysian mangrove ecosystem. The study assessed the concentrations of Cd, Cr, Cu, Pb and Zn in *Sonneratia caseolaris*. Results showed that both roots and leaves of *Sonneratia caseolaris* accumu-

lated and exceeded the general normal upper range of Cu and Pb in plants. In Iran, Parvaresh *et al.* [41] studied the bioavailability of different heavy metals (Ni, Cu, Cd, Pb and Zn) in the sediments of Sirik Azini creek. The outcome of their research revealed no heavy metal pollution was found in the area due to low geo-accumulation index of Pb in the sediment. They assessed that the concentration of heavy metals particularly Pb in the leaves were higher than the concentration of Pb in the sediment.

Qui *et al.* [42] studied the different accumulation and partitioning of seven trace metals, namely, As, Cd, Cr, Cu, Hg, Pb and Zn, in mangroves and sediments from three estuarine wetlands of Hainan Island, China. They analyzed the sediment samples and found out that the heavy metals present in the area were still at relatively low levels. Furthermore, Pb analysis of mangroves showed that this metal was found mostly in the branches of the different mangroves. Zhang *et al.* [43] investigated the physiological response of *Sonneratia apetala* (Buch) to the addition of wastewater nutrients and heavy metals (Pb, Cd, and Hg). They planted mangroves in four different treatments: 1) control, which has only salted water; 2) normal concentration of wastewater nutrients and heavy metals; 3) five times the normal treatment; and 4) ten times the normal treatment. Results revealed that growth of mangrove increased with increasing levels of wastewater pollution. The study showed that mangroves were potential phytoremediator in wetland ecosystem.

The research of Nirmal *et al.* [44] entitled, "An assessment of the accumulation potential of Pb, Zn, and Cd by *Avicennia marina* (Forssk) in Vamleshwar Mangroves, Gujarat, India," reported that sediments in the area are below critical soil concentration for heavy metals. *A. marina* possesses the capacity to uptake selected heavy metals, Pb, Zn and Cd, via its roots and storing them in their leaves without any sign of complications. The concentrations of heavy metals in the *A. marina* were in normal range except for Pb. The roots of mangrove contained the highest concentration of heavy metal except for Cd. Furthermore, *A. marina* had the capacity to uptake metals via its roots and accumulates them in their leaves without any sign of injury. The study showed *Avicennia marina* as a potential phytoremediation species for selected heavy metals in many mangrove ecosystems. Subramanian [45] cited that mangroves generally have low concentration of heavy metal. Kathiresan and Bingham [27] mentioned that mangroves can tolerate metal pollution because they were poor accumulators of heavy metals. A study on the metal uptake of *Rhizophora mangle* in Sepetiba Bay, Rio de Janeiro, Brazil showed that only one percent of the total heavy metals concentration in the sediment accumulated in the mangrove [27]. In their experiment, they used young *Bruguiera gymnorrhiza* and artificially synthesized wastewater treatments with different levels of Cu, Cd, Cr, Ni, and Zn. The control treatment showed higher biomass and growth than the plants treated with wastewater.

2.3. Phytoremediation of Soil Pollutants

Phytoremediation is a cleanup technology for metal contaminated soils, specifically Pb. In order for this type of remediation strategy to be successful, it is necessary to utilize metal accumulating plants to extract environmentally toxic metals from the soil, such as Pb, Ni, Cr, Cd and Zn. Certain plants have been identified not only to accumulate metals in the plant roots, but also to translocate the accumulated metals from the root to the leaf and to the shoot. While many plants performed this function, some plants, known as "hyperaccumulators", can accumulate extremely high concentrations of metals in their shoots (0.1% to 3% of their dry weight) [46]. The metal-rich plant material can then be harvested and removed from the site without extensive excavation, disposal costs, and loss of topsoil that is associated with traditional remediation practices.

Bioremediation process would be extremely slow because the rate of bioemediation is directly proportional to growth rate while the total amount of bioremediation is correlated with a plant's total biomass. No plant has been discovered yet capable of meeting all the ideal criteria of an effective phytoremediator. These criteria are fast growing, deep and extensive roots, high biomass, easy to harvest and hyperaccumulators of a wide range of toxic metals. A Pb absorption study by Huang and Cunningham [47] cited corn as a perfect phytoremediator due to its large biomass, fast rate of growth, and the existence of extensive genomic knowledge of this crop. Introduction of hyper accumulating genes as well as genetic information would better prepare these species to deal with diverse climatic conditions [19]. The mobilization of metal contaminants, both in the soil and the plant, is another important factor influencing the success of phytoremediation. The amount of soluble Pb^{2+} in the soil appears to be a key factor to the enhancement of Pb^{2+} uptake by plants [48].

Two main amendment techniques have been used to increase the bioavailability of Pb in soils and the mobility of Pb within plant tissue by lowering soil pH and adding synthetic chelates. Soil pH is a significant parameter in the uptake of metal contaminants because soil pH value is one of the principal soil factors controlling metal availability [49]. Maintaining a moderately acidic pH in the soil may be attained through the use of ammonium containing fertilizer or soil acidifiers. By this, Pb metal bioavailability and plant uptake can increase [50-52]. In a study performed by Cholpecka et al. [53] on metal contaminated soils in southwest Poland, reported that soil samples with pH of less than 5.6 contained relatively more metals in the exchangeable form than in soil samples with pH greater than 5.6. In addition, at lower pH, the Pb in soil has a greater potential to translocate from a plant's roots into its shoots. Synthetic chelates, such as ethylenediaminetetraacetic acid (EDTA), have been shown to aid in the accumulation of Pb2+ in the plant tissue. EDTA and other chelates have been used in soils and nutrient solutions to increase the solubility of metal cations and the translocation of Pb into shoots [54].

The physiological and biological mechanisms involved in Pb uptake of plants involving root to shoot transport of Pb may require some time to develop and become functional. Since plant species can differ significantly in Pb uptake and translocation, the success of using plants to extract Pb from contaminated soils requires the following: 1) the identification of Pb accumulating plants that can survive in the presence of contaminants; 2) the measurement of the concentration of pollutant in the soil, and 3) knowledge of chemistry (availability or speciation) of the metal in the soil matrix. The combination of soil amendment and foliar fertilizer application to plants capable of absorbing and translocation of Pb may be an effective means of remediating an area with varying levels of Pb concentrations.

Other model of phytoremediators includes various varieties of transgenic trees. Trees are ideal in the remediation of heavy metals because they can withstand higher concentrations of pollutants due to their large biomass. As such, they can accumulate large amounts of the contaminants in their systems because of their size capable of reaching huge area and great depths due to their extensive root systems. Furthermore, they can stabilize an area, prevent erosion, and minimize spread of contaminant because of their perennial presence. They can also be easily harvested and removed from the area with minimal risk, effectively taking with them a large quantity of the pollutants that were once present in the soil [19].

3. Phytoremediation, Is It Good or Bad?

Earlier discussion has illustrated many advantages and disadvantages of transgenic phytoremediation. The primary advantages of using plants in bioremediation are as follows: it is more cost-effective; more environmentally friendly; and more aesthetically pleasing than conventional methods. The conventional methods are usually expensive and environmentally disruptive [55]. Plants also offer a permanent, in situ, nonintrusive, self-sustaining method of soil contaminant removal. More importantly, contaminants can be removed much more easily through the harvest of plants than from the soil itself.

More benefits are derived through phytoextraction. It

enables scientists to reclaim and recycle usable materials, including a wide variety of precious metals from the soil [21]. Also, its potential benefits are extremely high and extremely attractive to scientists and businessmen alike [21]. Furthermore, phytoextraction is economical because only solar energy must be present to maintain the system [55]. Finally, the greatest advantage of this technology is that it utilizes the inherent agronomic benefits of plants [56]. These benefits include high biomass, extensive root systems that both stabilize the ecosystem by preventing contaminant to spread through leaching as well as reaching a large volume of contaminated soil and a greater ability to withstand adverse environmental conditions and interspecies competition than bacteria [56]. As extensive as these benefits are, the possible costs of using plants for bioremediation should not be ignored. Some concerns voiced out in response to phytoremediation include its slow speed in comparison to mechanical methods such as soil excavation and climatic restrictions that may limit growing many species of plants, and the unknown long-term environmental costs [19]. Also, potential danger might exist for animals that live in the areas in which phytoremediators are grown, especially if these animals typically feed on plants being used for phytoremediation [21].

Moreover, concerns have been raised regarding the potential for contaminants to move up the food chain more quickly. This problem may occur if toxic materials are sequestered in consumable sources such as plants [57]. Finally, issues with the disposal of these toxic materials still remain. Once contaminants have been extracted from the soil by the plants, we are still faced with the dilemma of what to do with these contaminants. It seems that the end result remains the same. This involves the removal of contaminants to a landfill location where the plants would eventually biodegrade and the contaminants could enter the soil system once again [57].

4. Case Study: Phytoremediation Research in the Tropics

4.1. Phytoremediation of Lead Contaminated Soils

The global problem concerning contamination of the environment as a consequence of human activities is increasing. Most of the environmental contaminants are chemical by-products such as Pb. Lead released into the environment makes its way into the air, soil and water. Lead contributes to a variety of health effects such as decline in mental, cognitive, and physical health of the individual. An alternative way of reducing Pb concentration from the soil is through phytoremediation. Phytoremediation is an alternative method that uses plants to clean up contaminated area. Hence, Paz-Alberto *et al.*

[58] conducted a study in the Philippines. The objectives of this study were 1) to determine the survival rate and vegetative characteristics of three grass species such as vetiver grass, cogon grass, and carabao grass grown in soils with different Pb levels; and 2) to determine and compare the ability of three grass species as potential phytoremediators in terms of Pb accumulation by plants. The three test plants: vetiver grass (*Vetiveria zizanioides* L.); cogon grass (*Imperata cylindrica* L.); and carabao grass (*Paspalum conjugatum* L.) were grown in different individual plastic bags containing soils with 75 mg·kg^{-1} (37.5 kg·ha^{-1}) and 150 mg·kg^{-1} (75 kg·ha^{-1}) of Pb, respectively. The Pb contents of the test plants and the soil were analyzed before and after experimental treatments using an atomic absorption spectrophotometer. This study was laid out following a 3×2 factorial experiment in a completely randomized design [58].

Results of the study (**Table 2**) revealed that on the vegetative characteristics of the test plants, vetiver grass registered the highest whole plant dry matter (33.85 - 39.39 Mg·ha^{-1}). Carabao grass had the lowest herbage mass production of 14.12 Mg·ha^{-1} and 5.72 Mg·ha^{-1} from soils added with 75 and 150 mg·Pb·kg^{-1}, respectively. Vetiver grass also had the highest percent plant survival which meant it best tolerated the Pb contamination in soils. Vetiver grass registered the highest rate of Pb absorption (10.16 ± 2.81 mg·kg^{-1}). This was followed by cogon grass (2.34 ± 0.52 mg·kg^{-1}) and carabao grass with the mean Pb level of 0.49 ± 0.56 mg·kg^{-1}. Levels of Pb among the three grasses (shoots + roots) did not vary significantly with the amount of Pb added (75 and 150 mg·kg^{-1}) to the soil. Vetiver grass yielded the highest biomass; it also has the greatest amount of Pb absorbed (roots + shoots). This can be attributed to the highly extensive root system of vetiver grass with the presence of an enormous amount of root hairs. Extensive root system denotes more contact to nutrients in soils, therefore more likelihood of nutrient absorption and Pb uptake. The efficiency of plants as phytoremediators (**Table 3**) could be correlated with the plants' total biomass. This implies that the higher the biomass, the greater the Pb uptake. Plants characteristically exhibit remarkable capacity to absorb what they need and exclude what they do not need. Some plants utilize exclusion mechanisms, where there is a reduced uptake by the roots or a restricted transport of the metals from roots to shoots. Combination of high metal accumulation and high biomass production results in the most metal removal in the soil [58]. The study indicated that vetiver grass possessed many beneficial characteristics to uptake Pb from contaminated soil. It was the most tolerant and could grow in soil contaminated with high Pb concentration. Cogon grass and carabao grass are also potential phytoremediators since they can absorb small amount of Pb in soils, although cogon

Table 2. Levels of Pb absorbed 1) by whole plants (roots + shoots) and estimated total uptake of Pb 2) by vetiver grass, cogon grass and carabao grass.

Grasses	Amount of Pb added (kg·ha⁻¹)		Mean ($LSD_{0.05}$ = 17.2)
	37.5	75	
1. Levels of Pb in whole plants	**(mg·kg⁻¹)**		**(mg·kg⁻¹)**
Vetiver grass	11.84 ± 2.94	8.47 ± 1.59	10.16 ± 2.81a[§]
Cogon grass	2.00 ± 0.19	2.68 ± 0.54	2.34 ± 0.52b
Carabao grass	0.40 ± 0.32	0.58 ± 0.25	0.49 ± 0.56c
Mean ($LSD_{0.05}$ = 1.5)	4.75 ± 2.5x[†]	3.91 ± 2.6x	($LSD_{0.05}$ = 1.8)
2. Plant uptake of Pb	**(kg·ha⁻¹)**		**(kg·ha⁻¹)**
Vetiver grass	29.71 ± 8.71	33.78 ± 10.02	31.74 ± 9.01a[§]
Cogon grass	1.93 ± 0.48	2.69 ± 0.19	2.33 ± 0.53b
Carabao grass	0.19 ± 0.06	0.34 ± 0.03	0.27 ± 0.03b
Mean ($LSD_{0.05}$ = 5.6)	13.95 ± 2.91x[†]	12.27 ± 3.32x	($LSD_{0.05}$ = 6.8)

[§]Means in respective columns (1 and 2) with the same letter(s) are not significantly different at 5% level of significance. [†]Means in respective rows (1 and 2) with the same letter(s) are not significantly different at 5% level of significance.

Table 3. Estimated removal (%) of Pb by three grasses from soils amended with varying levels of Pb.

Grasses	Amount of Pb added (kg·ha⁻¹)		Mean ($LSD_{0.05}$ = 17.2)
	37.5	75	
	(%)		
Vetiver grass	79.2	45.1	62.2
Cogon grass	5.1	3.6	4.4
Carabao grass	0.5	0.5	0.5

grass is more tolerant to Pb-contaminated soil compared with carabao grass. The important implication of the findings of this study is that vetiver grass can be used for phytoextraction on sites contaminated with high levels of heavy metals, particularly Pb [58].

A field survey was conducted by Bautista [59] to identify phytoremediators present in the selected cities in the province of Nueva Ecija, Philippines. The plants found in the heavy traffic area of Cabanatuan City were the "balite" (*Ficus bengalensis*) and the "espada" (*Sanasaviera trifasciata*). In the heavy traffic area of San Jose City the most common plants are the Bougainvillea (*Bougainvillea* sp.) and the Cherry Pink plant. The Indian tree (*Polyalthia longifolia*) and the bougainvillea (*Bougainvillea sp.*) were the most common plants found along the traffic islands of the Science City of Muñoz. In Cabanatuan City, the balite absorbed 2.822 ppm of Pb, while espada absorbed 2.352 ppm of Pb; in San Jose City, the cherry pink plant absorbed 4.803 ppm, while the bougainvillea absorbed 1.521 ppm of Pb; and in the Science City of Muñoz, the Indian tree absorbed 0.217 ppm, and

the bougainvillea absorbed 0.528 ppm, respectively. Results of the chemical analysis proved that all of the plants along the traffic islands of the three selected cities of Nueva Ecija were phytoremediators of Pb. They were the most effective phytoremediator of Pb among the plants in the traffic area within the three selected cities.

As discussed previously, there are several different methods through which phytoremediation can occur. However, in order to maximize the success of a phytoremediation strategy, it is critical to have significant metal bioavailability at a contaminated site as well as a large quantity of plant biomass with high rates of growth. Metal contaminants that are not soluble, may limit the success of phytoremediation. In most Pb contaminated soils usually less than 0.1% of the total Pb present is bioavailable for plant uptake. The plants grown in a contaminated soil accumulated less Pb in both the roots and shoots than the plants grown hydroponically in a solution with a similar Pb concentration. The difference in uptake was because the Pb in the solution was much more bioavailable to the plants.

It should be noted that while hydroponic tests do not reflect accurately the accumulation potential in terrestrial applications, these tests could be valuable in the screening for Pb accumulating plant species and tolerance levels. The second limitation in Pb phytoextraction is the poor translocation of the metal from the roots to the harvestable shoots. In the plants that do translocate Pb, translocation is less than 30% [22].

Research has been conducted in the field to improve both the uptake and translocation of Pb through induced hyperaccumulation, which involves soil pH adjustments or the application of synthetic chelates. In general, the more biomass that the plant has, the more metal can be accumulated since the metal uptake is a function of the overall biomass [60]. The use of fertilizers can help facilitate rapid plant establishment and growth. For most Pb-contaminated soil, P availability is very low due to the precipitation of Pb-P precipitation. Thus, a foliar P fertilizer spray applied topically to the plant's leaves and stem increases phosphorous content in the plant, while not confounding the Pb-P binding problem in the soil. In a study reported by Huang et al. [50], soil to which phosphate fertilizer was added directly showed diminished Pb bioavailability, presumably due to Pb-P precipitation, in contrast to hydroponic uptake. Furthermore, although the foliar P application decreases Pb^{2+} concentration in shoots by 55% and root-Pb^{2+} concentration by 20%, the total amount of Pb^{2+} accumulation increased by 115% in shoots and 300% in the roots. This is the result of the large increase in biomass production made possible by overcoming phosphate limitations. These results further emphasize the relationship between Pb^{2+} accumulation and plant biomass in Pb^{2+} phytoextraction.

4.2. Phytoremediation Potential of Some Plant Species from Mining Sites

The focus of this study were on the accumulation of heavy metals in plants most commonly found in mine tailings of Victoria, Manlayan, Benguet, Philippines and identification of the different plant species within the area of the study. These plant species were assumed to be potential phytoremediation species [61].

The heavy metals extracted from the plants in the mine tailing were Cu, Cd, Pb and Zn. The fourteen plant species that were identified within the study were: *Eleusine indica* L.; *Amaranthus spinosus* L.; *Alternathera sessilis* L.; *Portuluca oleracea* L.; *Fimbristylis meliacea* L., Vahl, *Mikania cordata* ((Burm. F.) B. l. Robins; *Polygonun barbatum* L.; *Achyranthes aspera* L., *Blumea* sp., *Cyperus alternifolus* L.; *Crassocephalum crepidioides* (Benth.) S. Moore; *Cyperus compactus* Retz.; *Desmodium* sp. and *Muntingia calabura* L. These plants absorbed certain metals at low and high levels. Among the plants species,

A. spinosus was found to have almost all the metals extracted in large amounts particularly Pb. The other plant species with high concentration of Pb were *A. sessilis*, *Desmodium* sp., *P. oleracea*, and *A. aspera*. *E. indica* has the highest concentration of Zn together with *M. cordata*, *C. compactus*, *F. maliacea* and *A. spinosus*. In contrast, Cd was found in trace amount in soil, but high in the following species: *C. crepidioides*, *P. oleracea*, *A. sessilis*, and *C. alternifolius*. Nickel was found high only in A. *sessilis* and *Blumea* sp. but trace amount in *Desmodium* sp. and *F. meliacea*. Also, high Cu concentrations were found in *A. spinosus* and *P. oleracea*.

In this study, the phytoremediation potential was dependent on population within species. The potential of the surveyed species mentioned for phytoremediation was remarkable and promising because of the presence of heavy metals suspected to have accumulated in the soil. Root system of these plants showed higher root to shoot ratios compared to other plants found in the area indicating high translocation of metals to the shoot. These species also plays an important role in the phytostabilization of metals to reduce leaching and run off. Also, these may be transformed to less toxic forms. These typical plants have dense root systems which can be effective for phytostabilization and elimination of contaminants such as Pb, Cd, Zn, As, Cu, and Ni in mine tailing sites.

A similar study conducted in Poland was worth including in this section. Wislocka et al. [62] studied the bioaccumulation of heavy metals by selected plants from uranium mining dumps in the Sudety Mountains, Poland. They found out that the investigated plants from the uranium dumps in the Sudety Mountains grew on acidic soils with an unfavorable C/N ratio. However, the nutrient status as well as relatively high CEC, and organic matter of the soil allowed the growth of spontaneous vegetation. Contamination by heavy metals (Pb, Zn, Cu, Cd and Ni), being associated with the mineral assemblage of the spoil material, was found to be significant within all dumps. All plants examined (*Salix caprea*, *Betula pendula and Rubus idaeus*) accumulated high amounts of heavy metals, but in general *R. idaeus* showed lower concentration of heavy metals (except Mn) in its leaves. However, Pb was accumulated to a similar degree in both trees and *R. idaeus*. Among all the heavy metals analyzed in the three species, Cd exhibited the greatest accumulation rate and the Cd accumulation ratio was several times higher for *S. caprea*, in comparison to the other two species. *B. pendula* and *R. idaeus* exhibited higher accumulation rates for Mn than *S. caprea*. However, the potential use of *R. idaeus* in monitoring metal concentration in the environment requires further investigation. The significant positive correlation between Pb in soil and leaves of the same tree suggest that *S. caprea* should be employed for monitoring Pb in the environment.

4.3. Phytoremediation Potential of Selected Plants for Mutagenic Agents

Research and development has its own benefits and inconveniences. One of the inconveniences is the generation of enormous quantity of diverse toxic and hazardous wastes and its eventual contamination to soil and groundwater resources. Ethidium Bromide (EtBr) is one of the commonly used substances in molecular biology experiments. It is highly mutagenic and moderately toxic substance in DNA-staining during electrophoresis. Interest in phytoremediation as method to solve chemical contamination has been growing rapidly in recent years. The technology has been utilized to clean up soil and groundwater from heavy metals and other toxic organic compounds in many countries like the United States, Russia and most of European countries. Phytoremediation requires somewhat limited resources and is very useful in treating a wide variety of environmental contaminants. It is in this context that Uera *et al.* [63] conducted a study aimed to assess the potential of selected tropical plants as phytoremediators of EtBr.

This study used tomato (*Solanum lycopersicum*), mustard (*Brassica alba*), vetiver grass (*Viteveria zizanioides*), cogon grass (*Imperata cylindrical*), carabao grass (*Paspalum conjugatum*) and talahib (*Saccharum spontaneum*) to remove EtBr from laboratory wastes. The six tropical plants were planted in individual plastic bags containing 10% EtBr-stained agarose gel. The plants were allowed to establish and grow in the soil for 30 days. Ethidium Bromide content of the test plant s and the soil were analyzed before and after soil treatment. Ethidium Bromide contents of the plants and soils were analyzed using an UV VIS spectrophotometer.

Results showed a highly significant ($p \leq 0.001$) difference in the ability of the tropical plants to absorb the EtBr from the soils. Mustard registered the highest absorption of EtBr (1.4 ± 0.12 $\mu g \cdot kg^{-1}$) followed by tomato

and vetiver grass with average uptake of 1.0 ± 0.23 and 0.7 ± 0.17 $\mu g \cdot kg^{-1}$ EtBr, respectively. Cogon grass, talahib, and carabao grass had the least amount of EtBr absorbed (0.2 ± 0.6 $\mu g \cdot kg^{-1}$). Ethidium bromide content of the soil planted with mustard was reduced by 10.7%. This was followed by tomato with an average reduction of 8.1%. Only 5.6% reduction was obtained from soils planted to vetiver grass. Soils planted to cogon grass, talahib and carabao grass had the least reduction of 1.52% from its initial EtBr content (**Table 4** and **Figure 1**). Mustard had the highest potential as phytoremediator of EtBr in soil. However, the absorption capabilities of the other test plant may also be considered in terms of period of maturity and productivity. Uera *et al.* [63] recommended that a more detailed and complete investigation of the phytoremediation properties of the different plants tested should be conducted in actual field experiments. Plants should be exposed until they reach maturity to establish their maximum response to the toxicity and mutagenecity of EtBr and their absorbing capabilities. Different plant parts should be analyzed individually to determine the movement and translocation of EtBr from soil to the tissues of the plants. Since this study has an increased amount of EtBr application should be explored in future studies. It is suggested therefore that a larger, more comprehensive exploration of phytoremediation application in the management of toxic and hazardous wastes emanating from biotechnology research activities should be considered especially on the use of vetiver grass, a very promising tropical perennial grass.

4.4. Phytoremediation Potential of Selected Tropical Plants for Acrylamide

Environmentally hazardous and health risk substances in animals and humans in the environment have increased as a result of continuing anthropogenic activities. Examples of these activities are food processing, laboratory, food production, industrial and other relative activities that use various forms of acrylamide. All acrylamide in

Table 4. Levels of EtBr in soils and relative reduction of EtBr in soils after 30 days.

Plants (Treatments)	Initial Level in Soil ($\mu g \cdot kg^{-1}$)	Final Level in Soil ($\mu g \cdot kg^{-1}$)	Percent Reduction in Soil
Tomato	19.7	18.1 ± 0.17	8.12b[§]
Mustard	19.7	17.6 ± 0.23	10.66a
Vetiver grass	19.7	18.6 ± 0.23	5.58b
Talahib	19.7	19.4 ± 0.15	1.52c
Carabao grass	19.7	19.4 ± 0.20	1.52c
Cogon grass	19.7	19.4 ± 0.21	1.52c

[§]Means in column followed by a common letter(s) are not significantly different from each other at $p \leq 0.05$.

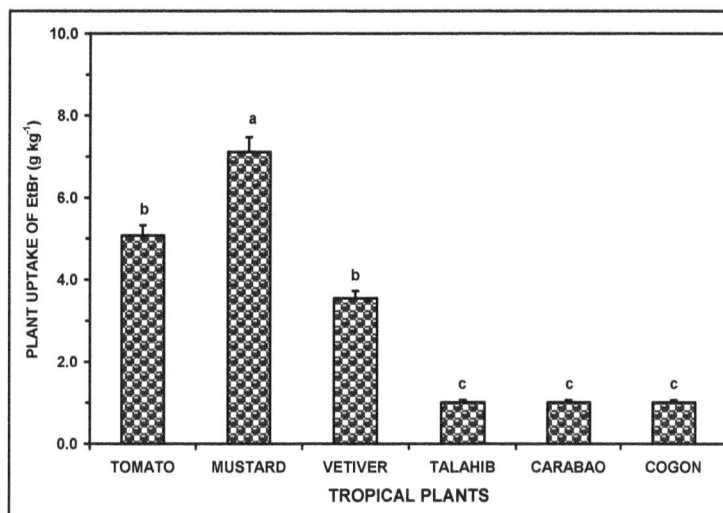

Figure 1. Average uptake of EtBr by the different tropical plants. Uptake of EtBr by different tropical plants are significantly different (p ≤ 0.05) when superscripts located at top of bars are different.

the environment are man-made. It is the building block for the polymer, polyacrylamide, which is considered to be a non-toxic additive. However, if the polymerization process is not perfect and complete, the polyacrylamide may still contain acrylamide which is toxic and may pose risks and hazards to the environment. Another form of acrylamide may pose danger as well in the environment is the acrylamide monomer, also a very toxic organic substance that could affect the central nervous system of humans and is likely to be carcinogenic.

Phytoremediation could be a tool to somehow absorb this neurotoxic agent and lessen the contamination in the soil. This technology could lessen the soil and water contamination by acrylamide thereby limiting the exposure of animals and humans. This technique may also help solve the problem of disposing of contaminated acrylamide waste materials. Thus, Paz-Alberto *et al.* [63] conducted a study 1) to evaluate phytoremediation potentials of some selected tropical plants in acrylamide contaminated soil; and 2) to compare the performance of tropical plants in absorbing acrylamide through accumulation in their roots, stems, and leaves. The 200 grams polyacrylamide gel (PAG) was poured in each pot and mixed thoroughly with the soil by stirring manually. The soil was watered with 1,000 ml water and the test plants were transplanted after three days. Plant samples were collected at 45 days and 60 days after being planted onto PAG contaminated soil. The mustard and pechay were collected after 45 days of exposure while vetiver grass, hogweeds, snake plant, and common sword fern were collected after 60 days of exposure.

Among the plants tested, the highest concentration of acrylamide was absorbed by the whole plant of mustard (6512.8 mg·kg^{-1}) compared with pechay (3482.7 mg·kg^{-1}), fern (2015.4 mg·kg^{-1}), hogweeds (1805.3 mg·kg^{-1}),

vetiver grass (1385.4 mg·kg^{-1}) and snake plants (887.5 mg·kg^{-1}). Results of the study regarding the acrylamide absorption of the whole plants of mustard and pechay conformed to previous findings of other studies (**Figure 2**). Two members of *Brassica* family, *Brassica juncea* L. (mustard) and *Brassica chinensis* L. (pechay) were found to be effective in removing wide ranges of contaminants. Mustard, pechay, and fern plants had 60% survival rate while hogweeds had 80% survival rate. Snake plant and vetiver grass had 100% survival rate.

All the test plants planted in soil without acrylamide had survival rate of 100%. The 100 percent survival rate of vetiver grass and snake plant was due to the tolerance of these plants to acrylamide (**Table 5**). These findings could be attributed to the extraordinary features of vetiver grass such as its massive and deep root system and heavy biomass including its highly tolerance to extreme soil conditions like heavy metal toxicities and high metal concentration.

Results of the study proved that all the test plants are potential phytoremediators of acrylamide. However, mustard and pechay were the most effective as they absorbed the highest acrylamide concentrations in their roots, shoots and the whole plants. On the other hand, vetiver grass and snake plant had the highest uptake of acrylamide even though these plants did not absorb the highest acrylamide concentration. Therefore, these two plants can be considered as the best phytoremediator of acrylamide because they are perennial plants with heavier biomass with long, dense and extended root system. As such, these plants are capable of absorbing acrylamide in the soil for a long period of time.

As preventive measures and for application purposes, vetiver grass and snake plants could be planted along and around the wastewater treatment ponds of laboratories

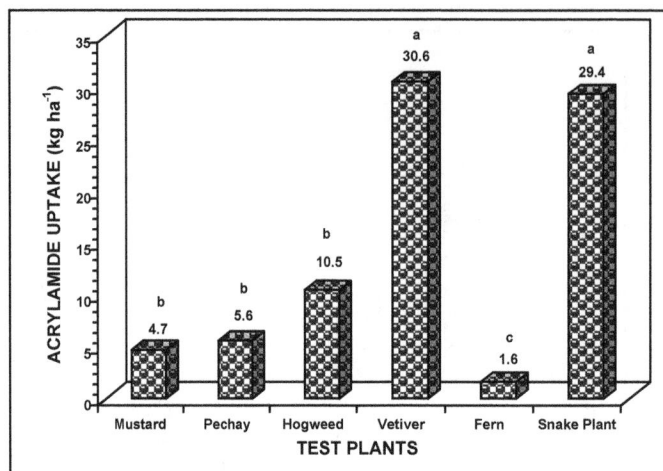

Figure 2. Comparative amount of acrylamide uptake among the different tropical plants. Acrylamide uptakes among the different tropical plants are significantly different ($p \leq 0.05$) when superscripts located at top of bars are different.

Table 5. Survival rate and weight of test plants at harvest.

Test Plants	Survival Rate (%)	Weight at Harvest (g)
Mustard (*Brassica juncea* L.)	60	1.3
Pechay (*Brassica chinensis* L.)	60	2.9
Hogweed (*Portulaca oleracea* L.)	80	10.5
Vetiver (*Vetiveria zizaniodes* L.)	100	39.7
Fern (*Nephrolepsis cordifolia* L.)	60	1.4
Snake plant (*Sanseviera trifasciata* Prain)	100	59.6

using polyacrylamide gel. These plants can prevent further migration of pollutants to the environment aside from making the ponds more resistant to soil erosion. Further studies are suggested to evaluate acrylamide contaminations from laboratory washing, primary treatment pond, and seepage ponds that have earth dikes. Vetiver grass and snake plants are recommended for further phytoremediation studies for longer period of time to test the reduction of acrylamide in soil. Moreover, the outcome of acrylamide accumulation in the plants is also recommended for further study in conjunction with labeled-carbon tracer to determine its effects on the plants.

5. Outlook

Phytoremediation using "green plants" has potential benefits in restoring a balance in stressed environment. It is an emerging low cost technology, non-intrusive, and aesthetically pleasing using the remarkable ability of green plants to metabolize various elements and compounds from the environment in their tissues. Phytoremediation technology is applicable to a broad range of contaminants, including metals and radionuclides, as well as organic compounds like chlorinated solvents, polycyclic aromatic hydrocarbons, pesticides, explosives,

and surfactants. However, phytoremediation technology is still in its youthful development stages and full scale application is still inadequate. As with all new technology, it is important to proceed with caution.

The largest barrier to the advancement of phytoremediation, however, may be public opposition to genetic modification in general. Because all natural hyperaccumulator species are small in size, genetic modification can be used to introduce this technology to other species or to increase the biomass of the natural hyperaccumulators in order to create effective phytoremediators. This public opposition was the same fears that surround the issue of genetic modification of crops, and includes concerns regarding decreased biodiversity, the entry of potentially harmful genes into products consumed by humans, and the slippery slope created by introducing and transferring novel, foreign DNA between non-related species. Nonetheless, the benefits of using phytoremediation to restore balance to a stressed environment seem to far outweigh the costs.

REFERENCES

[1]   N. Kumar, V. Dushenkov, H. Motto and I. Raskin, "Phytoextraction: The Use of Plants to Remove Heavy Metals

from Soils," *Environmental Science and Technology*, Vol. 29, No. 5, 1995, pp. 1232-1238.

[2] M. M. Lasat, "Phytoextraction of Metals from Contaminated Soil: A Review of Plant/Soil/Metal Interaction and Assessment of Pertinent Agronomic Issues," *Journal of Hazardous Substance Research*, Vol. 2, No. 5, 2000, pp. 1-25.

[3] J. A. Pollard and A. J. Baker, "Deterrence of Herbivory by Zinc Hyperaccumulation in *Thlaspi caerulescens*," *New Phytologist*, Vol. 135, No. 4, 1997, pp. 655-658.

[4] Freshwater Management Series No. 2, "Phytoremediation: An Environmentally Sound Technology for Pollution Prevention, Control and Remediation: An Introductory Guide to Decision-Makers. United Nation Environment Program 2000," 2000.

[5] "Sunflowers Bloom in Tests to Remove Radioactive Metals from Soil and Water," *Wall Street Journal*, Vol. 6, No. 2, 1996, pp. 23-28.

[6] National Research Council (NRC), "Lead in the Human Environment," National Academy of Sciences, Washington DC, 1980, p. 525.

[7] P. N. Truong and D. Baker, "Vetiver Grass System for Environmental Protection. Royal Development Projects Protection," Technical Bulletin No. 1998/1, Pacific Rim Vetiver, Office of the Royal Development Projects Board, Bangkok, 1996.

[8] A. Mekonnen, "Handbook on Vetiver Grass Technology—From Propagation to Utilization," GTZ IFSP S/GONDER, Ethiopia, 2000.

[9] C. Knoll, "Rehabilitation with Vetiver," *African Mining*, Vol. 2, 1997, p. 43.

[10] H. Chen, "Chemical Methods and Phytoremediation of Soil Contaminated with Heavy Metals," Vol. 41, American Chemical Society, Washington DC, 1998, pp. 229-234.

[11] P. N. Troung, "Stiffgrass Barrier with Vetiver Grass. A New Approach to Erosion and Sediment Control," *2nd International Vetiver Conference*, 18-22 January 1995, Phechaburi.

[12] N. Roongtanakiat and P. Chairoj, "Vetiver Grass for Remedying Soil Contaminated with Heavy Metals," 2010. http://www.google.com/ Roongtanakiat+ N+and+Chairoj+ R=2001&meta= &aq=o&aqi= &aql=&oq=&gs_rfai= & fp=da1b4ba80a870679

[13] E. R. R. L. Johnson and D. G. Shilling, "Cogon Grass. Plant Conservation Kirchner, A. 2001. Mine-Land Restoration: Phytoremediation of Heavy-Metal Contaminated Sites—A Critical View," International Ecological Engineering Society, 2002. http://www.iees.ch/EcoEng011/EcoEng011_R2.html

[14] L. Mannetje, "Paspalum Conjugatum," 2004. http://www.fao.org/ag/AGP/ AGP/doc/GBASE/DATA/PF000492.HTM

[15] M. Y. Sykes, B. J. Vina and S. Abubakr, "Biotechnology: Working with Nature to Improve Forest Resources and Products," *International Vetiver Conference*, 4-8 February 1999, Chiang Rai, pp. 631-637.

[16] C. Garbisu, "Phytoremediation: A Technology Using Green Plants to Remove Contaminants from Polluted Areas," *Reviews on Environmental Health*, Vol. 17, No. 3, 2002, pp. 173-188.

[17] T. Macek, *et al*, "Phytoremediation: Biological Cleaning of a Polluted Environment," *Reviews on Environmental Health*, Vol. 19, No. 1, 2004, pp. 63-82.

[18] United States Environmental Protection Agency (USEPA), "Hazard Summary. Lead Compounds," 2004. http://www.epa.gov/ttn/atw/hlthef/ lead.html

[19] S. Clemens, M. G. Palmgren and U. Kramer, "A Long Way Ahead: Understanding and Engineering Plant Metal Accumulation," *Trends in Plant Science*, Vol. 7, No. 7, 2002, pp. 309-314.

[20] A. S. Moffat, "Plants Proving Their Worth in Toxic Metal Cleanup," *Science*, Vol. 9, No. 2, 1995, pp. 302-303.

[21] J. W. Huang and S. D. Cunningham, "Lead Phytoextraction: Species Variation in Lead Uptake and Translocation," *The New Phytologist*, Vol. 134, No. 1, 1996, pp. 75-84.

[22] P. C. C. Cortez, "Assessment and Phytoremediation of Heavy Metals in the Panlasian Creek," An Unpublished High School Thesis, University Science High School, Central Luzon State University, Science City of Muñoz, Nueva Ecija, Philippines, 2005.

[23] H. Xia and X. Ma, "Phytremediation of ethion by water hyacinth (*Echhornia crassipes*) from water. Bioresour Technology. College of Food Sciences, Biotechnology and Environmental Engineering, Zhiejang Ghongshang University, Hangzou, 2005.

[24] B. Letachowicz, J. Krawczyk and A. Klink, "Accumulation of Heavy Metals in Organs of *Typha latifolia*," *Polish Journal of Environmental Studies*, Vol. 15, No. 2a, 2006, pp. 407-409.

[25] J. H. Primavera, R. B. Sadaba, M. J. H. L. Lebata and J. P. Altamirano, "Handbook of Mangroves in Philippines-Panay," SEAFDEC Aquaculture Department, Iloilo, 2004.

[26] K. Kathiresan and B. L. Bingham, "Biology of Mangroves and Mangrove Ecosystems," 2011. http://faculty.wwu.edu/ bingham/mangroves.pdf

[27] Philippines Environment Monitor, "Resources and Ecosystems," 2012. http://siteresources.worldbank.org/INTPHILIP PINES/Resources/PEM05-ch2.pdf

[28] J. B. Long and C. Giri, "Mapping the Philippines' Mangrove Forests Using Landsat Imagery," 2011. www.mdpi.com/1424-8220/11/3/2972/pdf

[29] P. J. Hogarth, "The Biology of Mangroves and Seagrasses," Oxford University Press, Oxford, New York, 2007, p. 217.

[30] W, Zheng, X. Chen and P. Lin, "Accumulation and Biological Cycling of Heavy Metal Elements in *Rhizophora stylosa* Mangroves in Yingluo Bay, China," 2012.

http://www.int-res.com/articles/meps/159/m159p293.pdf

[31] G. R. Macfarlane and M. D. Burchett, "Toxicity, Growth and Accumulation Relationships of Copper, Lead and Zinc in the Grey Mangrove *Avicennia marina* (Forsk.) Vierh," 2011.
http://www.mendeley.com/research/ toxicity-growth-and-accumulation-rel ationships-of-copper-lead-and-zinc-in-the-grey-mangrove-avicennia-mari na-forsk-vierh/

[32] G. R. Macfarlane and M. D. Burchett, "Cellular Distribution of Copper, Lead and Zinc in the Grey Mangrove, *Avicennia marina* (Forsk.) Vierh," 2011.
http://www.sciencedirect.com/science/article/pii/S030437 7000001054

[33] S. Cheng, "Heavy Metals in Plants and Phytoremediation," 2012.
http://ir.ihb.ac.cn/bitstream/152342/9658/1/Heavy% 20metals%20in% %20journals.pdf

[34] I. Sari, Z. Bin Din, and G. W. Khoon, "The Bioaccumulation of Heavy Metal Lead in Two Mangrove Species (*Rhizophora apiculata* and *Avicennia alba*) by the Hydroponics Culture," 2011.
http://eprints.usm.my/ 1490/1/ The_Bioaccumulation_Of_ Heavy_ Metal_Lead_In_Two Mangrove_Species_ (Rhizophora_apiculata_and_Aviicennia_alba)_by_The_Hydropo nics_Culture.pdf

[35] P. Saenger and D. McConchie, "Heavy Metals in Mangroves Methodology, Monitoring and Management," 2011.
http://www.frienvis.nic.in/bulletinwork/bulletin.htm

[36] A. Shete, V. R. Gunale and G. G. Pandit, "Bioaccumulation of Zn and Pb in *Avicennia marina* (Forsk.) Vierh. and *Sonneratia apetala* Buch. Ham. from urban Areas of Mumbai (Bombay), India," 2011.
http://www.ajol.info/index.php/jasem/article/viewFile/55 142/4 3614

[37] B. Y. Kamaruzzaman, M. C. Ong, K. C. A. Jalal, S. Shahbudin and O. Mohd Nor, "Accumulation of Lead and Copper in *Rhizophora apiculata* from Setiu Mangrove Forest, Terengganu, Malaysia," 2011.
www.jeb.co.in/journal_ issues/200909_sep09_supp/paper_ 08.pdf

[38] V. Pahalawattaarachchi, C. S. Purushothaman and A. Venilla, "Metal Phytoremediation Potential of *Rhizophora mucronata* (Lam.)," 2011.
http://nopr.niscair.res.in/bitstream/123456789/.../IJMS%2 038(2)%20178-183.pdf

[39] M. F. Nazli and N. R. Hashim, "Heavy Metal Concentrations in an Important Mangrove Species, *Sonneratia caseolaris*, in Peninsular Malaysia," 2011.
www.tshe.org/ea/pdf/vol3s%20p50-55.pdf

[40] H. Parvaresh, Z. Abedi, P. Farshchi, M. Karami, N. Khorasani and A. Karbassi, "Bioavailability and Concentration of Heavy Metals in the Sediments and Leaves of Grey Mangrove, *Avicennia marina* (Forsk.) Vierh, in Sirik Azini Creek, Iran," 2011.
http://www.mendeley.com/research/bioavailability-conce ntration-heavy-m etals-sediments-leaves-grey-mangrove-avicennia-marina-forsk-vierh-sirik-azini-creek-iran/

[41] Y. W. Qui, K. F. Yu, G. Zhang and W. X. Wang, "Ac-

cumulation and Partitioning of Seven Trace Metals in Mangroves and Sediment Cores from Three Estuarine Wetlands of Hainan Island, China," 2012.
http://www.sklog.labs.gov.cn/atticle/B11/B11012.pdf

[42] J. E. Zhang, J. L. Liu, Y. Ouyang B. W. Lia and B. L. Zhao, "Physiological Responses of Mangrove *Sonneratia apetala* Buch-Ham Plant to Wastewater Nutrients and Heavy Metals," 2011.
http://www.tandfonline.com/doi/abs/10.1080/1522651100 3671395#preview-physio

[43] I. J. Nirmal, P. R. Sajish, R. Nirmal, G. Basil, and V. Shailendra, "An Assessment of the Accumulation Potential of Pb, Zn and Cd by *Avicennia marina* (Forsk.) Vierh. in Vamleshwar Mangroves, Gujarat, India," 2011.
http://notulaebiologicae.ro/nsb/artic le/viewFile/5593/5343

[44] A. N. Subramanian, "Persistent Chemicals," 2011.
http://ocw.unu.edu/international-network-on-water-enviro nment-and-health/unu-inweh-course-1-mangroves/Persist ant-Chemicals.pdf

[45] US Environmental Protection Agency, "Lead-How Lead Affects the Way We Live and Breath. Office of Air Quality Planning and Standards," 2004.
http://www.epa.gov/air/urbanair/lead/index.html

[46] J. W. Huang and S. D. Cunningham, "Lead Phytoextraction: Species Variation in Lead Uptake and Translocation," *New Phytologist*, Vol. 134, No. 1, 1996, pp. 75-84.
http://www.lenntech.com/Pb-en.htm

[47] J. Wu, F. C. Hsu and S. D. Cunningham, "Chelate-Assisted Pb Phytoextraction: Pb Availability, Uptake, and Translocation Constraints," *Environmental Science and Technology*, Vol. 33, No. 11, 1999, pp. 1898-1904.

[48] B. J. Alloway, "Cadmium. Heavy Metals in Soil. London," Blackie and Son, London, 1990.

[49] J. W. Huang, J. Chen, W. R. Berti and S. D. Cunningham, "Phytoremediation of Lead-Contaminated Soils: Role of synthetic Chelates in Lead Phytoextraction," *Environmental Science and Technology*, Vol. 31, No. 3, 1997, pp. 800-805.

[50] D. E Salt, M. Blaylock, I. Chet, S. Dushenkov, B. Ensley, P. Nanda and I. Raskin, "Phytoremediation: A Novel Strategy for the Removal of Toxic Metals from the Environment Using Plants," *Biotechnology*, Vol. 13, No. 5, 1995, pp. 468-474.

[51] S. R. Smith, "Effect of Soil pH on Availability to Crops of Metals in Sewage Sludge Treated Soils, Cadmium Uptake by Crops and Implications for Human Dietary Intake," *Environmental Pollution*, Vol. 86, No. 1, 1994, pp. 5-13.

[52] A. Cholpecka, J. R. Bacon, M. J. Wilson and J. Kay, "Heavy Metals in the Environment. Forms of Cadmium, Lead, and Zinc in Contaminated Soils from Southwest Poland," *Journal of Environmental Quality*, Vol. 25, No. 1, 1996, pp. 69-79.

[53] M. Sadiq and G. Hussain, "Effect of Chelate Fertilizers

on Metal Concentrations and Growth of Corn in a Pot Experiment," *Journal of Plant Nutrition*, Vol. 16, No. 4, 1993, pp. 699-711.

[54] S. P. Bizily, C. L. Rugh, A. O. Summers and R. B. Meagher, "Phytoremediation of Methylmercury Pollution: merB Expression in Arabidopsis Thaliana Confers Resistance to Organomercurials," *Proceedings of the National Academy of Sciences of the United States of America*, Vol. 96, No. 12, 1999, pp. 6808-6813.

[55] S. Abdulla, "Tobacco Sucks up Explosives," *Nature*, 2002. http://www.nature.com/nsu/990429/990429-5.html

[56] A. Kirchner, "Mine-Land Restoration: Phytoremediation of Heavy-Metal Contaminated Sites—A Critical View," International Ecological Engineering Society, 2002. http://www.iees.ch/EcoEng011/EcoEng011_R2.html

[57] A. M. Paz-Alberto, G. C. Sigua, B. G. Baui and J. A. Prudente, "Phytoextraction of Lead-Contaminated Soil Using Vetiver grass (*Vetiveria zizanioides* L.), Cogon grass (*Imperata cylindrica* L.) and Carabao grass (*Paspalum conjugatum* L.)," *Environmental Science and Pollution Research*, Vol. 14, No. 7, 2007, pp. 498-504.

[58] M. B. Bautista, "Phytoremediation Potential of Selected Plants on Lead in Nueva Ecija," An Unpublished High School Thesis, University Science High School, Central

Luzon State University, Science City of Muñoz, Nueva Ecija, 2006.

[59] K. A. Gray, "The Phytoremediation of Lead in Urban, Residential Soils," Northwestern University, Evanston, 2000.

[60] J. R. Undan, R. T Alberto, A. M. Paz-Alberto and C. T. Galvez, "Heavy Metals in Plant Species in Mine Tailings of Victoria in Manlayan, Benguet Province," *CAS Faculty Journal*, Vol. 2, No. 1.

[61] M. Wislocka, J. Krawczyk, A. Klink and L. Morrison, "Bioaccumulation of Heavy Metals by Selected Plant Species from Uraniom Mining Dumps in the Sudety Mountains, Poland," *Polish Journal of Environmental Studies*, Vol. 15, No. 5, 2006, pp. 811-818.

[62] R. B. Uera, A. M. Paz-Alberto and G. C. Sigua, "Phytoremediation Potentials of Selected Tropical Plants for Ethidium Bromide," *Environmental Science and Pollution Research*, Vol. 14, No. 7, 2007, pp. 505-509.

[63] A. M. Paz-Alberto, M. J. J. De Dios, R. T. Alberto and G. C. Sigua, "Assessing Phytoremediation Potentials of Selected Tropical Plants for Acrylamide," *Journal of Soils and Sediments*, Vol. 11 No. 7, 2011, pp. 1190-1198.

11

Modelling Weather and Climate Related Fire Risk in Africa

Flávio Justino[1], F. Stordal[2], A. Clement[3], E. Coppola[4], A. Setzer[5], D. Brumatti[1]

[1]Departamento de Engenharia Agrícola, Universidade Federal de Viçosa, Viçosa, Brazil
[2]Department of Geosciences, University of Oslo, Oslo, Norway
[3]Department of Earth and Planetary Sciences, Johns Hopkins University, Baltimore, USA
[4]The Adbus Salam International Centre for Theoretical Physics, Trieste, Italy
[5]National Institute of Space Research, S. J. Campos, Brazil

ABSTRACT

Based on regional climate model simulations conducted with RegCM3 and NCEP Reanalyses, the impact of anomalous climate forcing on environmental vulnerability to wildfire occurrence in Africa is analyzed by applying the Potential Fire Index (*PFI*). Three different model-based vegetation distributions were analyzed for a present day simulation (1980-2000) and for the end of the twenty-first century (2080-2100). It was demonstrated that under current climate and vegetation conditions the *PFI* is able to reproduce the principal fire risk areas which are concentrated in the Sahelian region from December to March, and in subtropical Africa from July to October. Predicted future changes in vegetation lead to substantial modifications in magnitude of the *PFI*, particularly for the southern and subtropical region of Africa. The impact of climate changes other than through vegetation, was found to induce more moderate changes in the fire risk, and increase the area vulnerable to fire occurrence in particular in sub-Saharan. The *PFI* reproduces areas with high fire activity, indicating that this index is a useful tool for forecasting fire occurrence worldwide, because it is based on regionally dependent vegetation and climate factors.

Keywords: Climate Changes; Soybeans; Maize; Amazon

1. Introduction

Vegetation fires are extensive in Africa, and any changes in the risk of fires in a changing climate will have important impacts on societies in several African countries. Burning of biomass plays an important role in global emissions of carbon and other trace gases [1,2]. Vegetation fires on a global scale are the second largest anthropogenic source of greenhouse gas emissions [3]. Reference [4] reported that the amount of carbon dioxide release from Indonesian fires in 1997 and 1998, which was equivalent to 25% of total CO_2 emissions resultant from combustion of fossil fuels worldwide.

As discussed by [5], Africa is a continent highly prone to lightning storms and fire distribution is associated with both dry and wet periods. In the past two million years, lightning was the primary ignition source of fires in the Africa savannas, however, at present humans are playing a more important role in starting fires [6]. In this sense, in the year 2000 the most extensively burned areas were located in sub-Saharan Africa with the total area burned estimated at 959,480 km^2 [7].

Based on satellite data [8] demonstrated that in northern hemisphere Africa between 2001 and 2004, over 10^6 km^2 burned. It should be noted that the total burned area during this period in the Northern and Southern Hemisphere Africa represents 70% of the total area burned globally. Nevertheless, due to the coarse resolution of satellite detected-fire and scarcity of spatial distribution, a systematic evaluation of long term interannual variability of fire and burned area remain a matter of discussion. However, based on modeling results [9] claim that, by the year 2060 Africa may experience a strong decrease of burned area of ca. 20% - 25%. This is attributed to be linked to changes in social conditions and land use. There is a lively debate on the importance of anthropogenic and climate factors contributing to the ignition of vegetation fires [e.g. 10]. Under current condition, [11,12] found the best agreement between simulation and observations for the fire by explicitly considering human caused ignition and fire suppression as a function of population density [12], however, based on temperature, precipitation, relative humidity, lightning activity, land cover and population density datasets, argued that future climate conditions will play the major role in driving global fire trends, overcoming the human effect on fire ignition. It has also been demonstrated that increased surface soil moisture conditions limit the extent of burned area [9,13].

Fire occurrence is determined by factors which start the combustion reaction and that allows its continuation depending on the potential energy stored in the combustible material [e.g. 14-17]. It has also been demonstrated that changes induced in biome distribution may also alter atmospheric/landscape susceptibility to fire occurrence [18,19]. Reference [20] argued that the clearing of tropical savannas increases temperatures and wind speeds while reducing precipitation and relative humidity. This is a situation which may occur in a climate influenced by anthropogenic factors and may substantially increase fire frequency.

Despite the relevance of vegetation fires in determining global vegetation patterns and the atmospheric concentration of greenhouse gases [4,15,21], there is a lack of systematic investigations focusing on future fire risk in Africa based on models which include interaction between climatic variables and vegetation patterns. Africa is subject to a majority of number of fires detected globally (50% of all detected) mainly occurring in savanna regions [9,22].

Distinct changes in vegetation patterns associated to future climate changes have been found to increase potential free activity [2,23,24]. Using the Potential Fire Index (*PFI*), [18] found that under greenhouse warming conditions the *PFI* indicates an increase in the fire risk area, particularly for the Amazon region. Model-based investigations have also been conducted to understand the interaction between fire, the CO_2 fertilization effect and Earth's ecosystems and dominant plant communities [e.g. 2,15,25].

It should be noted that these more advanced methods based on process-based fire regime models coupled with ecosystem dynamics models [e.g. 15,26] have been successfully used to investigate the spatial and temporal evolution of fires, and their interaction with vegetation dynamics. Several parameters are used for simulating fire processes involving soil characteristics, carbon allocation, fuel loads and the moisture content of litter. One limitation of using these complex models is the need for several parameters that must be included to simulate the link between fire and ecosystem dynamics [15]. Reference [27] argued that these models may also include a considerable level of uncertainties due to difficulties in properly simulating the fuel biomass and plant competition.

This study aims to investigate how changes in climate and vegetation impact the risk of fire in Africa (vulnerability of a region to fire initiation). The method proposed here to investigate climate and vegetation factors which may induce wildfire, is built on the Potential Fire Index (*PFI*), in a simpler approach than estimating fires in e.g. by using dynamic global vegetation models (DGVMs) [e.g. 15,28]. This is carried out through climate simulations using the RegCM3 climate model associated with three vegetation biome distributions derived from the present day (1980-2000) and global warming induced climate (2080-2100). Moreover, the *PFI* methodology provides the possibility of calculation of fire risk as a result of computation of equations which are function of four parameters only, allowing for calculation of fire risk on regional and local basis. It must be noted that the ignition of fires is not taken into account, so that the risk is only potential in this respect.

2. Data and Methodology

2.1. Climate Data

The African climate is affected by regional and global climate processes [e.g. 29-31] which involves the interaction between sea surface temperatures [32,33], vegetation/land cover and soil moisture [34].

To evaluate the fire risk in Africa climate data provided by ENSEMBLES Regional climate models and future projections were utilized [35]. We have used results from an experiment with the RegCM3 which was integrated in a continuous transient scenario simulation for a 120-year period from 1980-2100 [35]. RegCM3 is a 3-dimensional, sigma-coordinate, primitive equation regional climate model. Initial and 6-hourly lateral boundary conditions and SSTs necessary to run the model were obtained from a corresponding simulation with the ECHAM5 AOGCM (European Centre Hamburg Model—Atmosphere Ocean Global Climate Model) [36]. The experiment was based on the SRESAIB scenario (Special Report on Emission Scenario A1B) [37].

As demonstrated by [35,38], analysis of the present day period (PD, 1980-2000) shows that the regional model is able to capture the basic climate characteristics of the African continent. For instance, the main African circulation such as tropical jet streams, monsoon flows and the mid-tropospheric African easterly jet (AEJ) stream are well reproduced, although the model has a tendency to produce a weaker AEJ. The position and strength of the core of the AEJ and monsoon flow are located correctly as well as the strength of the tropical easterly jet stream core and the height of the AEJ.

The December-January-February (DJF) maximum temperature bias is low and confined between ± 2 degrees except over the Sahara desert were the negative bias is slightly larger. Positive biases are mainly located over the tropical forest regions of the Congo Basin, in the southwestern coasts of the continent. JJA temperature biases are similar in magnitude to that of the DJF. Generally the cold bias is found in the central Sahara and a strong positive bias over the Southern Saudi Arabian peninsula. Over the rest of the continent the RegCM3 presents a more mixed distribution of positive (e.g. over the Congo Basin) and negative (e.g. over the Ethiopian Plateau) biases. The precipitation bias in DJF shows an area of negative bias extending in a southwestward direction from the Congo Basin to the southwestern African coast. A positive bias is found over eastern equatorial Africa and the southeastern African coast and a dry bias over the African monsoon region. An overestimation of precipitation is evident over the Ethiopian highlands. These biases do not exceed 2 mm/day during the rainy season.

We also use climate data from the RegCM3 for a future climate to access changes in fire risk. In this case, data for the period 2080-2100 have utilized in the A1B experiment described above. We denote this a Greenhouse Warming (GW) simulation.

2.2. Formulation of the Potential Fire Index (*PFI*)

The *PFI* is formulated on the principle that the vegetation fire risk increases with the increasing duration of dry periods. Type and natural cycle of vegetation phenology, maximum temperature and relative humidity of the air are also required to compute the *PFI*. The reference of the calculations is the *Number of Dry Days* or *Days of Drought* (*DD*), which is the number of days without precipitation during the proceeding 120 days [18].

For the *PFI* calculation, a set of parameterization equations applied for Brazil are utilized for Africa. It is briefly described below, and is presented in details by [18]. A set of modification are conducted, however, to compute the basic risk (*BR*) for the African vegetation types considered:

$$BR_i = f\left[1 + \sin\left(A_i \times DD\right)\right] \quad (1)$$

i = 1, 7, *DD* must be in radians. The 7 values of "*A*" vary as a function of vegetation flammability, and $f = 0.45$ and are given in **Table 1**. The resulting 7 values for *BR* are showing in **Figure 1**.

The *A* and *f* values are fitted based on 30 years of fire observation in Brazil from satellite. Specifically, the presence of fire has been taken in association with the amount of precipitation occurrence in the last 120 days in 11 discreet periods. Afterwards, empirical equations have been set to correlate the fire, the amount of precipitation (precipitation factor) and the vegetation type. The Basic

Table 1. Vegetation types and respectives *A* values used to define the basic risk (*BR*). African vegetation is in italic.

Vegetation	Vegetation Class	*A* Value
0	Water	-x-
1	Evergreen Needleleaf Forest	2
2	Evergreen Broadleaf Forest	1.5
3	Deciduous Needleleaf Forest	2
4	Deciduous Broadleaf Forest	1.72
5	Mixed Forest	2
6	Closed Shrublands	2.4
7	Open Shrublands	3
8	Woody Savannas	2.4
9	Savannas	3
10	Grasslands	6
11	Permanent Wetlands	1.5
12	Croplands	4
13	Urban and Built-Up	-x-
14	Cropland/Natural Vegetation Mosaic	4

Risk (*BR*) increases as a sine curve over time. This pattern has been chosen because the variation of intensity and duration of sunlight along the year is also sinusoidal and phenology of vegetation naturally follows the same pace.

The "*A*" values may represent the amount of the combustible material during periods without precipitation or drought. In this sense the dense evergreen broadleaf forests attain "*A*" value of 15, savannas of 3 and cropland and grassland (4 and 6) which are more susceptible to fire. **Figure 1(f)** shows that 30 consecutive days without precipitation results in a *BR* maximum (0.9) assuming grassland. It important to note that soil moisture, which is tightly linked to the precipitation amount, and will increase leaf litter is not explicitly included in the parameterization.

Two other factors are considered for calculation of the potential fire risk (*PFI*): the minimum relative humidity and the maximum temperature of the air:

$$PFI = BR \times \left(a \times RH_{\min} + b\right) \times \left(c \times T_{\max} + d\right) \quad (2)$$

where the values $a = -0.006$, $b = 1.3$, $c = 0.02$ and $d = 0.4$ are taken from [18]. $RH_{\min}$ is the minimum daily relative humidity and $T_{\max}$ is the maximum daily temperature. It should be noted that the *PFI* is computed on daily basis but results presented here are shown based on the monthly average.

The *PFI* has been extensively used in South America and Cuba to evaluate the atmospheric susceptibility to fire development [39,40]. It has been demonstrated that among 290 thousand fires in the continent in 2003, 94% occurred in areas of high and critical risk (*PFI*) in the fire season.

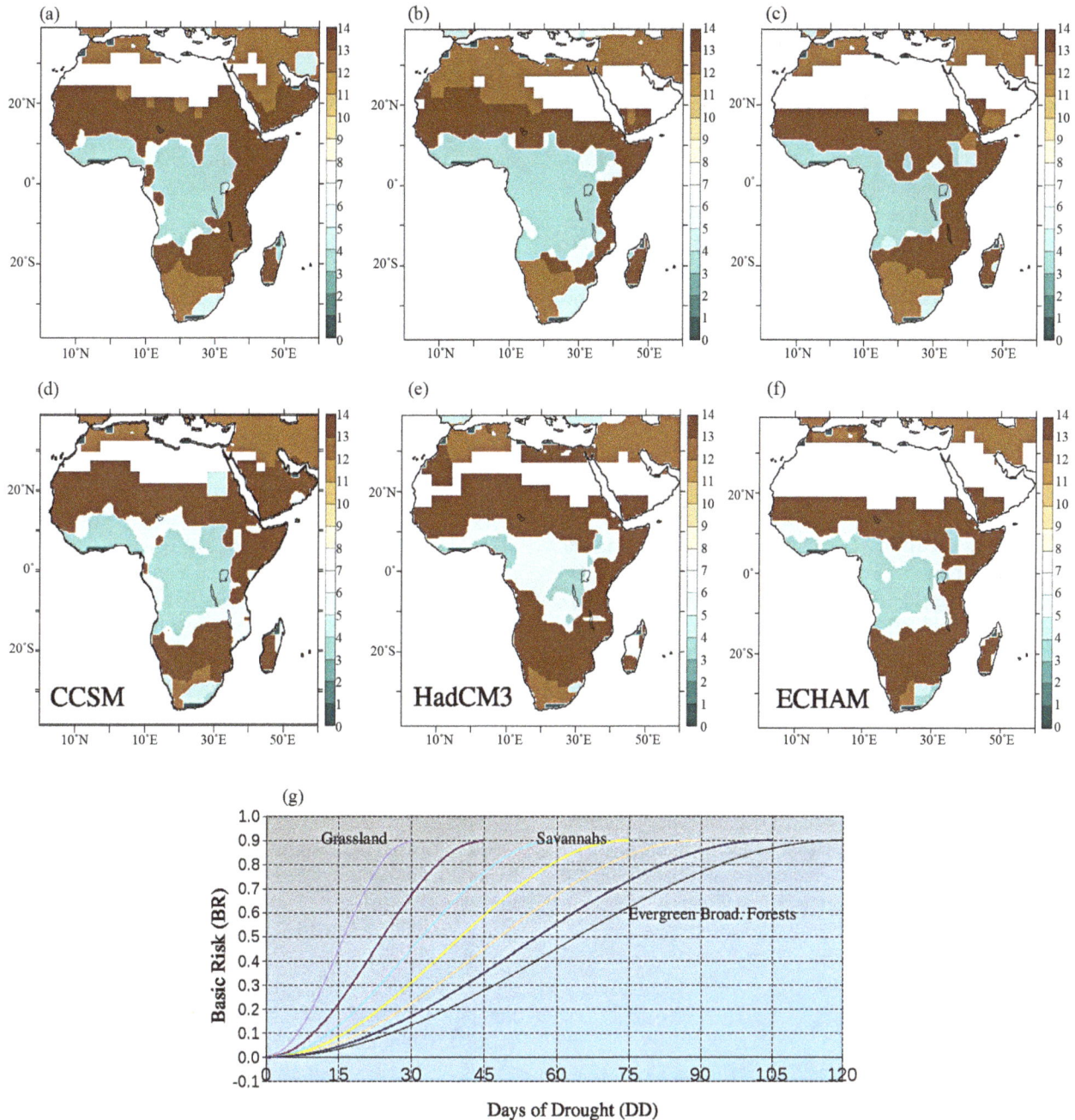

Figure 1. Vegetation distribution simulated by CLM-DGVM driven by climate forcing from CCSM (left), HadCM3 (middle) and ECHAM (right) for pre-industrial conditions (a)-(c) and for greenhouse warming conditions (d)-(f) (Alo and Wang, 2008). See Table 1 for the biomes denomination, (g) Time evolution of the basic risk as a function of the days of drought.

It has been considered to derive new parameters for Africa but the ones derived for Brazil are shown to work well in Africa, which is reasonable as they are fitted in the tropics and subtropics. In fact, the main point here is that changes in vegetation types play a major role. It should be noted that the *PFI* parameterization as proposed here does not work properly for extra-tropical climates with different precipitation regimes and lower temperatures.

2.3. Vegetation Data

We base our vegetation data on the results from the study of [19] which used CLM-DGVM to derive vegetation distributions in equilibrium with climates of several global climate models. In this paper we use the results based on the CCSM, HadCM and ECHAM models. The CLM-DGVM is composed of a land surface scheme CLM3.0, a phenology module, and biogeochemistry and vegetation dynamic modules based on CLM-DGVM [41]. Biogeophysical and biogeochemical processes are simulated with a 20-min time step while plant phenology is evaluated daily. Vegetation structure and distribution are updated yearly based on knowledge of the integrated processes for shorter time steps.

CLM-DGVM models 10 plant functional types (PFTs), namely, needleleaf evergreen temperate trees, needleleaf evergreen boreal trees, broadleaf evergreen tropical trees, broadleaf evergreen temperate trees, broadleaf deciduous tropical trees, broadleaf deciduous temperate trees, broadleaf deciduous boreal trees, C3 artic grasses, C3 nonartic grasses and C4 grasses. Up to 10 PFTs may coexist in each grid cell. The fractional coverage of the PFTs was used to assign the grid cells to the biome-type vegetation classes in **Table 1** (following the biome types and associated PFTs) [42].

Pertaining to each of the three GCMs employed (*i.e.*, CCSM, HadCM and ECHAM), control vegetation was simulated with CLM-DGVM driven by the present day mean climate (derived based on a 30-year period of model integration from the present day (PD) control GCM (CO_2 concentration held at 275 ppm experiment), and simulated future vegetation with the vegetation model based on the 2100 mean climate data derived for the period 2071-2100 from the GCMs SRESA1B stabilization denoted as the global warming (GW) case. In this case CO_2 concentration stabilizes at 720 ppm beyond 2100. In all the vegetation simulations, the CLM-DGVM was initiated considering bare ground and run for 200 years in order to attain vegetative equilibrium with the specified climate forcing (*i.e.*, with respect to leaf area index (LAI) and vegetation coverage). A detailed description of the CLM-DGVM and vegetation simulations is given in [19].

3. Results and Discussion

3.1. Present Day Analysis

In order to investigate the impact of climate and vegetation changes on fire susceptibility in Africa, three evaluations were performed: the first is based on the present day (PD) climate (1980-2000) as predicted by the RegCM3 and simulated pre-industrial vegetation from [19] (**Fig-**

ures 1(a)-(c)). The second calculation uses results of the RegCM3 greenhouse warming simulation (GW simulation) and the greenhouse vegetation distribution (GW, **Figures 1(d)-(f)**). One should note that we have used equilibrium vegetation for both epochs. The third analysis was performed for future climate conditions (2080 to 2100) and disregarded changes in vegetation, using the current biome distribution.

In what follows the monthly evolution of the *PFI* as well as some related parameters are shown in **Figure 2**. It should be noted that the largest number of vegetation fires in Africa occurs between December and February (DJF) in the equatorial/Sahelian region (10°N - 20°N) and between August and October (ASO) in the subtropical region of the Southern Hemisphere [43,44]. Therefore, the analyses conducted herein are restricted to these months. **Figure 3(a)** shows the averaged daily maximum air temperature at 2 m averaged from December to February for the PD simulation (1980-2000). There are evidently higher temperatures along the equatorial belt and in subtropical Africa with values as high as 37°C. Lower values are predominant in Northern Africa.

As expected, analyzes for relative humidity demonstrated that areas with high temperatures are also dominated by low relative humidity (**Figure 3(b)**). These conditions intuitively may favor intensified fire occurrence. However, climate conditions in accordance with the days of drought (*DD*) shed some light on the role of precipitation. For instance, the equatorial region (5°N - 15°N) exhibits larger *DD* values or long periods without rain (up to 1), whereas the sub-tropical region is dominated by small *DD* values which indicates frequent precipitation (**Figure 3(c)**).

The combination of these three factors (maximum temperature, relative humidity and days of drought) associated with the vegetation pattern allows for simulation of the potential fire risk (*PFI*). As shown in **Figures 3(d)-(f)**, the atmospheric susceptibility to fire or higher *PFI* is located between 10°N and 20°N (Western Africa) in the area covered by non-forest, with high temperature and low relative humidity. Impact of the *DD* on the *PFI* is highlighted by the decreasing fire risk between 0 - 20°S latitude. Despite high temperatures and moderate relative

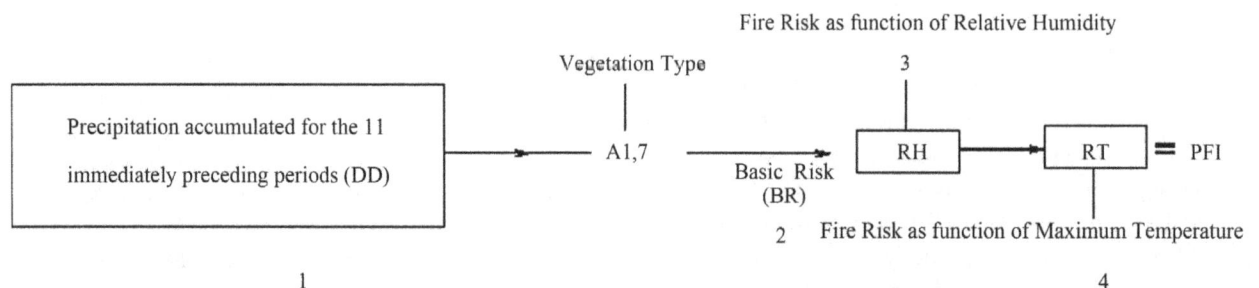

Figure 2. Flowchart presenting the sequence (as represented by numbers) of calculation for the *PFI*.

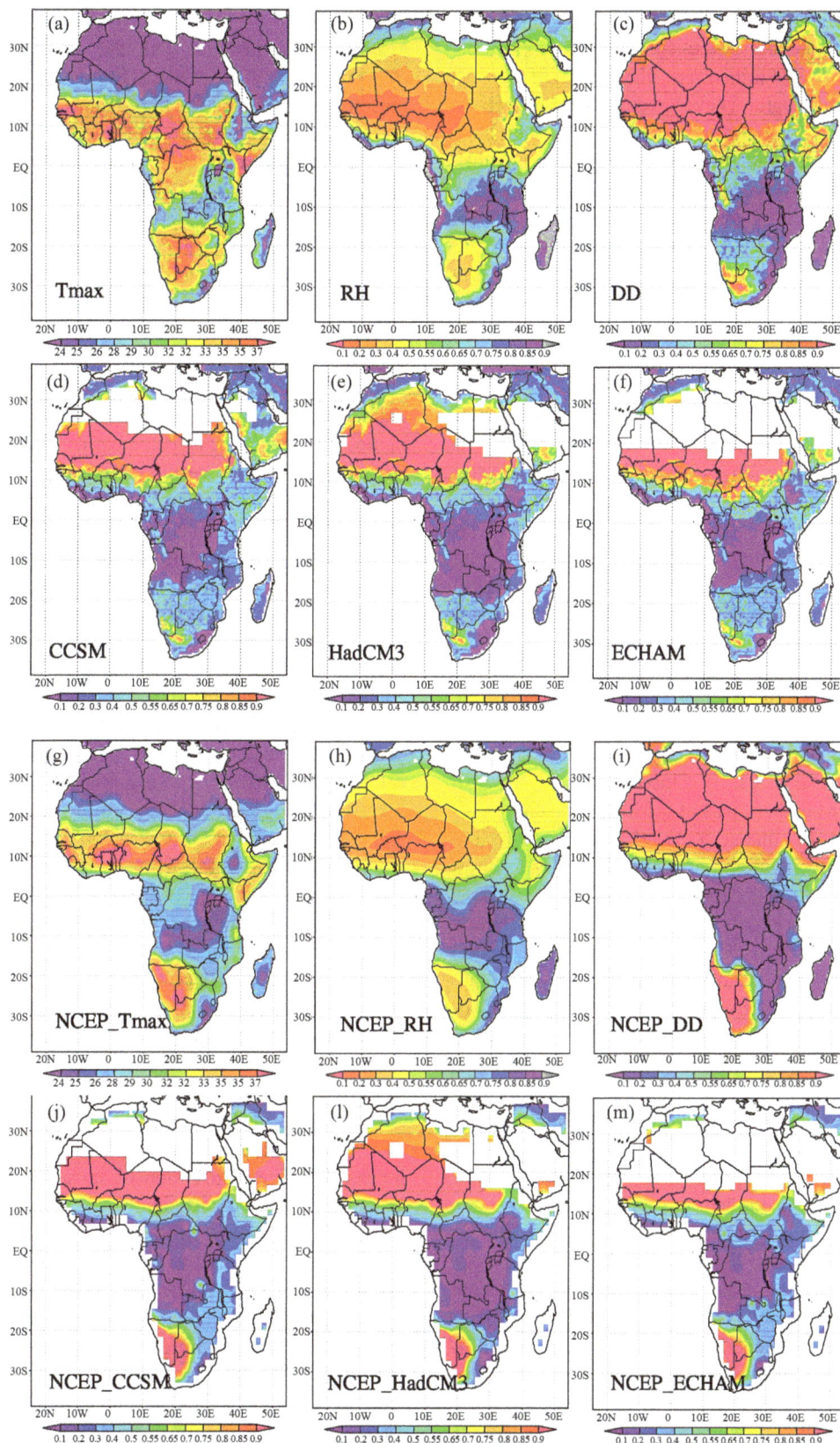

Figure 3. Present day maximum temperature ((a), °C), relative humidity ((b), %) and number of days of drought (c) averaged for December-January-February (DJF). (d), (c) and (f) is the potential fire index (*PFI*) based on CCSM, HadCM3 and ECHAM forced-vegetation. Figures from (g) to (m) are the same for NNR2.

humidity this region does not show high *PFI* due to the small *DD* which indicates frequent precipitation associated with the ITCZ. Furthermore, the dominant vegetation pattern in this region is evergreen or semi-evergreen forests, where the fuel consists of densely packed litter layers. It is interesting to note that the vegetation pattern based on by the CCSM, HadCM3 and ECHAM disagree on the Saharan desert extension but they do show similar biomes in central Africa. The CLM-DGVM forced by the CCSM and ECHAM simulates the African tropical forests more confined as compared to the simulation with the HadCM3 forcing (**Figures 1(a)-(c)**).

Because the *PFI* is based on RegCM3 model data, it is important to identify possible model biases in the predicted present-day maximum temperature, relative humidity and days of drought, and consequently in the *PFI* as represented by RegCM3. This is done by comparing the modeled data and the daily NCEP Reanalyses 2 (NNR2) for the period of 1980-2000 [45]. **Figures 3(a)-(l)** demonstrated that the two datasets, the RegCM3 and NNR2 yield many similarities, particularly in the Sahelian region and southern Africa (10°N and 10°S, **Figures 1(a)-(c) and (g)-(i)**).

Regarding to evaluation of the *PFI*, it may be noted that the NNR2 exhibits higher *PFI* as compared to the modeled output, over the Sahelian region due to lower *DD* values and higher (lower) air temperature (relative humidity). Differences between the two datasets are also identified in the southern part of Africa where the NNR2 shows higher *PFI* is linked to the *DD* values or long periods without precipitations. In general, the *PFI* shows many similarities in both datasets.

Vegetation fires primarily follow seasonal climate shifts [e.g. 10,46,47]. During ASO (August-September-October) maximum temperature values are observed in the Saharan region accompanied by lower relative humidity. Contrary to DJF climate features, ASO exhibits moderate temperatures and higher relative humidity along the equatorial belt (**Figures 4(a)-(c)**). Influence of the ITCZ is noted by the extremely low value of *DD* which extends from 10°N to 5°S.

A second interval with substantial fire activity in Africa is in agreement with the onset of the dry season due to the northward displacement of the ITCZ, which is located in the Northern Hemisphere. During ASO, the highest *PFI* occurs in Central and Southern Africa, which include Angola, Zambia, Zimbawe, Namibia and Botswana, due to high Tmax, low *RH* and critical *DD* (**Figure 4**), whereas Malawi and central Africa, for instance, have relatively low *PFI* magnitudes. Higher *PFI* in parts of the African continent is associated with the dominance of deciduous forest-woodland savanna and brush-grass savanna. According to the formulation of the *PFI* these biomes are associated with increased susceptibility to fire

activity due to the vulnerability of drought-induced fuel properties. Comparison with the climate variables and *PFI* provided by the NNR2 (**Figures 4(g)-(i)**), shows a reasonable agreement for maximum temperature, relative humidity and days of drought over the vast majority of Africa. Some discrepancies may be found in southern Africa. In this region the *PFI* attains critical values in the NNR2 model that extends for a larger area, whereas in the RegCM3 the critical *PFI* is confined in particular to the western part (**Figures 4(c)**, **(d)** and **(j)-(m)**).

Comparison with previous results based on models with higher complexity [e.g. 11,15], reveals that the *PFI* is able to reproduce areas dominated by frequent fire activity (see **Figure 1** in [15]). According to [15], moderate fire activity is shown in central Africa, whereas the highest fire incidence is located in the Sahel and southern Africa, similar to the results proposed in the present study. It should be noted that [11] shows present time averaged data.

In order to further validate the *PFI*, **Figure 5** shows the satellite based hot spots/fires from the World Fire Atlas. This data forms a unique long time series of global fire location and timing (http://dup.esrin.esa.int/ionia/wfa/index.asp), and may be compared with satellite data obtained from MODIS [43, 44], http://rapidfire.sci.gsfc.nasa.gov/firen. This may be used to represent fire activity during DJF and ASO because fire is detected with brightness temperature higher than 308 K. **Figure 5** includes all fire activity during DJF and ASO detected in 2004 and 2008. The year 2004 was characterized by a weak El Niño and 2008 experienced a recovery from La Niña conditions which may provide a range of climate-induced vegetation fires. It should be noted that the spatial resolution of the RegCM3 model (40 km) makes it difficult to compare the *PFI* with small scale observed hot spots. By showing hot spots it is only intended to give the preferential location of fire occurrence in Africa, to provide inferences about the validity of the presented *PFI*.

One may argue that there is no substantial inter-annual modification in the climate-driven preferable area of fire incidence, since distribution of savannas, shrubland and grasslands must be sufficiently dry to ignite (**Figure 5**). Since this study aims to identify areas with fire risk the comparison of the *PFI* with hot spots is a reasonable assumption. The close correspondence between higher *PFI* (0.7 - 0.9) and the detected fires is remarkable. One should keep in mind, however, that the presented *PFI* is smoothened by utilizing the seasonal average contrary to what is shown for fires which accumulates on a daily basis.

The satellite-based wildfires reveal the high number of fires in the Sahelian region which are more recurrent in DJF, and in the subtropical region of Africa which mostly

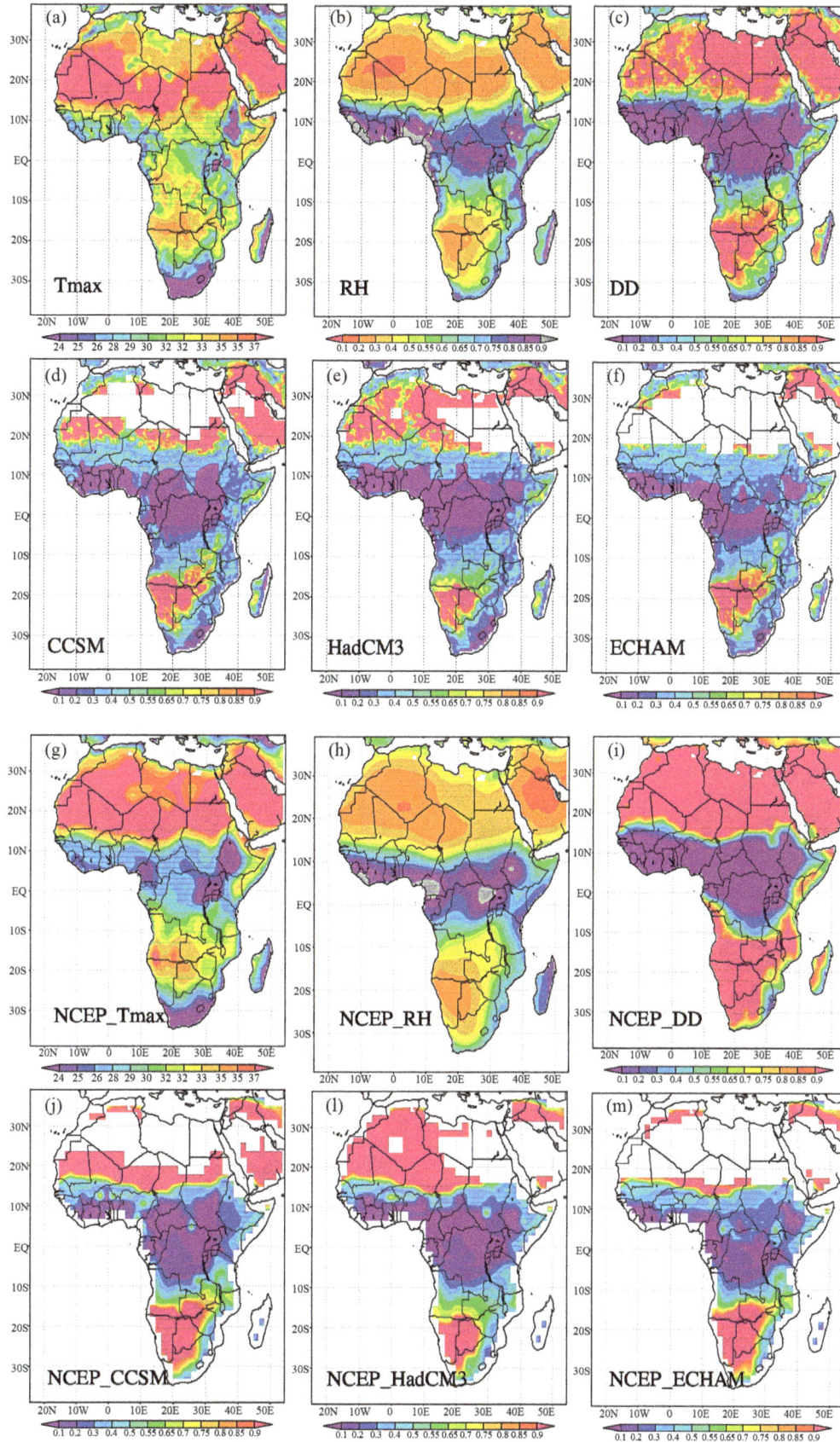

Figure 4. Same as Figure 3 but for August-September-October (ASO).

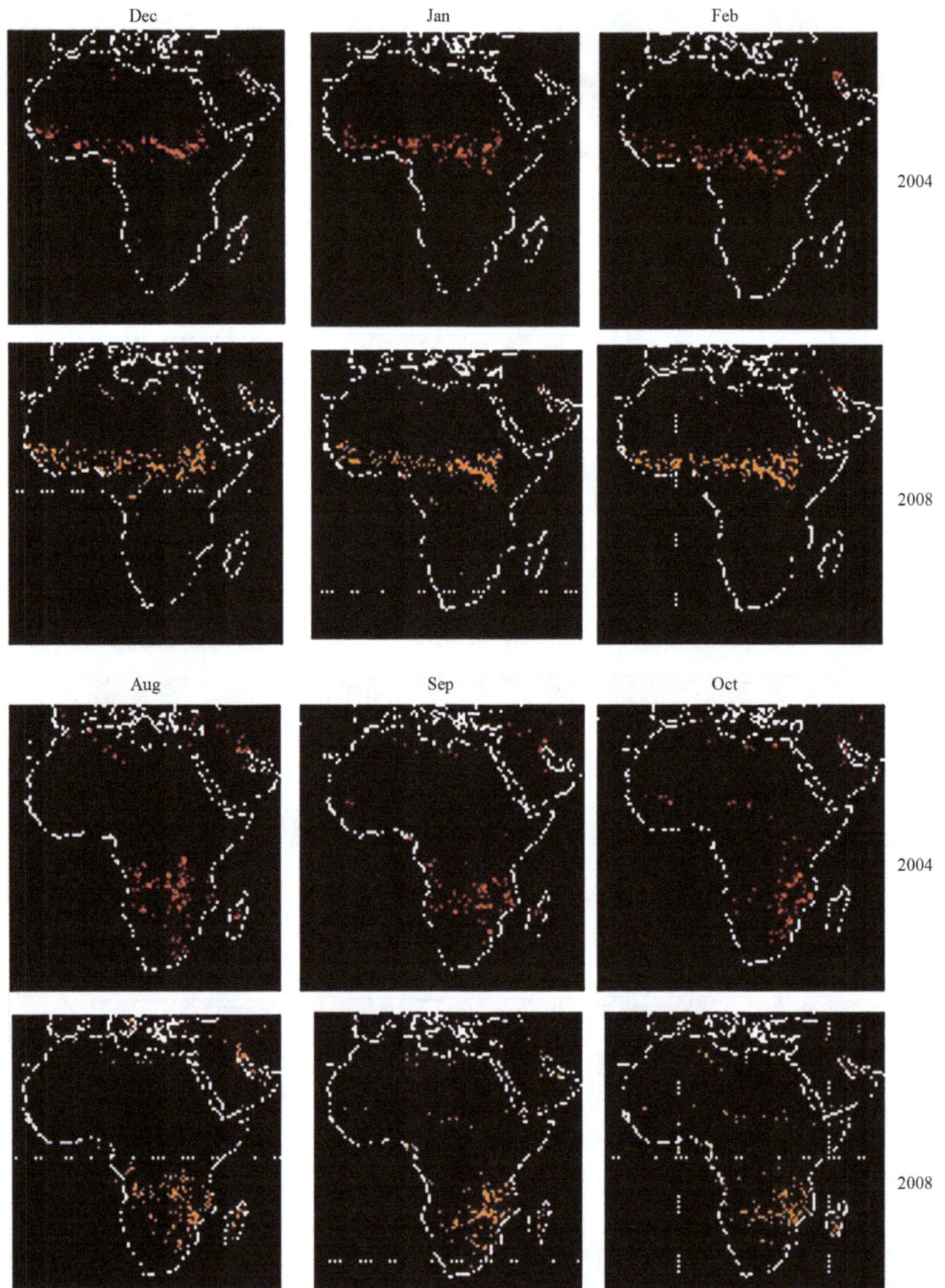

Figure 5. Satellite-detected hot spots with brightness temperature higher than 308 K in the years 2004 and 2008 during December-January-February and August-September-October. Source: http://dup.esrin.esa.int/ionia/sfa/index.asp.

occur during ASO on the west and east coasts. Fires occuring on the east coast are not entirely supported by the *PFI* since these countries show low *PFI*, except in Madagascar where the *PFI* is reasonable. Similarly, fires identified in South Africa are in areas dominated by high *PFI* in particular based on the NNR2 results. The high agreement between the fire index and the satellite-based fires in Angola, Zambia, Zimbabwe and Namibia may be stressed.

It should be emphasized that anthropogenic factors contribute to the discrepancy between the *PFI* and active fire observations. For example, in Namibia the climatic conditions (and *PFI*) appear conducive to biomass burning. However, the active fire observations are quite sparse which may be due to low fuel loads (*i.e.* that is unable to sustain a fire under real conditions). Secondly, this area is sparsely populated and may be affected with less anthropogenic burning. Reference [48] found relationships between precipitation (drought extent) to be weakly correlated with active fire observations over parts of Africa, suggesting that other factors (such as land management and precipitation quantity in the proceeding wet season) may play an important role. It should be noted that despite these drawbacks, the *PFI* is able to reproduce the most susceptible area for vegetation fire development in response to climate (**Figures 4(d)-(g)**). Results presented here are very similar in terms of area with high *PFI* with those discussed in [9,12]. It should be noted however that [9] focused on fraction of area burned.

In relation to natural and anthropogenic fire occurrence based on climatic factors, the *PFI* indicates risk and vulnerability to erratic fire. It is crucial to assess the potential for fire development in order to avoid the occurrence of erratic wildfires which may cause substantial changes to the plant community, animal habitats and human health. The *PFI* methodology has been successfully applied in Brazil by political and economic decision markers in order to reduce hazards.

3.2. Greenhouse Warming Analyzes

In the following section the *PFI* response to climate and vegetation changes is evaluated. As previously discussed, forcing the vegetation model with outputs based on ECHAM, HadCM3 and CCSM model leads to a reduction in the evergreen forest and enlargement of savannas by the end of the 21st Century. The most pronounced changes occur when the vegetation model is forced with HadCM3 input data (**Figure 1(e)**).

These modifications play an important role in defining the fire risk level by increasing the basic potential risk (*BR*, Equation (1)). **Figures 6(a)-(c)** show the climate anomalies between the GW and PD experiments for averaged maximum temperature, relative humidity and days of drought for DJF. Positive T_{max} anomalies of 5° Celsius (C)

are evident in southern Africa and along 10°N. This region also exhibits lower relative humidity by up to 10%.

It is interesting to observe that no remarkable changes are predicted to occur in terms of the periods of dry days. Comparison between the current and simulated number of drought days (*DD*) in the GW simulation in general shows values ± 0.1. This demonstrates that in the RegCM3 the daily frequency of precipitation is not as distinct as could be expected between the two climate regimes. These results agree with previous investigations for the southern Africa continent [49-51]. It might be noted, however, that parts of equatorial Africa in the GW simulation experiences lower *DD* as compared to the PD simulation. In this region the number of dry days was already quite low in the PD. It should be stressed that close to the south coast of Mozambique there is presently a low number of dry days, but this is predicted to increase substantially in future scenarios (**Figure 6(c)**).

Figures 6(d)-(f) illustrate the fire risk (*PFI*) in response to future climate changes. Throughout the African continent the *PFI* increases as a result of simulated warmer climate expected for the end of the 21st Century. During DJF the dominant area of high susceptibility to wild-fire activity is the Sahelian region, which corroborates with present day analyzes (**Figures 6(d)-(f)**). Small changes are predicted to occur in the central part of Africa despite a shrinkage in the area covered by evergreen forests (**Figures 1(d)-(f)**). This feature is reasonable since the *DD* plays a leading role in defining the fire risk, as expressed by Equation (1) (**Figure 6(c)**). As previously shown *DD* does not experiences large changes between the two epochs. The *PFI* in DJF shows an additional region with high vulnerability to wildfire located between 10 - 30°S (**Figures 6(d)-(f)**). This has not been identified under current conditions (PD simulation). This feature is closely linked to positive (negative) air temperature (relative humidity) anomalies (**Figures 6(a)** and **(b)**).

However, the role of vegetation clearly stands out when comparing the *PFI* anomalies between PD and GW simulation (**Figures 6(g)-(i)**). Since the HadCM3 forcing leads to the most severe changes in vegetation (**Figure 1(b)**), the corresponding *PFI* configuration is also associated with larger values as compared to anomalies resulting from the ECHAM and CCSM. The effect of climate change, however, reduces the fire risk in areas which experience increasingly more frequent precipitation such as the majority of equatorial Africa (0 - 10°S). For instance, Sudan which exhibits substantial number of fires under PD conditions is expected to show a decrease in the future.

Regarding evaluation of the *PFI* in ASO (August-September-October) (**Figure 7**), more favorable climate conditions for vegetation fire occurrence can be observed. Under PD conditions the most susceptible region for fire

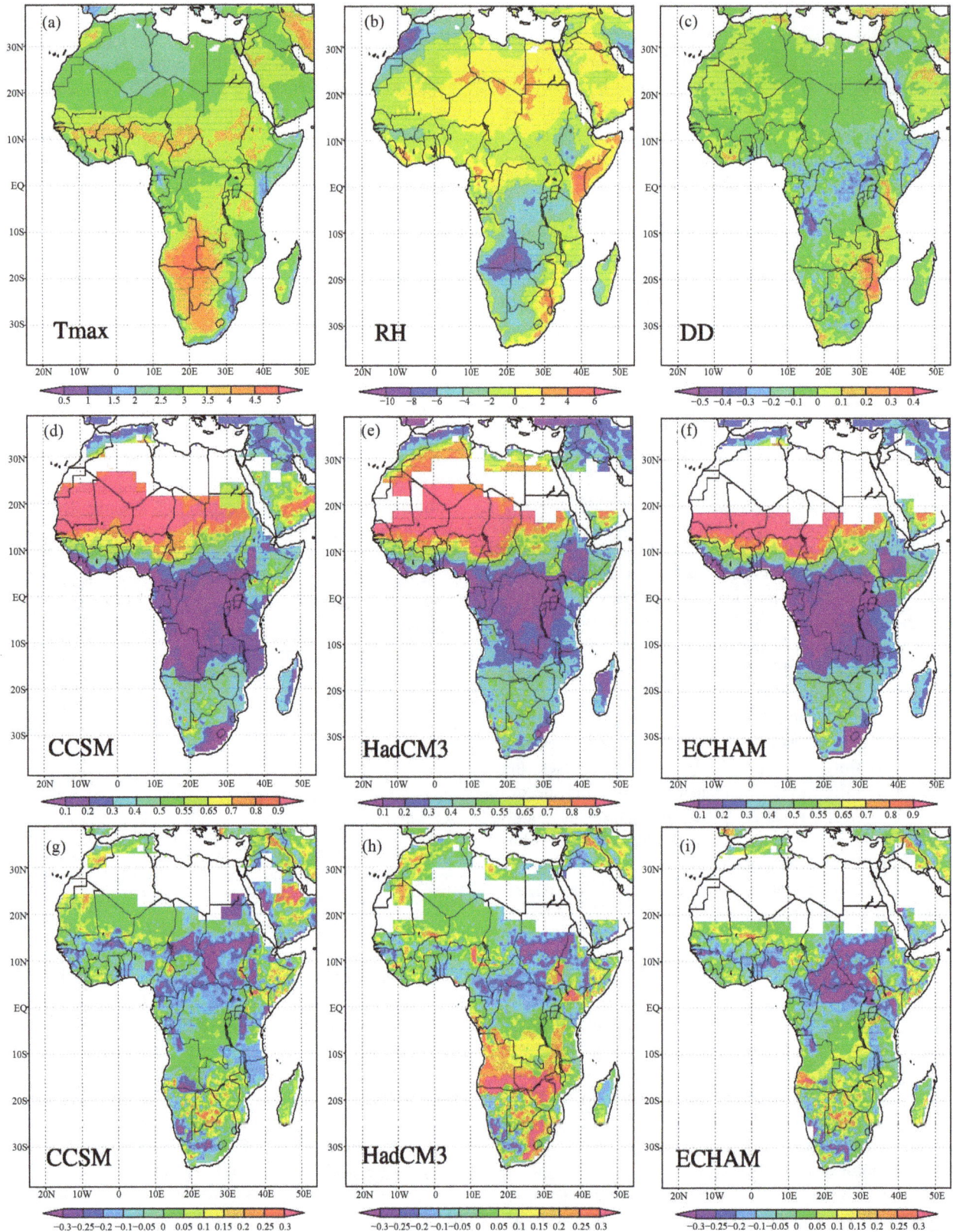

Figure 6. Greenhouse warming anomalies as compared to present Day in DJF. Maximum temperature ((a), °C), relative humidity ((b), %) and days of drought (c). (d)-(f) show the *PFI* estimated from the greenhouse warming simulation based on CCSM, HadCM3 and ECHAM forced-vegetation. (g)-(i) show the *PFI* anomalies between greenhouse warming and present day simulation.

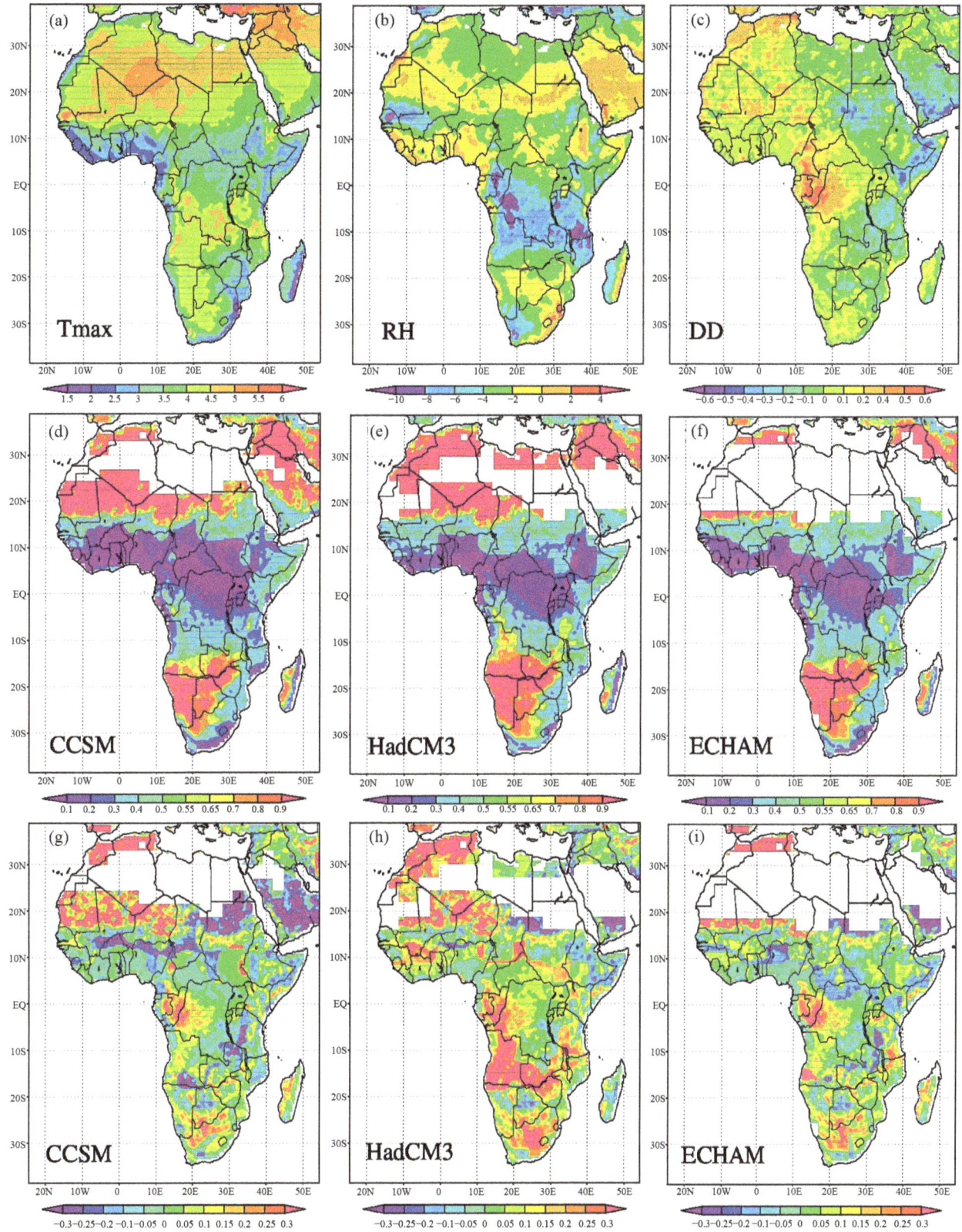

Figure 7. Same as Figure 6 but for August-September-October (ASO).

development is found in southern Africa (**Figures 4(d)-(f)**). In the GW simulation there exists an enlargement of this region as well as high fire risk between 10 - 20°N and on the Island of Madagascar (**Figures 6(g)-(i)**). These results are similar to those found by [12] (their **Figure 3**) who found increased (reduced) projected fire activities in Sahel, East Africa and South Africa (Central Africa). Fire risk is expected to increase also on the west coast from the equator to 20°S. This is more evident in the HadCM3 case (**Figure 6(h)**).

To further evaluate the fire risk in Africa, **Figure 8** shows the zonally averaged *PFI* computed at latitude belts between 5°N - 20°N, 10°S - 5°N, 25°S - 10°S and 40°S - 25°S throughout the year. This aids in identifying changes in the amplitude of the seasonal cycle under both climate regimes and in determining the individual month with the highest fire risk. According to the figure, the region confined between 10°S - 20°N shows higher *PFI* in March and gradually attains its minimum fire risk in November (**Figures 8(a)** and **(b)**). This is closely linked with the meridional migration of the ITCZ and its associated precipitation over the region. In this region there is an increase in the fire risk in 10 out of 12 months when future climate and vegetation conditions are applied. Vegetation distribution as predicted based on HadCM3 leads to the highest values of the *PFI*. It should be noted,

however, that under current conditions the *PFI* associated with the HadCM3 model attains the smallest value.

A similar evaluation for the region between 40°S - 10°S shows that, in comparison to current conditions, the fire risk in the future increases substantially and it is high from July to October, especially between 25°S - 10°S (**Figure 8(c)**). Under PD conditions this is critical in September only. This suggests that an increase in the length of the fire season in sub-Saharan Africa is likely in the future. The same is also predicted in the Amazon region [18,52]. Moreover, in subtropical Africa (**Figure 8(d)**), the *PFI* increases from low in the PD simulation to moderate in the GW simulation.

The interannual variability of the *PFI* as represented by the standard deviation (STD, not shown), for present day and greenhouse warming show that higher values of the STD are located in Sahel and in the southern part of Africa. Lower values of the STD are, however, found in the equatorial belt where the Congo forest has a moderate vulnerability to fire occurrence. These findings are common features in CCSM, HadCM3 and ECHAM based estimates.

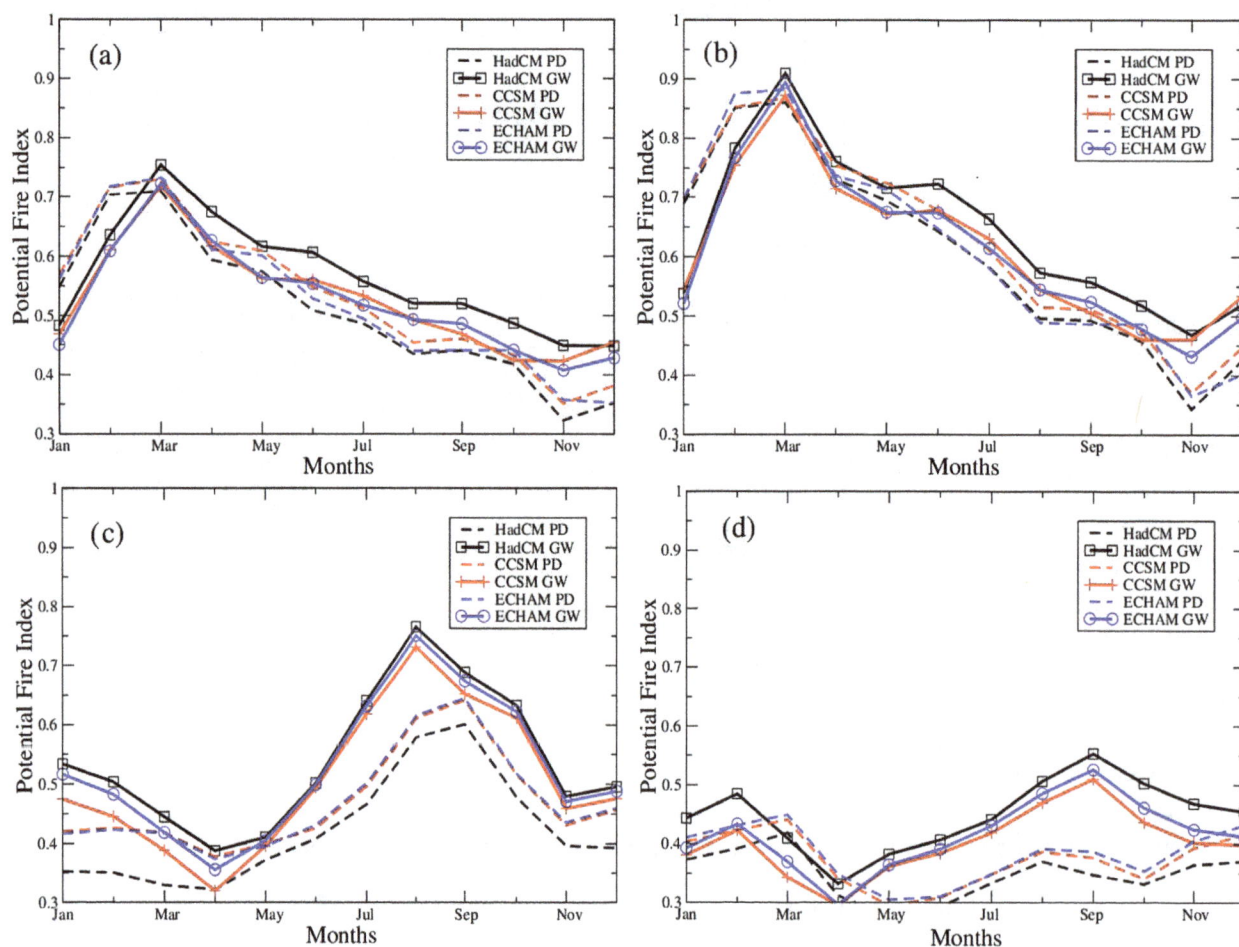

Figure 8. Monthly zonally averaged current and future *PFI* for (a) 5°N - 20°N; (b) 10°S - 5°N; (c) 25°S - 10°S and (d) 40°S - 25°S.

It should be mentioned that the regions experiencing higher standard deviation are larger than those with maximum or critical *PFI* (**Figures 6** and **7**), which demonstrates the north-south shift of the *PFI* throughout the year. An interesting picture emerges by comparing present day and greenhouse warming simulations. For instance, the interannual variability is partially reduced in the Northern Hemisphere tropical region but increases in the Southern Hemisphere counterpart.

Therefore, one may argue that climate changes increase the area vulnerable to fire occurrence in particular in sub-Sahara. Previous studies have shown that substantial changes are predicted to occur in precipitation extremes [e.g. 53], in particular an increase in consecutive dry days.

The investigation of fire risk under GW climate conditions but with PD vegetation (figure not shown) yields *PFI* similar to those under present day conditions (**Figures 3(c)**, **(d)**, **(f)** and **4(c)**, **(d)**, **(f)**). In this sense, taking into account a warmer and drier climate, as predicted to occur in the future, does not substantially change the fire risk. This reveals the dominant role of vegetation changes in determining environmental susceptibility to fire occurence.

4. Concluding Remarks

Despite advances in the understanding of climate and its link with fire activities, studies evaluating the association of climate variables with fire occurrence are still scarce, particularly for future climate scenarios. Based on the Potential Fire Index (*PFI*) computed from regional climate simulations performed by the RegCM3, this study evaluated the impacts of climate changes on vegetation and vegetation-fire risk in Africa under current and future global warming conditions. Changes in vegetation due to climate changes have been taken from the calculations by [19], who used a dynamic vegetation model (CLM-DVGM) to estimate vegetation distributions in the current and future climate.

It was herein demonstrated that the *PFI* was able to identify the principal fire risk areas which are concentrated in the Sahelian region from December to March, and in subtropical Africa from July to October. By applying three different vegetation distributions for current conditions and those proposed for a global warming scenario, it was found that the future *PFI* is extremely sensitive to the conditions imposed by the vegetation. For instance, when forest was substituted for savannas a greater fire risk was projected. Calculations of the *PFI* indicate that with the predicted future vegetation distribution, Africa will be more exposed to large scale vegetation fires. We showed that the impacts on *PFI* from climate change itself, in terms of temperature, humidity and precipitation were less pronounced.

Limitation of our modeling approach may be noted since the *PFI* depends strongly on the number of dry days and thus on precipitation. Precipitation is a very uncertain parameter in future and present day climate predictions and significantly differs even in sign between different climate models. It is not unlikely that using a different global climate model as boundary condition will lead to a different change in fire risk over Africa. Taking into account the full range of future climate predictions is beyond the scope of this study. A final remark is that our *PFI* concept does not take ignition of fire into account.

5. Acknowledgements

This work has been supported by the SoCoCA project funded by the Research Council of Norway (contract 190159).

REFERENCES

[1]   D. M. J. S. Bowman, *et al.*, "Fire in the Earth System," *Science*, Vol. 324, No. 5926, 2009, pp. 481-484.

[2]   K. Thonicke, I. C. Prentice and C. Hewitt, "Modelling Glacial-Interglacial Changes in Global Fire Regimes and Trace Gas Emissions," *Global Biogeochemical Cycles*, Vol. 19, No. 3, 2005.

[3]   B. Langmann, B. Duncan, C. Textor, J. Trentmann and G. R. van der Werf, "Vegetation Fire Emissions and Their Impact on Air Pollution and Climate," *Atmospheric Environment*, Vol. 43, No. 1, 2009, pp. 107-116.

[4]   S. E. Page, F. Siegert, J. O. Rieley, H. D. V. Boehm, A. Jaya and S. Limin, "The Amount of Carbon Released from Peat and Forest Fires in Indonesia during 1997," *Nature*, Vol. 420, 2002, pp. 61-65.

[5]   J. G. Goldammer, "Fire in the Environment: The Ecological, Atmospheric and Climatic Importance of Vegetation Fires," 13th Edition, John Wiley & Sons, Chichester, 1993.

[6]   A. Aubréville, "Ancienneté la Destruction de la Couverture Forestiére Primitive de L'afrique Tropicale," *Bulletin Agricole de Congo Belge*, Vol. 40, No. 2, 1949, pp. 1347-1352.

[7]   J. M. N. Silva, J. M. C. Pereira, A. I. Cabral, M. J. P. Vasconcelos, B. Mota and J. M. Gregoire, "An Estimate of the Area Burned in Southern Africa during the 2000 Dry Season Using SPOT-VEGETATION Satellite Data," *Journal of Geophysical Research*, Vol. 108, No. D13, 2003.

[8]   L. Giglio, G. R. van der Werf, J. T. Randerson, G. J. Collatz and P. Kasibhatla, "Global Estimation of Burned Arca Using MODIS Active Fire Observations," *Atmospheric Chemistry & Physics Discussions*, Vol. 5, No. 6, 2005, pp. 11091-11141.

[9]   V. Lehsten, P. Harmand, I. Palumbo and A. Arneth, "Mo-

delling Burned Area in Africa," *Biogeosciences*, Vol. 7, 2010, pp. 3199-3214.

[10] J. R. Marlon, *et al.*, "Climate and Human Influences on Global Biomass Burning over the Past Two Millennia," *Nature Geoscience*, Vol. 1, No. 10, 2008, pp. 697-702.

[11] S. Kloster, *et al.*, "Fire Dyamics During the 20th Century Simulated by the Community Land Model," *Biodigeosciences*, Vol. 7, No. 6, 2010, pp. 1877-1902.

[12] O. Pechony and D. T. Shindell, "Driving Forces of Global Wildfires over the Past Millennium and the Forthcoming Century," *PNAS*, 2010.

[13] A. Bartsch, H. Balzter and C. George, "The Influence of Regional Surface Soil Moisture Anomalies on Forest Fires in Siberia Observed from Satellites," *Environmental Research Letters*, Vol. 4, No. 4, 2009, Article ID: 045021.

[14] S. J. Goetz, M. C. Mack, K. R. Gurney, J. T. Randerson and R. A. Houghton, "Ecosystem Responses to Recent Climate Change and Fire Disturbance at Northern High Latitudes: Observations and Model Results Contrasting Northern Eurasia and North America," *Environmental Research Letters*, Vol. 2, No. 4, 2007, Article ID: 045031.

[15] S. Scheiter and S. I. Higgins, "Impacts of Climate Change on the Vegetation of Africa: An Adaptive Dynamic Vegetation Modelling Approach," *Global Change Biology*, Vol. 15, No. 9, 2009, pp. 2224-2246.

[16] C. Gouveia, C. C. DaCamara and R. M. Trigo, "Post Fire Vegetation Recovery in Portugal Based on Spot-Vegetation Data," *Natural Hazards and Earth System Sciences*, Vol. 10, No. 4, 2010, pp. 673-684.

[17] A. H. Lynch, *et al.*, "Using the Paleorecord to Evaluate Climate and Fire Interactions in Australia," *Annual Review of Earth and Planetary Sciences*, Vol. 35, 2007, pp. 215-239.

[18] F. Justino, A. S. de Melo, A. Setzer, R. Sismanoglu, G. C. Sediyama, G. A. Ribeiro, J. P. Machado and A. Sterl, "Greenhouse Gas Induced Changes in the Fire Risk in Brazil in ECHAM5/MPI-OM Coupled Climate Model," *Climatic Changes*, Vol. 106, No. 2, 2010, pp. 285-302.

[19] C. A. Alo and G. Wang, "Potential Future Changes of the Terrestrial Ecosystem Based on Climate Projections by Eight General Circulation Models," *Journal of Geophysical Research*, Vol. 113, No. G1, 2008.

[20] W. A. Hoffmann, W. Schoeder and R. Jackson, "Positive Feedbacks of Fire, Climate and Vegetation and the Conversion of Tropical Savanna," *Geophysical Research Letters*, Vol. 29, No. 22, 2002, pp. 9-1-9-4.

[21] W. J. Bond, F. I. Woodward and G. F. Midgley, "The Global Distribution of Ecosystems in a World without Fire," *New Phytologist*, Vol. 165, No. 2, 2005, pp. 525-537.

[22] E. Dwyer, S. Pinnock, J. M. Gregoire and J. Pereira, "Global Spatial and Temporal Distribution of Vegetation Fire as Determined from Satellite Observations," *International Journal of Remote Sensing*, Vol. 21, 2000, pp. 1289-1302.

[23] A. Kilpeläinen, S. Kellomäki, H. Strandman and A. Venäläinen, "Climate Change Impacts on Forest Fire Potential in Boreal Conditions in Finland," *Climatic Change*, Vol. 103, No. 3-4, 2010, pp. 383-398.

[24] M. Flannigan, B. Amiro, K. Logan, B. Stocks and B. Wotton, "Forest Fires and Climate Change in the 21st Century," *Mitigation and Adaptation Strategies for Global Change*, Vol. 11, No. 4, 2006, pp. 847-859.

[25] S. J. Pyne, P. L. Andrews and R. D. Laven, "Introduction to Wildland Fire," Wiley & Sons, New York, 1996.

[26] K. Thonicke, A. Spessa, I. C. Prentice, S. P. Harrison, L. Dong and C. Carmona-Moreno, "The Influence of Vegetation, Fire Spread and Fire Behaviour on Biomass Burning and Trace Gas Emissions: Results from a Prtocee-Based Model," *Biogeosciences Discussions*, Vol. 7, No. 1, 2010, pp. 697-743.

[27] T. Hickler, *et al.*, "Implementing Plant Hydraulic Architecture within the LPJ Dynamic Global Vegetation Model," *Global Ecology and Biogeography*, Vol. 15, No. 6, 2006, pp. 567-577.

[28] H. Sato, *et al.*, "SEIB-DGVM: A New Dynamic Global Vegetation Model Using a Spatially Explicit Individual-Based Approach," *Ecological Modelling*, Vol. 200, No. 3, 2007, pp. 279-307.

[29] A. Giannini, M. Biasutti and M. M. Verstraete, "A Climate Model-Based Review of Drought in the Sahel: Desertification, the Re-Greening and Climate Change," *Global and Planetary Change*, Vol. 64, No. 3-4, 2008, pp. 119-128.

[30] S. Nicholson and J. C. Selato, "The Influence of La Nina Events on African Rainfall," *International Journal of Climatology*, Vol. 20, No. 14, 2000, pp. 1761-1777.

[31] K. H. Cook, "Large-Scale Atmospheric Dynamics and Sahelian Precipitation," *Journal of Climate*, Vol. 10, No. 6, 1997, pp. 1137-1152.

[32] C. J. C. Reason and M. Rouault, "Links between the Antarctic Oscillation and Winter Rainfall over Western South Africa," *Geophysical Research Letters*, Vol. 32, No. 7, 2005, Article ID: L07705.

[33] K. H. Cook, "A Southern Hemisphere Wave Response to ENSO with Implications for Southern Africa Precipita-

tion," *Journal of Climate*, Vol. 58, No. 15, 2001, pp. 2146-2162.

[34] I. Held, T. L. Delworth, J. Lu, K. L. Findell and T. R. Knutson, "Simulation and Sahel Drought in the 20th and 21st Centuries," *PNAS*, Vol. 102, No. 50, 2005, pp. 17891-17896.

[35] L. Mariotti, E. Coppola, M. Sylla, F. Giorgi and C. Piani, "Regional Climate Model Simulation of Projected 21st Century Climate Change over an All-Africa Domain: Comparison Analysis on Nested and Driving Model Results," *Journal of Geophysical Research*, Vol. 116, No. D15, 2011.

[36] E. Roeckner, *et al.*, "The Atmospheric General Circulation Model ECHAM5. Part I: Model Description," Max Planck Institute for Meteorology, 2003.

[37] N. Nakicenovic and R. Swart, "Special Report on Emissions Scenarios," Cambridge University Press, Cambridge, 2000.

[38] M. B. Sylla, E. Coppola, L. Mariotti, F. Giorgi, P. M. Ruti, A. Dell'Aquila and X. Bi, "Multiyear Simulation of the African Climate Using a Regional Climate Model (RegCM3) with the High Resolution ERA—Interim Reanalysis," *Climate Dynamics*, Vol. 35, No. 1, 2010, pp. 231-247.

[39] R. Sismanoglu, A. Setzer, A. L. Lopes and E. M. Sedeno, "Risco de Fogo Para a Vegetação de Cuba: Comparação Entre Duas Versões Para 2010 Utilizando Dados do Satélite TRMM e SYNOP," *Proceedings of* 15*th Simpósio Brasileiro de Sensoriamento Remoto*, 2011.

[40] R. Sismanoglu and A. Setzer, "Risco de Fogo da Vegetação na América do Sul: Comparação de Três Versões na Estiagem de 2004," *Proceedings of* 12*th Simpósio Brasileiro de Sensoriamento Remoto*, 2005.

[41] S. Sitch, *et al.*, "Evaluation of Ecosystem Dynamics, Plant Geography and Terrestrial Carbon Cyclin in the LPJ Dynamic Global Vegetation Model," *Global Change Biology*, Vol. 9, No. 2, 2003, pp. 161-185.

[42] G. Bonan, S. Levis, L. Kergoat and K. Oleson, "Landscapes as Patches of Plant Functional Types: An Integrating Concept for Climate and Ecosystem Models," *Global Biogeochemical Cycles*, Vol. 16, 2002, pp. 5.1-5.23.

[43] D. Davies, S. Kumar and J. Descloitres, "Global Fire Monitoring Using MODIS Near-Real-Time Satellite Data," *GIM International*, Vol. 18, No. 4, 2004, pp. 41-43.

[44] L. Giglio, J. Descloitres, C. O. Justice and Y. J. Kaufman,

"An Enhanced Contextual Fire Detection Algorithm for MODIS," *Remote Sensing of Environment*, Vol. 87, No. 2-3, 2003, pp. 273-282.

[45] E. Kalnay, *et al.*, "The NCEP-NCAR 40 Year Reanalysis Project," *Bulletin of the American Meteorological Society*, Vol. 77, No. 3, 1996, pp. 437-471.

[46] V. Markgraf, C. Whitlock and S. Haberle, "Vegetation and Fire History during the Last 18,000 cal yr B.P. in Southern Patagonia: Mallin Pollux, Coyhaique, Province Aisen (45°41'30"S, 71°50'30"W, 640 m Elevation)," *Palaeogeography, Palaeoclimatology, Palaeoecology*, 2007.

[47] C. Whitlock, M. M. Bianchi, P. J. Bartlein, V. Markgraf, J. Marlon, M. Walsh and N. McCoy, "Postglacial Vegetation, Climate, and Fire History along the East Side of the Andes (lat 41-42.5S), Argentina," *Quaternaly Research*, Vol. 66, No. 2, 2006, pp. 187-201.

[48] G. R. van der Werf, J. T. Randerson, L. Giglio, N. Gobron and A. J. Dolman, "Climate Controls on the Variability of Fires in the Tropics and Subtropics," *Global Biogeochemical Cycles*, Vol. 22, No. 3, 2008.

[49] B. Lyon, "Southern Africa Summer Drought and Heat Waves: Observations and Coupled Model Behavior," *Journal of Climate*, Vol. 22, No. 22, 2009, pp. 6033-6046.

[50] P. Rocha, A. Melo-Gonçalves, C. Martques, J. Ferreira and J. M. Castanheira, "High-Frequency Precipitation Changes in Southeastern Africa Due to Anthropogenic Forcing," *International Journal of Climatology*, Vol. 28, No. 9, 2008.

[51] M. Tadross, C. Jack and B. Hewitson, "On RCM-Based Projections of Change in Southern African Summer Climate," *Geophysical Research Letters*, Vol. 32, No. 23, 2005.

[52] M. Hirota, C. Nobre, M. D. Oyama and M. M. Bustamante, "The Climatic Sensitivity of the Florest, Savanna and Forest-Savanna Transition in Tropical South America," *New Phytologist*, Vol. 187, No. 3, 2010, pp. 707-719.

[53] M. E. Shongwe, G. J. van Oldenborgh, B. J. J. M. van den Hurk, B. de Boer, C. A. S. Coelho and M. K. van Aalst, "Projected Changes in Mean and Extreme Precipitation in Africa Under Global Warming. Part I: Southern Africa," *Journal of Climate*, Vol. 22, No. 13, 2009, pp. 3819-3837.

12

Detection and Projections of Climate Change in Rio de Janeiro, Brazil

Claudine Dereczynski[1], Wanderson Luiz Silva[1], Jose Marengo[2]
[1]Department of Meteorology, Federal University of Rio de Janeiro, Rio de Janeiro, Brazil
[2]Center for Earth System Science, National Institute for Space Research, Cachoeira Paulista, Brazil

ABSTRACT

A study on the detection and future projection of climate change in the city of Rio de Janeiro is here presented, based on the analysis of indices of temperature and precipitation extremes. The aim of this study is to provide information on observed and projected extremes in support of studies on impacts and vulnerability assessments required for adaptation strategies to climate change. Observational data from INMET's weather stations and projections from INPE's Eta-HadCM3 regional model are used. The observational analyses indicate that rainfall amount associated with heavy rain events is increasing in recent years in the forest region of Rio de Janeiro. An increase in both the frequency of occurrence and in the rainfall amount associated with heavy precipitation are projected until the end of the 21st Century, as are longer dry periods and shorter wet seasons. In regards to temperature, a warming trend is noted (both in past observations and future projections), with higher maximum air temperature and extremes. The average change in annual maximum (minimum) air temperatures may range between 2°C and 5°C (2°C and 4°C) above the current weather values in the late 21st Century. The warm (cold) days and nights are becoming more (less) frequent each year, and for the future climate (2100) it has been projected that about 40% to 70% of the days and 55% to 85% of the nights will be hot. Additionally, it can be foreseen that there will be no longer cold days and nights.

Keywords: Climate Change; Extreme Event Indices; Air Temperature; Precipitation; Eta-HadCM3 Model; Rio de Janeiro

1. Introduction

The changes occurring in the climate extremes, such as an increase in hot days, decrease in cold days, longer heat waves or more frequent intense rains or severe drought, have a major impact on ecosystems and on the society at large [1-4]. Thus, climate change detection and projections based on the analysis of extreme event indices are of great importance to support studies on impacts and vulnerability and to prepare strategies to adapt to climate change. Among others, such studies have been undertaken at global and regional level by [4-12]. This detection type of study requires high-quality and long-term observed data, which are not always available.

There is consensus in most of the climate change detection studies for South America [5-7,11] concerning warming, that there is an increase (decrease) in the frequency of warm (cold) nights, and more frequent days with heavy rain in Southeast South America. However, in the states of São Paulo, Rio de Janeiro and Minas Gerais, the results concerning precipitation are not uniform, with nearby locations showing opposite trends, as shown in [9]. Thus, a detailed investigation of the behavior of

extreme climate event indicators for the city of Rio de Janeiro, in the state of the same name, will be tremendously useful.

Several groups have projected future climate change by nesting regional climate models within general circulation coupled atmosphere-ocean models. The advantage of this technique, which is known as dynamic downscaling [13-18], is that it captures regional scale aspects appropriately. Important future climate change projection results with regional models for South America can be found in [10,18-25]. As is the case with detection studies, future climate change projections using different models and CO_2 emissions scenarios point to consensus insofar as the warming projected for the late 21st century is concerned, but are divergent in respect to rainfall outlooks.

This research project undertakes climate change detection and investigates climate change projections for the city of Rio de Janeiro by analyzing precipitation and air temperature-related observed extremes. The aim is to support vulnerability studies and adapt them to climate change scenarios. Observed extreme events trends are detected using observational data coming from two weather stations located in the city of Rio de Janeiro. The simulations of the Eta regional model of the National Institute for Space Research (INPE) driven by the UK Met Office Hadley Centre (HadCM3) global model for the current climate (1961-1990) are compared with observational data as a prior step for the assessment of the projections generated by the same regional model for the 2011 to 2100 period.

2. Methodology and Data

The observational data used in this study comprise a series of maximum and minimum air temperatures and total daily rainfall of two weather stations from the Brazilian National Institute of Meteorology (INMET), located in the city of Rio de Janeiro. Despite the limitations imposed by the use of only two weather stations, these stations did not suffer from change in location. Furthermore, they are located in distinct environments, allowing for comparisons to be made of two extreme urban conditions: the Alto da Boa Vista station (22°57'57.50"S/43°16'46.20"W), which is located within a tropical forest at the Tijuca Forest National Park at an altitude of 347.1 m; and the Santa Cruz station (22°55'19.59"S/43°41'12.90"W), positioned in the western zone of the city at an altitude of 63.0 m and in a region where significant urbanization has taken place (**Figure 1**). Unfortunately, several flaws took place in daily data collection at each station during the analyzed data period. For the Alto da Boa Vista (Santa Cruz) station, the numerous gaps in the daily rainfall data precluded the calculation of total annual rainfall in 11 (13) years, and only 30 (33) years remained for analysis, within the period ranging from 01/01/1967 to 12/31/2007 (01/01/1964 to 12/31/2009). At Alto da Boa Vista (Santa Cruz) station the maximum number of consecutive missing years was four (six), from 1988-1991 (1993-1998).

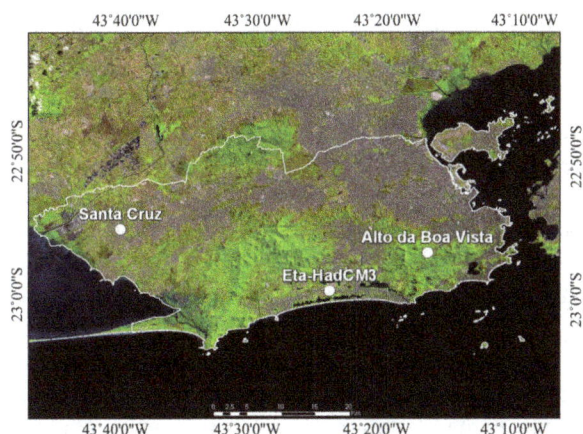

Figure 1. Map of the city of Rio de Janeiro (city border in white) showing the location of Alto da Boa Vista and Santa Cruz stations and the Eta-HadCM3 model grid point.

At INPE, the Eta-CPTEC numerical model (with a horizontal resolution of 40 km), derived from the Eta regional model developed at the University of Belgrade [26], was driven by the lateral boundary conditions of the global coupled HadCM3 model [27]. For the present climate conditions (1961-1990), a CO_2 concentration equal to 330 ppm was considered, while the Special Report on Emission Scenarios-SRES [28] A1B emission scenario was used for future climate (2011-2100). This study uses simulations which were generated for a set of 4 members: the control member plus 3 integrations with physical disturbances, and members with different climate sensitivities, herein referred as Ctrl, Low, Mid and High. Although the 3 members correspond to the same A1B scenario, each member's temperature increase projections are similar to the increases characteristic of scenarios of low, medium and high emissions from the IPCC SRES. Henceforth, the regional Eta model simulations integrated with the boundary conditions from the HadCM3 global model will be called Eta-HadCM3. The evaluation of the current climate and future projections of the Eta-HadCM3 model are described in [24,25], respectively. The grid point of the Eta-HadCM3 model from which the series of rainfall and air temperature data were extracted lies between the two weather stations (23.00°S/43.40°W and at an altitude of 62.9 m). In **Figure 1** the grid point is located in 23.00°S/43.40°W, representing the area 22.8°S - 23.2°S/ 43.20°W - 43.60°W.

The RClimdex software [29], developed by the Canadian Meteorological Service, was used to generate extreme climate event indicators [4] from observational data and simulations of the Eta-HadCM3 model, for the 1961-1990 base period. **Table 1** presents the definition of the indices used in this study.

The Mann-Kendall non-parametric statistical test was used to assess the climate trends [30]. This is the most appropriate method to analyze the significance of possible climate change in climatological series [31]. One of the benefits afforded by this test is that the data doesn't have to belong to a particular distribution. Another benefit is that its result is less affected by outlier values because the calculation is based on the sign of the differences, and not directly on the values of the variables. The confidence level adopted in the Mann-Kendall statistical test was 95% for all data series. The non-parametric method called the Sen Curvature [32] was used to estimate the magnitude of the trends found in the data series.

3. Rio de Janeiro City Climate

The study area, the city of Rio de Janeiro, located in Southeastern Brazil, between the 22.8° and 23.1°S parallels and the meridians of 43.1° and 43.8°W, features a warm, rainy climate in the summer and cold, dry weather in the

Table 1. Definitions and magnitudes of the climate extremes indicators related to precipitation, maximum and minimum air temperature observed at Alto da Boa Vista and Santa Cruz. The values in boldface are considered statistically significant at a confidence level of 95%. RR is the daily rainfall rate. A wet day has RR ≥ 1 mm. A dry day has RR < 1mm. TX and TN are daily maximum and minimum air temperature, respectively.

Indicator	Definition and unity	Alto da Boa Vista	Santa Cruz
PRCPTOT	Annual total precipitation from wet days (mm)	↗ +7.83 mm/year	↗ +2.54 mm/year
R95 p	Annual total precipitation on the days when RR > 95th percentile of the wet days (mm)	↗ **+11.77 mm/year**	No trend
R99 p	Annual total precipitation on the days when RR > 99th percentile of the wet days (mm)	↗ +3.40 mm/year	No trend
R30 mm	Annual count of days when RR ≥ 30 mm (days)	↗ +0.07 day/year	↗ +0.03 day/year
RX1 day	Annual max 1-day precipitation (mm)	↗ +1.04 mm/year	↘ −0.86 mm/year
RX5 day	Annual max consecutive 5-day precipitation (mm)	↗ +1.54 mm/year	↘ −0.47 mm/year
CDD	Max number of consecutive dry days in the year (days)	No trend	No trend
CWD	Max number of consecutive wet days in the year (days)	No trend	No trend
SU25	Number of days with TX > 25°C (days)	↗ **+1.42 day/year**	↗ +0.44 day/year
TMAXmean	Annual mean of TX (°C)	↗ **+0.04°C/year**	↗ **+0.03°C/year**
TX90 p	Percentage of days with TX > 90th percentile (%)	↗ **+0.15% day/year**	↗ **+0.15% day/year**
TX10 p	Percentage of days with TX < 10th percentile (%)	↘ **−0.20% day/year**	↘ **−0.11% day/year**
WSDI	Number of days in the year with at least 6 consecutive days of TX > 90th percentile	↗ **+0.17 day/year**	↗ +0.02 day/year
TMINmean	Annual mean of TN (°C)	No trend	↗ **+0.01°C/year**
TN90 p	Percentage of days with TN > 90th percentile (%)	↗ +0.03% day/year	↗ **+0.17% day/year**
TN10 p	Percentage of days with TN < 10th percentile (%)	↘ −0.03% day/year	↘ −0.06% day/year
DTR	Mean of the difference between TX and TN (°C)	↗ **+0.05°C/year**	↗ +0.01°C/year

winter. The city is influenced by the presence of the Atlantic Ocean to the south, the Guanabara Bay to the east, and the Sepetiba Bay to the west. The city's topography is marked by the formation of three massifs: Gericinó-Mendanha, to the north; the Tijuca massif to the east, where the Alto da Boa Vista station is located; and the Pedra Branca massif, to the west. The other areas of the city are lowlands with an average altitude of about 20 m. The Santa Cruz station is in the West Zone lowland (**Figure 1**).

Although the study area is under the influence of the South Atlantic Subtropical Anticyclone during most of the year, the most frequent winds are not from the northeast. In the city, dominant winds are southerly during the afternoon and night (sea breeze) and northerly at late night and in the morning (land breeze). In spring and summer, when land surface heating is more intense, the temperature gradient between continent and ocean intensifies the sea breeze circulation and increases the frequency of winds blowing from the south quadrants compared with the annual pattern. In fall and winter, winds coming from the southerly direction are less frequent, while the northerly ones increase. In Santa Cruz, the prevailing winds are from the northeast at dawn and in the morning, while from the southwest in the afternoon and evening [33].

The spatial distribution of air temperature is influenced by the topography and by the action of sea and land breezes. The Tijuca and Pedra Branca massifs, located near the waterfront, hamper the penetration of the sea breeze and transient systems into the city, making the Northern and Western regions of the city much warmer and drier than the Southern region and the downtown area. The stations located in the Western Zone register the highest temperatures, while the Alto da Boa Vista station registers the lowest ones.

The same regional geographic features affect rainfall, with maximums concentrated along the three massifs in the city. In the Western Zone, the total annual rainfall adds up to less than 1200 mm, while in Tijuca (where the Alto da Boa Vista station is located), it is almost double, averaging 2200 mm. This spatial rainfall distribution pattern is explained by the displacement of weather systems, mainly from south to north, producing maximum (minimum) rainfall to the windward (leeward) of the mountains. The orographic lifting of moist air advected by the sea breeze further enhances this process. The analysis of heavy rainfall events (greater than 30.0 mm/day) indicates that in 77% of cases, the rains are caused by frontal systems that occur year-round, although less frequently in winter [34].

Based on what was stated above, it is clear that the two stations selected for this study represent very different climates, with the Alto da Boa Vista station located in a high region with low temperatures and the highest rainfall in the area, while the Santa Cruz station is in lowland, where temperatures are higher and total rainfall low. Moreover, near the Alto da Boa Vista station, as it is situated within the Tijuca Forest, the vegetation has not undergone major changes and urban sprawl has been limited in recent decades, while in areas near the Santa Cruz station there has been significant urban area densification and expansion, with possible reductions in the vegetation cover.

4. Climate Extremes: Present and Future Climates

The results of climate extremes on the present and future climates are presented first for rainfall, then for air temperature. Although the Eta-HadCM3 model simulations for present and future climate are on the same graph, its trends and significances are calculated separately. The ensemble of the four members of the model is used in the model's simulations for the present climate. For future climate, member averages are used for indicators based on precipitation, since no large dispersion is seen among members; meanwhile, each of the members is analyzed for temperature (Ctrl, Low, Mid and High) separately in order to explore the difference between them.

4.1. Rainfall

Figure 2 features the total annual rainfall (PRCPTOT) of wet days (days when rainfall is greater than or equal to 1 mm) and the total annual rainfall of days when it is above the 95th percentile of wet days (R95p). Initially, a high interannual variability is noted in the observational data, especially in Alto da Boa Vista, and it is worthy of note that 1967 and 1998 were the wettest (El Niño years). There is also an increasing PRCPTOT trend for the two INMET stations. Despite the rate of +7.83 mm/year for Alto da Boa Vista and of +2.54 mm/year for Santa Cruz, these trends are not statistically significant at a confidence level of 95%. These results agree with those obtained for Southeastern South America by [7], in which PRCPTOT shows an increase. On the other hand, negative trends in

total annual precipitation were found for the State of Rio de Janeiro by [9] for Nova Friburgo in the period ranging from 1951 to 2001; [35], for Resende in the period of 1932 to 2000, and even [36] for the Capela Mayrink station (1977-2002), also located in the Tijuca massif, less than 2 km away from the Alto da Boa Vista station. This shows how rainfall is considerably dependent on local conditions, such as the land and sea breeze effects. As it will be discussed later, the Eta-HadCM3 model doesn't reproduce the strong interannual variability of PRCPTOT, mainly at Alto da Boa Vista station, located in a higher altitude, with heavy rainfall events. The model simulates, in the present climate, a reduction in PRCPTOT (−6.17 mm/year), and continues showing a slight reduction for this index in the future (−0.39 mm/year).

The R95p index (**Figure 2**) shows a statistically significant increase for Alto da Boa Vista (+11.77 mm/year) and a null trend for Santa Cruz in the current climate. This implies that at Alto da Boa Vista either heavy rainfall events have become stronger, *i.e.*, associated with a higher volume of rainfall each year, or they are becoming more frequent, or both. In the present climate, the regional model does not capture this increase in R95p; however, it reverts in the future and shows an increasing trend (+1.17 mm/year), which is statistically significant. The R99p index shows the same R95p behavior, *i.e.* positive trend for Alto da Boa Vista (+3.40 mm/year) and no trend for Santa Cruz (both without statistical significance). The regional model, however, indicates a slight positive trend for R99p in the current climate (+0.19 mm/year-not statistically significant) and in the future scenario (+0.73 mm/year), with statistical significance.

The RX5day and RX1day indicators that show the accumulated maximum rainfall in 1 day and in 5 days, respectively, are rising only for Alto da Boa Vista and without statistical significance. In Santa Cruz, both indi-

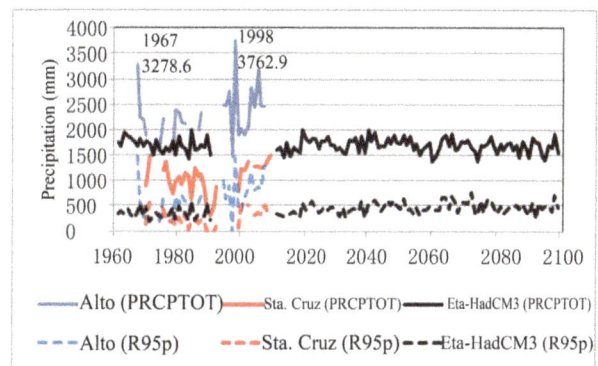

Figure 2. Temporal evolution of the PRCPTOT (solid lines) and R95 p (dashed lines) indices, both in mm, during the present climate, observed for Alto da Boa Vista in blue and in red for Santa Cruz. The black line refers to the Eta-HadCM3 model simulations (mean of the 4 members of the ensemble) for present and future climates.

ces are in a slight decline. In general, it is noted that the current rainfall climate positive trends are more marked in Alto da Boa Vista than in Santa Cruz.

The R30 mm indicator, illustrating the number of days in the year on which the total daily rainfall was above 30 mm, shows no statistically significant positive trends for Alto da Boa Vista (+0.07 day/year) and Santa Cruz (+0.03 day/year). In the latter, the frequency of R30 mm that occurred on average 6 times a year from 1975-1984, became on average 10 times a year in the 1998-2007 period. [37] shows that the frequency of extreme daily rainfall events (>100 mm/day) increased at the Capela Mayrink Station from 1977 to 2002. This confirms that moderate to heavy rain events are becoming more frequent every year. In the present climate, the regional model presents a statistically non-significant negative trend for R30 mm (−0.05 day/year); however, this reverts in the future, with a statistically significant trend of +0.02 day/year.

The CDD and CWD indices that illustrate the maximum number of consecutive dry and wet days, respectively, show null trends, both in Alto da Boa Vista and in Santa Cruz, and this characteristic is well represented by the Eta-HadCM3 model. For CDD, the values are close to those noticed in Alto da Boa Vista (between 12 and 33 days). A high interannual variability is noted in Santa Cruz, where CDD reaches 59 days (occurring in the 01/24 to 03/23/1990 period). The regional model shows an upward trend in this index (+0.04 day/year) for the future, and it is statistically significant. For CWD in the future, the model projects a statistically significant trend of decline (−0.01 day/year), i.e., reducing the duration of consecutive wet days.

The regional model simulations, with a horizontal resolution of 40 km, cannot reproduce the high interannual variability of the extreme rainfall-related indices. Besides the patchy nature of rainfall and the influence of the topography, in the global and even in regional models, precipitation is mediated in the area and, therefore, the magnitude of the extremes is reduced compared to the values observed at the stations [38]. It has been recognized that the comparison between model grid output and station data is not straightforward [39-41] and that calculations of precipitations extremes indices could be sensitive to model resolution [42-44]. [43] notes that precipitation extremes obtained from individual station records are essentially point estimates and are not directly comparable to gridded model output that presumably represents precipitation variability on much coarse spatial scales. Nevertheless, for most of the indices that were calculated, the magnitude of the indicators was limited between the values observed in Alto da Boa Vista and in Santa Cruz, increasing the confidence in the usefulness of their future projections. Regarding the simulated trends, although most of them exhibit trends that are opposite to those

that have been observed, it should be noted that there are differences even in series of observational data coming from stations that are relatively close to each other, as was the case of PRCPTOT in Alto da Boa Vista and Capela Mayrink and also of RX1 day and RX5 day, with positive trends for Alto da Boa Vista and negative ones for Santa Cruz.

4.2. Air Temperature

For Rio de Janeiro, the extreme event indices associated with maximum air temperature trends show more significant augmentation than those associated with the minimum air temperature, contrary to most of the studies undertaken to detect climate change in the globe [4,6,8] and in South America [9-11,45]. Thus, the DTR index, that represents the daily thermal amplitude, which should be decreasing as in most locations, is rising in Rio de Janeiro, i.e., the maximum air temperature increases at a rate higher than the minimum temperature. This index is statistically significant for Alto da Boa Vista (+0.05°C/year), but not for Santa Cruz (+0.01°C/year). At the Alto da Boa Vista station the maximum air temperature is increasing but the minimum temperature has no trend, while at Santa Cruz both maximum and minimum air temperature are increasing.

The TMAXmean and TMINmean indices that depict the annual average of the maximum and minimum daily air temperatures, respectively, are presented in **Figure 3**, with the top curves being for TMAXmean and lower ones for TMINmean. At both stations, one can notice a warming trend statistically significant in the current climate, but the upward trend is more significant for TMAXmean (+0.04°C/year in Alto da Boa Vista and +0.03°C/year in Santa Cruz) than for TMINmean (no trend in Alto da

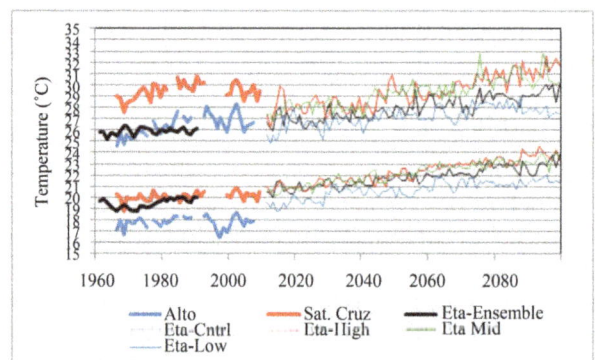

Figure 3. Temporal evolution of the TMAXmean (upper lines) and TMINmean (lower lines) indices (°C) during the present climate, observed for Alto da Boa Vista in blue, in red for Santa Cruz, and simulated by the Eta-HadCM3 model (average of the 4 members of the ensemble) in black. For the period of 2011-2099, the fine black, blue, green and red lines refer to the Eta-Cntrl, Eta-Low, Eta-Low, Eta-Mid and Eta-High members, respectively.

Boa Vista and +0.01°C/year in Santa Cruz). The Eta-HadCM3 model simulates a mild climate for Rio de Janeiro in the current climate scenario, with maximum (minimum) air temperatures near the values seen in Alto da Boa Vista (in Santa Cruz), with a positive trend of +0.03°C/year for TMINmean and +0.01°C/year for TMAXmean. For the late 21st century, the model projects annual average values for maximum (minimum) air temperatures between 2°C and 5°C (between 2°C and 4°C) above their current climatology (1961-1990).

The TX90p (warm days), shown in **Figure 4**, and TX10p (cold days) indices, that show the percentage of days in the year on which the maximum air temperature (TX) is, respectively, above the 90th percentile (TX > 32°C in Alto da Boa Vista and TX > 35°C in Santa Cruz) and below the 10th percentile (TX < 20.3°C in Alto da Boa Vista, and TX < 23.4°C in Santa Cruz), show statistically significant trends in both locations. Hot (cold) days are becoming more (less) frequent, at a rate of +0.15% days/year for both locations (−0.22% days/year) for Alto da Boa Vista and −0.11% days/year for Santa Cruz. The Eta-HadCM3 model managed to adequately capture the behaviors of such indicators in the current climate. For the late 21st century, it has been projected that about 40% (with Eta-Low) to 70% (with Eta-High) of the days of the year will be hot and there will be no cold days for all members of the ensemble.

Although the TN90p indices (warm nights: TN > 21.5°C in Alto da Boa Vista and TN > 23.4°C in Santa Cruz) and TN10p (cold nights: TN < 14.0°C in Alto da Boa Vista and TN < 16.2°C in Santa Cruz) show trends of an increase and a reduction, respectively, based on the observations made in Alto da Boa Vista and in Santa Cruz, they are not statistically significant (except for TN90p in Santa Cruz), as is the case for the simulation of the Eta-

HadCM3 model in the current climate. For the future climate scenario, it is projected there will be about 55% of warm nights in the year with Eta-Low and 85% with Eta-High. There will also be no cold nights in the late 21st century.

The WSDI indicator (heat waves), which represents the maximum number of consecutive days in the year when the maximum air temperature is above the 90th percentile, shows a statistically significant positive trend for Alto da Boa Vista (+0.17 day/year) and a weak positive trend for Santa Cruz (+0.02 day/year) not statistically significant. These heat waves tend to be more frequent and to last longer by the late 21st century, as per the four members of the Eta-HadCM3 model. The control member shows that the city of Rio de Janeiro could face more than 90 consecutive days of temperatures above the 95th percentile by 2100.

Table 1 shows a summary of the trends observed for the climate extremes indicators related to rainfall, maximum air temperature and minimum air temperature that were analyzed earlier. It is noted that trends are significant to air temperature but not for precipitation, except for R95p at Alto da Boa Vista. Besides, as interannual variability at the present climate is higher for precipitation than for temperature, the magnitudes of the differences between future and present are stronger for temperature than precipitation. According to [6] precipitation changes around the world are much less spatially coherent compared with temperature change, with large areas showing both increasing and decreasing trends.

Sea surface temperature near Rio de Janeiro state had increased 1 to 1.5 degree in the past few decades [47]. Warmer sea surface tends to evaporate more water, and sea breeze will bring more rainfalls to the city. [46] comments that generally, precipitation extremes are expected to increase in severity with climate change, and these will have adverse impacts on metropolitan region of Rio de Janeiro, given that such area already experiences extreme flooding and landslides on a roughly 20 year basis. This is consistent with the discovery of this paper for Alto da Boa Vista, located in a forested region, near the waterfront, windward of Tijuca massif. In terms of temperatures, results at **Table 1** show that warm (cold) days and nights are more (less) frequent, consistent with a global warming scenario.

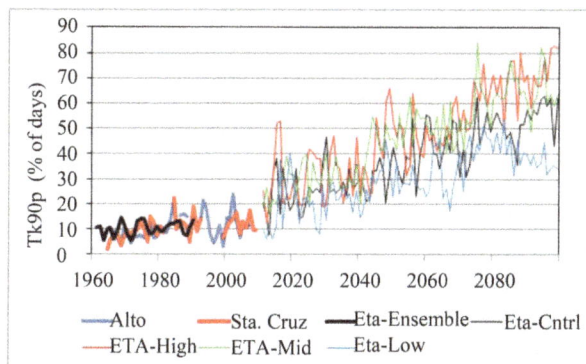

Figure 4. Temporal evolution of TX90p in a percentage of days per year, during the present climate, observed for Alto da Boa Vista in blue, for Santa Cruz in red, and simulated by the Eta-HadCM3 model (average between the 4 members of the set) in black. For the period of 2011-2099, the fine black, blue, green and red lines refer to the Eta-Cntrl, Eta-Low, Eta-Mid and Eta-High members, respectively.

5. Conclusions

This study undertakes a climate change detection analysis for the city of Rio de Janeiro based on a series of observational data on the total daily rainfall and the maximum and minimum daily air temperatures for two INMET stations: Alto da Boa Vista and Santa Cruz. The climate extremes indicators are calculated as per [4]. The simulations of the regional Eta-HadCM3 model are used to investigate future projections (2011-2100) of such indices, considering the SRES-A1B scenario. The model's assessment in the current climate (1961-1990) is performed as a prior step to investigate its future projections.

The results indicate that the climate in the city of Rio de Janeiro is becoming more humid in the forested region, with heavy rainfall events producing a greater amount of rainfall. The differences in rainfall indicators in Alto da Boa Vista and in Santa Cruz, with more marked trends in the first weather station than in the second one, suggests changes in wind patterns, the circulation of sea and land breezes, and moisture transport into the city, which needs to be investigated in more detail. This trend in extreme precipitation could perhaps be related to changes in the frequency and duration of cold fronts, since most of rainfall in this region is produced by cold front penetrations. If the frequency of cold fronts increases, it will affect both places (Alto da Boa Vista and Santa Cruz), although as Alto da Boa Vista is more humid, nearest the waterfront, within a tropical forest and in a higher altitude, rainfall should be stronger there. More investigation is needed on the assessments of trends of cold fronts in the region, helping to explain changes in rainfall extremes. For the future, it is projected an increase in both the frequency and intensity of heavy rainfall events and, on the other hand, a statistically significant increase in the duration of dry periods and a reduction in the duration of humid periods. This indicates a trend of unequally distributed rainfall for the future, with longer dry periods, but, on the other hand, more frequent heavy rainfall events.

In terms of temperature, as expected one can note a warming trend in the city; however, the most remarkable trends occur for indices associated with the maximum air temperature, in opposition to the results for most of the globe, in which the minimum air temperature increases at a higher rate than maximum air temperature. Hot (cold) days are more (less) frequent. Heat waves, i.e., periods with high maximum air temperatures, are becoming longer. The trends associated with the minimum temperature are more significant in Santa Cruz than in Alto da Boa Vista. The differences in the two regions (urban and forested) may be associated with circulation changes at the synoptic scale and also on a local scale due to the urban heat island effect. Warm evenings are also more frequent, while cold nights are in decline. According to the regional model projections, the average annual maximum (minimum) air temperature anomalies can reach between 2°C and 5°C (between 2°C and 4°C) above the current values. It is projected that about 40% to 70% of the days in the year will be hot and there will be no cold days. Similar behavior is seen for hot (cold) nights, with a frequency between 55% and 85% of warm nights in the year and without any cold nights.

In the present climate, the regional model shows air temperature results that are quite consistent with the observations made at Alto da Boa Vista and in Santa Cruz, especially for the indices based on percentiles (TN10p, TN90p, TX10p, TX90p). The model's TMAXmean (TMINmean) is close to that observed in Alto da Boa Vista (Santa Cruz), suggesting that the model simulates a milder climate for Rio de Janeiro, generating DTR values about 2°C (3°C) lower than those observed in Alto da Boa Vista (Santa Cruz). With respect to the precipitation simulated by the model in the current climate, most of the trends are opposite to what has been observed. However there is no consensus even among the observations expressed in this study, possibly due to the rainfall depending on local systems, such as land and sea breezes, or to the limitation of the observational data itself, with countless flaws.

With regard to the model's future projections, it is important to emphasize that the simulations only take into account the increased concentration of greenhouse gases, and not changes in land use and the urban heat island effect resulting from urban sprawl. Moreover, when it comes to regional models, the reliability of the high resolution projections depends on the quality of the lateral boundary condition, which in this study is provided by only one global model, and also on the ability of the regional model to develop realistic regional characteristics for the present climate. We have also to consider that regional model simulations, with a horizontal resolution of 40 km, cannot reproduce the high interannual variability of the extreme rainfall-related indices [7]. It is suggested in here new downscaling experiments using more than one global model as boundary conditions, to allow for detailed uncertainty analyses.

More detailed studies to detect the possible intensification (weakening) of sea (land) breezes, in addition to analyses on the seasonal level, using other sets of observational data could contribute to increase the knowledge about climate change in the city of Rio de Janeiro.

5. Acknowledgements

Jose Marengo and Wanderson Luiz Silva were supported by the Brazilian National Council for Scientific and Technological Development (CNPq). Additional funding was provided by Rede-CLIMA, the National Institute of Science and Technology for Climate Change (INCT-CC), the FAPESP-Assessment of Impacts and Vulnerability to Climate Change in Brazil and strategies for Adaptation options project (Ref. 2008/58161-1), and the European Community's Seventh Framework Programme (FP7/ 2007-2013) under Grant N0. 212492: CLARIS-LPB project. We acknowledge the National Meteorological Institute (INMET) for providing their observed data.

REFERENCES

[1] T. Karl, N. Nicholls and J. Gregory, "The Coming Climate," *Scientific American*, Vol. 276, 1997, pp. 54-59.

[2] D. Easterling, *et al.*, "Climate Extremes: Observations, Modeling and Impacts," *Science*, Vol. 289, No. 5487, 2000, pp. 2068-2074.

[3] G. Meehl, *et al.*, "An Introduction to Trends in Extreme Weather and Climate Events: Observations, Socioeconomic Impacts, Terrestrial Ecological Impacts and Model Projections," *Bulletin of the American Meteorological Society*, Vol. 81, No. 3, 2000, pp. 413-416.

[4] P. Frich, *et al.*, "Observed Coherent Changes in Climatic Extremes during the Second Half of the Twentieth Century," *Climate Research*, Vol. 19, No. 3, 2002, pp. 193-212.

[5] L. A. Vincent, *et al.*, "Observed Trends in Indices of Daily Temperature Extremes in South America 1960-2000," *Bulletin of the American Meteorological Society*, Vol. 18, No. 23, 2005, pp. 5011-5023.

[6] L. V. Alexander, *et al.*, "Global Observed Changes in Daily Climate Extremes of Temperature and Precipitation," *Journal of Geophysical Research*, Vol. 111, No. D5, 2006, pp. 1-22.

[7] M. R. Haylock, *et al.*, "Trends in Total and Extreme South American Rainfall in 1960-2000 and Links with Sea Surface Temperature," *Journal of Climate*, Vol. 19, No. 8, 2006, pp. 1490-1512.

[8] C. Tebaldi, K. Hayhoe, J. Arblaster and G. Meehl, "Going to the Extremes: An Intercomparison of Model-simulated Historical and Future Changes in Extreme Events," *Climatic Change*, Vol. 79, No. 3-4, 2006, pp. 185-211.

[9] G. Obregón and J. A. Marengo, "Mudanças Climáticas Globais e Seus Efeitos Sobre a Biodiversidade: Caracterização do Clima no Século XX no Brasil: Tendências de Chuvas e Temperaturas Médias e Extremas," Relatório, 2007.

[10] J. A. Marengo, R. Jones, L. M. Alves and M. C. Valverde, "Future Change of Temperature and Precipitation Extremes in South America as Derived from the PRECIS Regional Climate Modeling System," *International Journal of Climatology*, Vol. 29, No. 15, 2009, pp. 2241-2255.

[11] J. A. Marengo, M. Rusticucci, O. Penalba and M. Renom, "An Intercomparison of Observed and Simulated Extreme Rainfall and Temperature Events during the Last Half of the Twentieth Century: Part 2: Historical Trends," *Climatic Change*, Vol. 98, No. 3-4, 2010, pp. 509-529.

[12] M. Rusticucci, J. A. Marengo, O. Penalba and M. Renom, "An Intercomparison of Model-Simulated in Extreme Rainfall and Temperature Events during the Last Half of the Twentieth Century. Part 1: Mean Values and Variability," *Climatic Change*, Vol. 98, No. 3-4, 2010, pp. 493-508.

[13] R. E. Dickinson, R. M. Errico, F. Giorgi and G. T. Bates, "A Regional Climate Model for the Western US," *Climate Change*, Vol. 15, No. 3, 1989, pp. 383-422.

[14] F. Giorgi and T. Bates, "The Climatological Skill of a Regional Model over Complex Terrain," *Monthly Weather Review*, Vol. 117, No. 11, 1989, pp. 2325-2347.

[15] F. Giorgi, "On the Simulation of Regional Climate Using a Limited Area Model Nested in a General Circulation Model," *Journal of Climate*, Vol. 3, No. 9, 1990, pp. 941-963.

[16] H. Kida, T. Koide, H. Sasaki and M. Chiba, "A New Approach to Coupling a Limited Area Model with a GCM for Regional Climate Simulations," *Journal of the Meteorological Society of Japan*, Vol. 69, No. 6, 1991, pp. 723-728.

[17] R. G. Jones, *et al.*, "Generating High Resolution Climate Change Scenarios Using PRECIS," Report, Met Office Hadley Centre, Exeter, 2004.

[18] J. A. Marengo, *et al.*, "Future Change of Climate in South America in the Late Twenty-First Century: Intercomparison of Scenarios from Three Regional Climate Models," *Climate Dynamics*, Vol. 35, No. 6, 2009, pp. 1073-1097.

[19] R. D. Garreaud and M. Falvey, "The Coastal Winds off Western Subtropical South America in Future Climate Scenarios," *International Journal of Climatology*, Vol. 29, No. 4, 2008, pp. 543-554.

[20] W. R. Soares and J. A. Marengo, "Assessments of Moisture Fluxes East of the Andes in South America in a Global Warming Scenario," *International Journal of Climatology*, Vol. 29, No. 10, 2008, pp. 1395-1414.

[21] S. A. Solman, M. N. Nuñez and M. F. Cabré, "Regional Climate Change Experiments over Southern South America. I: Present Climate," *Climate Dynamics*, Vol. 30, No. 5, 2008, pp. 533-552.

[22] M. N. Nuñez, S. A. Solman and M. F. Cabré, "Regional Climate Change Experiments over Southern South America. II: Climate Change Scenarios in the Late Twenty-First Century," *Climate Dynamics*, Vol. 32, No. 7-8, 2009, pp. 1081-1095.

[23] L. M. Alves and J. A. Marengo, "Assessment of Regional Seasonal Predictability Using the PRECIS Regional Climate Modeling System over South America," *Theoretical and Applied Climatology*, Vol. 100, No. 3-4, 2010, pp. 337-350.

[24] S. C. Chou, *et al.*, "Downscaling of South America Present Climate Driven by 4-Member HadCM3 Runs," *Climate Dynamics*, Vol. 38, No. 3-4, 2011, pp. 635-653.

[25] J. A. Marengo, *et al.*, "Development of Regional Future Climate Change Scenarios in South America Using the Eta CPTEC/HadCM3 Climate Change Projections: Climatology and Regional Analyses for the Amazon, São Francisco and the Parana River Basins," *Submitted to Climate Dynamics*, Vol. 38, No. 9-10, 2011, pp. 1829-1848.

[26] F. Mesinger, *et al.*, "An Upgraded Version of the Eta Model," *Meteorology and Atmospheric Physics*, Vol. 116, No. 3-4, 2012, pp. 63-79.

[27] C. Gordon, *et al.*, "Simulation of SST, Sea Ice Extents and Ocean Heat Transport in a Version of the Hadley Centre Coupled Model without Flux Adjustments," *Climate Dynamics*, Vol. 16, No. ,2-3 2000, pp. 47-168.

[28] N. Nakicenovic, *et al.*, "Special Report on Emission Scenarios," Cambridge University Press, Cambridge, 2000, p. 599.

[29] X. Zhang and F. Yang, "RClimDex (1.0)—User Manual," Climate Research Branch Environment Canada Downsview, Ontario, 2004.

[30] R. Sneyers, "Sur l'Analyse Statistique des Series Dóbservations," Vol. 192, Organisation Méteorologique Mondial, Gênevè, 1975.

[31] C. Goossens and A. Berger, "Annual and Seasonal Climatic Variations over the Northern Hemisphere and Europe during the Last Century," *Annales Geophysicae*, Vol. 4 No. B4, 1986, pp. 385-400.

[32] P. K. Sen, "Estimates of the Regression Coefficient Based on Kendall's Tau," *Journal of American Statistics Association*, Vol. 63, No. 324, 1968, pp. 379-1389.

[33] P. Jourdan, "Caracterização do Regime de Ventos Próximo à Superfície na Região Metropolitana do Rio de Janeiro," Course Completion Paper at the Geoscience Institute Department of Meteorology, Rio de Janeiro, Federal University of Rio de Janeiro, 2007.

[34] C. P. Dereczynski, J. S. Oliveira and C. O. Machado, "Climatologia da Precipitação no Município do Rio de Janeiro," *Revista Brasileira de Meteorologia*, Vol. 24, No. 1, 2009, pp. 24-38.

[35] A. S. Figueiró, "Mudanças Ambientais na Interface Floresta-Cidade e Propagação de Efeito de Borda no Maciço da Tijuca, Rio de Janeiro: Um Modelo de Vizinhança," Doctoral Thesis in Geography, Federal University of Rio de Janeiro, 2005.

[36] A. S. Figueiró and A. L. Coelho Netto, "Do Local ao Regional: Análise Comparativa de Transectos Pluviométricos em Diferentes Escalas," V Encontro Nacional da Associação Nacional de Pós Graduação em Geografia/ ANPEGE, Florianópolis, 2003.

[37] A. L. Coelho Netto, A. Avelar and R. D'orsi, "Domínio do Ecossistema da Floresta Atlântica de Encostas," In: Rio Próximos 100 Anos, O Aquecimento Global e a Cidade, Rio de Janeiro, Instituto Municipal de Urbanismo Pereira Passos, 2008.

[38] C. Chen and T. Knutson, "On the Verification and Comparison of Extreme Rainfall Indices from Climate Models," *Journal of Climate, Notes and Correspondence*, Vol. 21, No. 7, 2008, pp. 1605-1621.

[39] D. Kiktev, *et al.*, "Comparison of Modeled and Observed Trends in Indices of Daily Climate Extremes," *Journal of Climate*, Vol. 16, No. 22, 2008, pp. 3560-3571.

[40] M. Wehner, "Predicted Twenty-First-Century Changes in Seasonal Extreme Precipitation Events in Parallel Climate Model," *Journal of Climate*, Vol. 17, No. 21, 2004, pp. 4281-4290.

[41] G. C. Hegerl, *et al.*, "Detectability of Anthropogenic Changes in Annual Temperature and Precipitation Extremes," *Journal of Climate*, Vol. 17, No. 19, 2004, pp. 3683-3700.

[42] J. Iorio, *et al.*, "Effect of Model Resolution and Subgrid-Scale Physics on the Simulation of Precipitation in the Continental United States," *Climate Dynamics*, Vol. 23, No. 3-4, 2004, pp. 243-258.

[43] V. V. Kharin, *et al.*, "Intercomparison of Near-Surface Temperature and Precipitation Extrems in AMIP-2 Simulations, Reanalyses, and Observations," *Journal of Climate*, Vol. 18, No. 24, 2005, pp. 5201-5223.

[44] S. Emori, *et al.*, "Validation, Parameterization Dependence, and Future Projection of Daily Precipitation Simulated with a High-Resolution Atmospheric GCM," *Geophysical Research Letters*, Vol. 32, No. 6, 2005, pp. 1-4.

[45] J. A. Marengo and M. C. Valverde, "Caracterização do Clima no Século XX e Cenário de Mudanças de Clima para o Brasil no século XXI usando Modelos do IPCC-AR4," Revista Multiciência, Campinas, Edição n. 8, 2007, Mudanças Climáticas.

[46] R. Blake, *et al.*, "Urban Climate: Processes, Trends and Projections. Climate Change and Cities: First Assessment Report of the Urban Climate Change Research Network," Cambridge University Press, Cambridge, 2011, pp. 43-81.

[47] I. Zurbenko and M. Luo, "Restoration of Time-Spatial Scale in Global Temperature Data," *American Journal of Climate Change*, Vol. 1, No. 3, 2012, pp. 154-163.

13

Climate Change: Concerns, Beliefs and Emotions in Residents, Experts, Decision Makers, Tourists, and Tourist Industry

Igor Knez[1], Sofia Thorsson[2], Ingegärd Eliasson[3]
[1]Department of Social Work and Psychology, University of Gävle, Gävle, Sweden
[2]Department of Earth Sciences, University of Gothenburg, Göteborg, Sweden
[3]Department of Conservation, University of Gothenburg, Göteborg, Sweden

ABSTRACT

The aim was to investigate effects of different groups of individuals (residents, tourists, experts, decision makers and members of tourist industry) and demographic variables (gender, age, education) on climate change-related concerns, beliefs and emotions. In line with the predictions: 1) Experts were shown to be least concerned for and afraid of climate change impact; 2) Youngest participants were found to be most, and oldest least, concerned for their future; 3) Women were shown to be more concerned for and afraid of the consequences of climate change; and 4) Men and the least educated participants believed their jobs to be more threatened by the environmental laws and protection, and the latter ones believed moreover that the claims about climate change are exaggerated. Implications of these findings for value orientations and their relationships to environmental concerns, beliefs and emotions are discussed.

Keywords: Climate Change; Demographic Variables; Concerns; Beliefs; Emotions

1. Introduction

Climate change is a continuing present day issue. News about floods, heat waves and storms and their impact on society reach us almost every day. These reports imply a threat to our present way of living, urging for responsebility, pro-environmental behaviour and ecologically sustainable progress. There is an increasing awareness that climate change is not only an ecological and economic dilemma, but also a social and psychological one [1-3], meaning that drastic policies are necessary to prevent a serious lack of natural resources by promoting a sustainable behaviour [4].

It is difficult to conceptualize and frame the climate change problem as well as to temporally and psychologically foresee its consequences, meaning that human processes of climate change related perception are for the most part uncertain [5,6]. Due to the embedded conflict between an individual level of short-term self-interest (e.g. accumulating damaging gases by driving the car a lot) and long-term collectivistic natural resource management (to decrease urban air pollution), the climate change issues can also be conceptualized as resource dilemmas, involving a conflict between individual and collective interests [7].

Climate change and its consequences are disseminated to the general public by the media, scientists and politicians [8]; information that can be vague, is often "scientifically uncertain" [9], and is due to this misunderstanding [10]. This relation between knowledge of climate change per se and confidence in that knowledge on an individual level has recently been found to vary across types of people [11], showing that both knowledge and confidence in one's own knowledge were highest among scientists followed by journalists, politicians and laypersons. It is also indicated that better knowledge of [4,12], and greater concern for [13,14], climate change may promote pro-environmental behaviour.

In accordance with the work of Schwartz [15,16] on structure and contents of human values, Stern and colleagues [17-21] and Schultz and colleagues [22-26] have extended this account to comprise environmental-related concerns (affect associated with environmental issues), attitudes (beliefs, values and behavioural intentions associated with environmental issues) and worldviews (our relationships to environment). More precisely, it is assumed that awareness of climate change has an impact on:

Oneself (*belief* about consequences "for oneself") is grounded on the person's *egoistic value* orientation (comprising dimensions of power, social power, authority and wealth vs. achievement, ambition, capability, success, influence);

Others (*belief* about consequences "for others") are grounded on the person's *benevolence value* orientation (comprising dimensions of being helpful, forgiving, loyal, responsible);

Biosphere (*belief* about consequences "for biosphere") is grounded on the person's *universalism value* orientation (comprising dimensions of broad-mindedness, equality, social justice, a world of peace).

Each of the beliefs about consequences for oneself, others and the biosphere is assumed to correlate positively with each of the environment-related *concerns* for: 1) "myself, my lifestyle, my health, my future"; 2) "all people, people in my country, children, my children"; and 3) "plants, marine life, birds, animals" respectively; see also [27,28]. Hansla *et al* [29] however, have, pointed out that the relationship between environmental-related values, beliefs and concerns may vary across different groups of individuals. In addition, Olofsson & Öhman [30] have indicated that demographic variables such as age, gender, education, political affiliation and location may influence people's environmental concerns; see also [17,31,32].

It has also been shown that people experience and express different types of emotion, such as worry and hope, related to global environmental problems [33]. For example, Garcia-Mira *et al.* [34] have indicated that we may estimate local (e.g. increased number of cars) compared to global (e.g. increased pollution of atmosphere) environmental problems as less worrying; and Ojala [35] reported that an emotional reaction/expression of worriment may vary with gender, indicating that women embrace environmental-related altruistic values to a higher degree than do men.

Present Study

Consequences of climate change on tourism are in particular difficult to predict [36-38]. This societal and economic sector will, in general terms, be deeply affected, because tourism by definition is related to climate and weather [39-43]. It is, for example, suggested that warmer summers may decrease tourism in the Mediterranean region [44], but expand it in the Northern and Western Europe [45], as well as enhance and develop spring and fall tourism in Southern Europe [44].

Despite the fact that tourism as a phenomenon consti-

tutes a considerable part of the global economy and that tourist streams may be radically modified in step with the climate change [46,47], there still appears to be a lack of responsiveness and commitment among politicians, stakeholders and the tourist industry [48-50].

Based on the findings reviewed above the aim was to investigate effects of different groups of individuals (residents, tourists, experts, decision makers and members of tourist industry) and demographic variables (gender, age and education) on participants' climate-change-related concerns, beliefs and emotions, including the following two hypotheses:

Hypothesis 1. In line with some previous findings which suggest differences in environment-related socio-psychological constructs between different groups of individuals [8-11,29,49,51], the present study included participants from five types of group (residents, tourists, experts, decision makers and members of the tourist industry) defining the independent variable of Type of Group. The hypothesis was that the impact of climate change measured as environmental-related concerns, beliefs ("awareness" of the climate change impact), and emotions would vary across these groups of individuals, due to differences in their underlying environmental issue-related value orientations [18,21,52].

Hypothesis 2. In addition to Hypothesis 1 and across the Type of Group variable, a second analysis included demographic (independent) variables of Gender, Age and Education previously indicated to influence environmental concerns [30-32,53-55] and emotions [33,35,56]. The hypothesis was that older vs. younger, men vs. women, and participants with lower vs. higher education would show less environment-related concerns and beliefs; and that men would be found to be less worried about climate change.

2. Method

2.1. Study Area and Climate

The city of Gothenburg (Göteborg, Sweden, 57°42'N, 11°58'E—with a population of approx. 500,000, approx. 900,000 in the region). Being located on the west coast of Sweden, Gothenburg is one of the country's warmest cities thanks to the Gulf Stream and its warming waters. Summers are predictably warm, with an average maximum daily air temperature from June to August of about 20°C/68°F or more. Summer evenings in Gothenburg are warm and balmy at times, with night-time temperatures rarely falling below 13°C/55°F.

2.2. Sample

A total of 1000 households located within the City of Gothenburg were sent a "climate survey" during the summer 2009. They were randomly identified from a population register. The questionnaire was also handed out at different tourist locations to 1000 tourists visiting the town in June to August 2009. The survey was also sent to 30 experts (scientists working with climate and environment issues); 67 decision makers (politicians, stakeholders); and 156 members of the tourist industry (hotels, museums, theatres, amusement parks, hauliers, conference centres, event companies, etc.). The survey comprised a number of sections including questions about demographic variables, climate, climate change-related behaviours and attitudes, etc. Data on climate change-related concerns, beliefs and emotions will be reported in the present study.

Procedures and response rates. After two contacts (After one week a reminder was sent to all those that had not answered. A week later a second reminder was sent.) 1257 responses were achieved, distributed across five groups of participants: Residents (528); Tourists (576); Experts (18); Decision Makers (33); and Tourist Industry (102).

Tourists. According to the World Tourism Organisation (UNWTO, 2008), a tourist is either a person who for *pleasure* has travelled at least 100 km. from their domicile, for a stay of at least 24 hours, or a person who for *various reasons* (e.g. pleasure, shopping, official journey) has left their domicile for another place. In the present study a definition of a tourist was a person living *outside* the Gothenburg region (comprising 13 municipalities) visiting Gothenburg city for pleasure and/or other reasons. For each tourist who agreed to participate, name, e-mail and telephone number were documented. Each person was given a questionnaire (in English, Germany or Swedish), stamped and addressed reply envelope, and an e-mail login code. The majority (75%) of tourists came from northern Europe (Sweden, Norway, Finland, and Denmark), 20% from other European countries and 5% from outside Europe. This is in accordance with local tourist statistics (from the Swedish Agency for Economic and Regional Growth) for accommodation showing that domestic tourists dominate.

2.3. Measures

Concerns. Climate change-related concerns [29] were measured with 12 self-report items responding to the statement, "I am concerned about climate change because of the consequences for: myself, my lifestyle, my health, my future, all people, people in my country, children, my children, plants, marine life, birds, other animals." Responses were made using a 7-point Likert scale ranging from 1 (completely disagree) to 7 (completely agree).

Beliefs. Climate change-related beliefs [29] were meas-

ured with 6 self-report statements: 1) Laws that protect the environment limit my choices and personal freedom; 2) Protecting the environment will threaten jobs for people like me; 3) Effects of climate change on public health are worse than people realize; 4) Pollution generated in one country harms people all over the world; 5) Over the next several decades, thousands of species will vanish; 6) Claims that there is climate change are exaggerated. Participants were asked to respond to these statements on a 7-point Likert scale ranging from 1 (completely disagree) to 7 (completely agree).

Emotions. Affect states related to climate change [57] were measured with 4 self-report items answering the questions of: 1) How *hopeful* do you feel about the place where you live (Item 1) and the world (Item 2) respectively when you think about climate change risks? and 2) How *afraid* do you feel about the place where you live 1) and the world 2) respectively when you think about climate change risks? Participants were asked to respond to these questions on a 7-point Likert scale ranging from 1 (not at all) to 7 (very much).

2.4. Design

A non-equivalent, comparison-group, quasi-experimental design [58] was used. Compared with a "true experiment" [59], this means that inferences drawn about the causal relationships between independent and dependent variables are considered to be weaker.

Independent variables. Five groups of participants: Residents, Tourists, Experts, Decision Makers, Tourist Industry. These were the independent variables involved in "Analyses 1" (see Result section below). In "Analysis 2", the independent variables of Gender (males vs. females; 590, 578 participants per category), Age (−25, 26 - 35, 36 - 45, 46 - 55, 56 - 65, 66+; 147, 211, 256, 233, 208, 193 participants per category) and Education (Primary, High School or equivalent, Bachelor's Degree, Master's Degree, Doctoral Degree; 140, 542, 285, 203, 52 participants per category) across the five groups of participants were used as the independent variables (see Result below).

Dependent variables. 22 items (statements, questions) distributed across three measures of climate change-related concerns (12 items), beliefs (6 items) and emotions (4 items).

3. Results

All data were subjected to MANOVAs (multivariate analyses of variance) due to the several items involved in each dependent variable. This section is divided into two main types of analyses: 1) "Effects of five groups of participants"; and 2) "Effects of gender, age and education" (demographic variables across the five groups). Accordingly, the first type of analyses involved one independent variable, that of Type of Group (related to *Hypothesis* 1, see Introduction), and the second type of analyses involved three independent variables, that of Gender, Age and Education (related to *Hypothesis* 2, see Introduction).

3.1. Effects of Five Groups of Participants

Concerns. A main significant effect of Type of Group was obtained, Wilk's Lambda = 0.89, $F(48, 3892) = 2.49$, $p < 0.001$, associated with all 12 statements (p values < from 0.05 to 0.01). As can be seen in **Figure 1**, the grand mean for the concerns across the five groups of participants varied between values of 3.55 (I am concerned about climate change because of the consequences for *my lifestyle*) and 5.53 (I am concerned about climate change because of the consequences for *children*) showing that participants in general were least concerned about the consequences for their lifestyle and most for future generations, $t(1061) = 27.99, p < 0.01$.

According to **Figure 2** and in general terms, Experts (similarly with Decision Makers and Tourist Industry) were shown to be least, and Tourists most, concerned for the consequences of climate change, $t(582) = 2.11, p < 0.05$.

Beliefs. A main significant effect of Type of Group was obtained, Wilk's Lambda = 0.95, $F(24, 3988) = 2.56$, $p < 0.01$, associated with the statements of: 1) The effects of climate change on public health are worse than people realize ($p = 0.06$); 2) Pollution generated in one country harms people all over the world ($p < 0.01$); and 3) Claims that there is climate change are exaggerated ($p < 0.01$).

As can be seen in **Table 1**, Tourists were the group that most believed that the effects of climate change on public health are worse than people realize and that pollution generated in one country harms people all over the world. In addition, Residents considered the claims about consequences of a climate change as more exaggerated than did the other four groups.

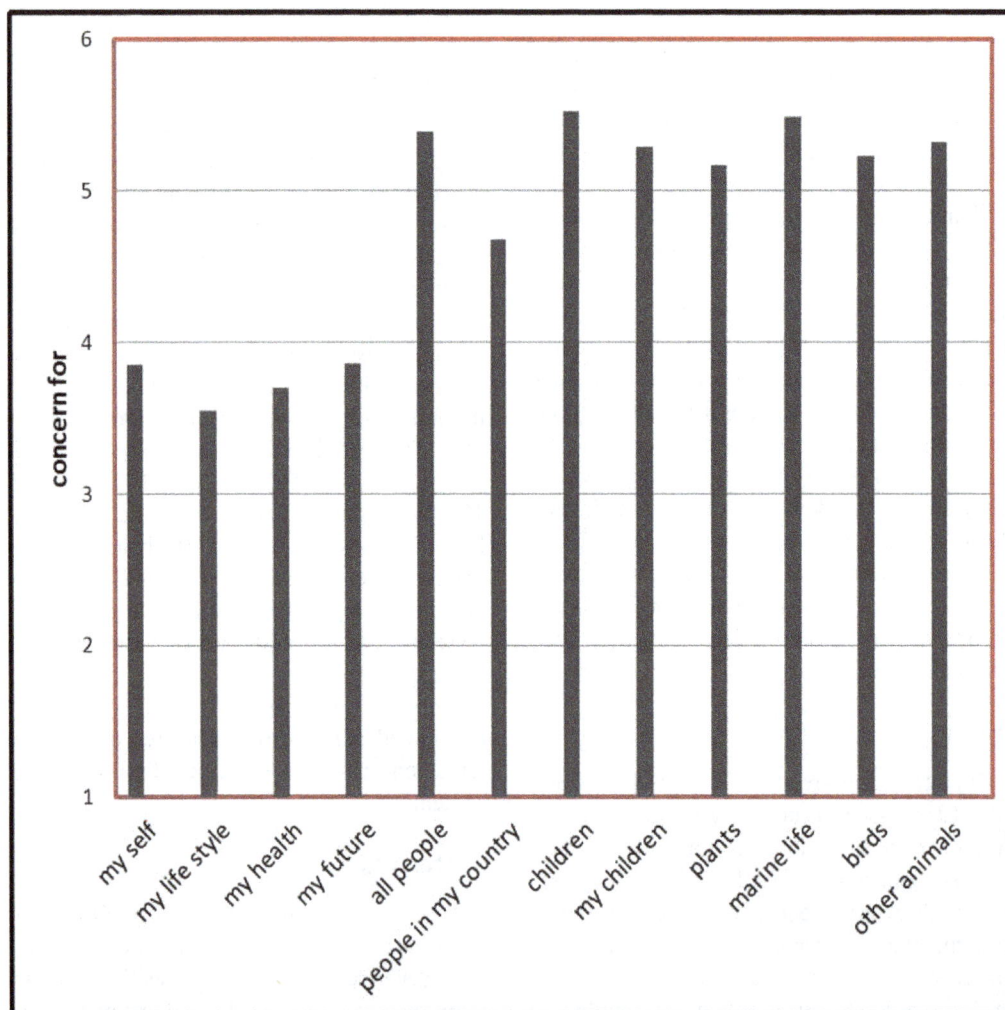

Figure 1. Mean concern for type of concern (concerns for: myself, my lifestyle, my health, my future, all people, people in my country, children, my children, plants, marine life, birds, other animals) across five groups of participants.

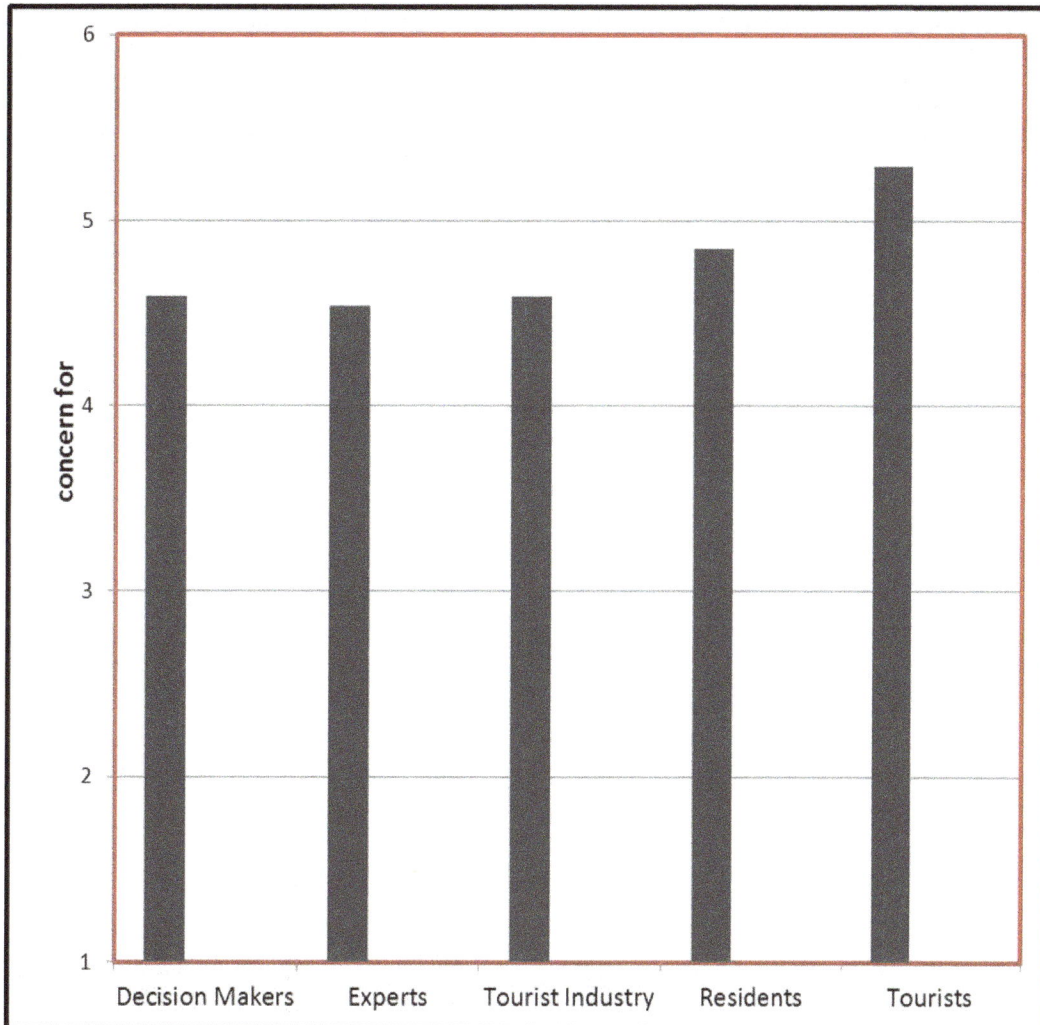

Figure 2. Mean concern (across type of concern) for Type of Group (Decision Makers, Experts, Tourist Industry, Residents, Tourists).

Furthermore, all participants (independently of group) believed for the most part that pollution generated in one country harms people all over the world and least that protecting the environment will threaten jobs for people like them (see **Figure 3**), $t(1185) = 58.71$, $p < 0.01$.

Emotions. A main significant effect of Type of Group was obtained, Wilk's Lambda = 0.93, $F(16, 3507) = 5,11$, $p < 0.001$, associated with the questions: 1) How hopeful do you feel when you think about climate change risks, in the world? ($p < 0.01$); and How afraid do you feel when you think about climate change risks, where you live 2) and in the world (3)? (p values < 0.01). The former showed Residents as most, and Decision Makers as least, hopeful (see **Table 2**) and the latter indicated that Residents and Tourists feared most, and Decision Makers and Experts feared least, climate change consequences for the place where they lived. Tourists feared most and Experts least the impact of climate change for the whole world (see **Table 3**).

However, and in general, all participants (independently of the group) were found to be most hopeful for the environment where they live compared to the whole world and, consistently, more afraid for the whole world than for their neighboring environment concerning climate change consequences (see **Figure 4**).

3.2. Effects of Gender, Age and Education

Concerns. Main significant effects of Age, Wilk's Lambda = 0.90, $F(60, 4054) = 1.49$, $p < 0.01$, and Education, Wilk's Lambda = 0.92, $F(48, 3334) = 1.55$, $p < 0.01$, were shown. The first effect was associated with the concerns for *my future* ($p < 0.01$) and the second with the concerns for *my lifestyle* ($p < 0.01$) and *my health* ($p < 0.05$). A tendency towards a main significant effect of Gender was also indicated, Wilk's Lambda = 0.98, $F(12, 865) = 1.52$, $p = 0.11$, significantly associated with all 12 statements (p values $<$ from 0.05 to 0.01).

Table 1. Post hoc, multiple comparisons (LSD, least significant difference) between the type of group (Decision Makers, DM; Experts, E; Tourist Industry, TI; Residents, R; Tourists, T) associated with the statements of (1) Effects of climate change on public health are worse than people realize; (2) Pollution generated in one country harms people all over the world; and (3) Claims that there is a climate change are exaggerated. Only significant differences are reported with statistics of mean difference (M.D.), standard error (S.E.), p value and lower and upper bound of 95% confidence interval (L.B. and U.B.).

Statements	Groups	M.D.	S.E.	p	L.B.	U.B.
Effects...	R vs. T	−0.30	0.11	0.01	−0.51	−0.09
Pollution...	R vs. T	−0.29	0.09	0.00	−0.47	−0.09
Claims...	DM vs. R	−0.87	0.34	0.01	−1.54	−0.21
	E vs. R	−0.92	0.46	0.05	−1.83	−0.02
	TI vs. R	−0.78	0.20	0.00	−1.17	−0.39
	R vs. T	0.51	0.11	0.00	−0.29	0.73

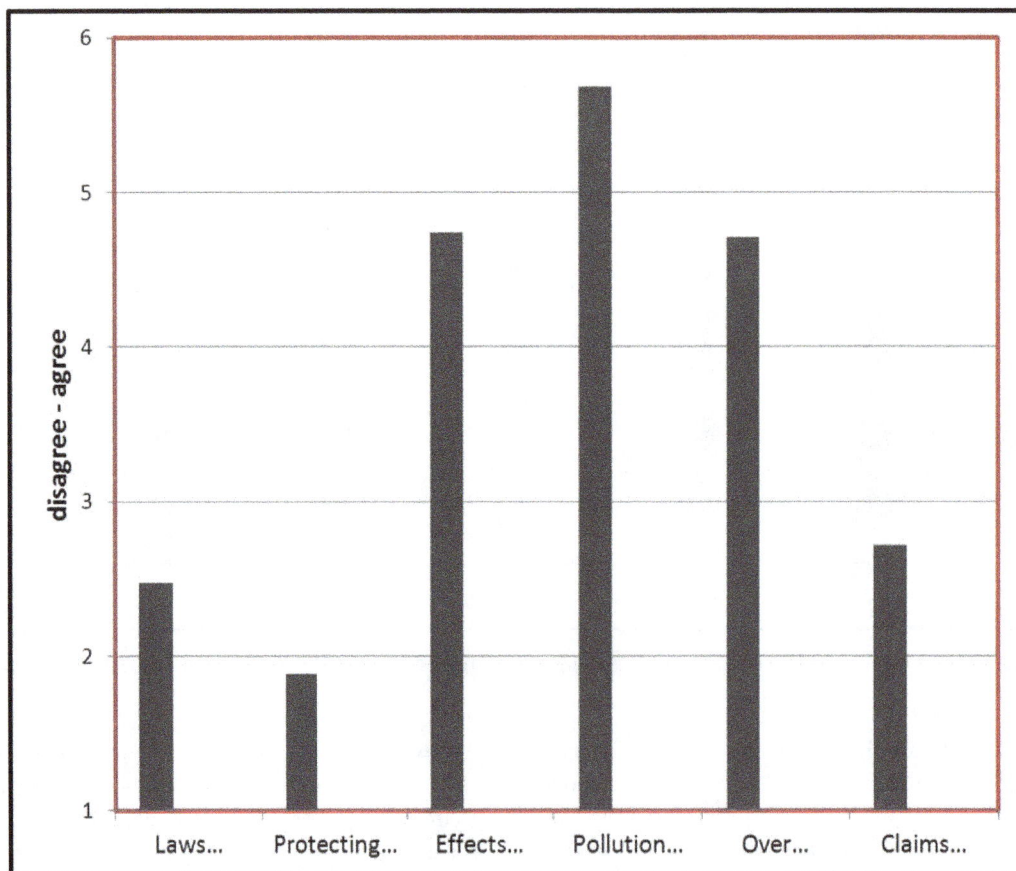

Figure 3. Mean disagreement-agreement across five groups of participants associated with statements of (1) Laws that protect the environment limit my choices and personal freedom; (2) Protecting the environment will threat jobs for people like me; (3) Effects of climate change on public health are worse than people realize; (4) Pollution generated in one country harms people all over the world; (5) Over the next several decades, thousands of species will vanish; (6) Claims that there is a climate change is exaggerated.

Table 2. Post hoc, multiple comparisons (LSD, least significant difference) between the Type of Group (Decision Makers, DM; Experts, E; Tourist Industry, TI; Residents, R; Tourists, T) associated with the question of "How much do you feel hopeful for the world when you think about the climate change risks?" Only significant differences are reported with statistics of mean difference (M.D.), standard error (S.E.), p value and lower and upper bound of 95% confidence interval (L.B. and U.B.)

Question	Groups	M.D.	S.E.	p	L.B.	U.B.
Hope for the world	DM vs. R	−0.66	0.28	0.02	−1.21	−0.11
	TI vs. R	−0.42	0.17	0.02	−0.76	−0.08
	R vs. T	0.40	0.09	0.00	0.21	0.58

Table 3. Post hoc, multiple comparisons (LSD, least significant difference) between the Type of Group (Decision Makers, DM; Experts, E; Tourist Industry, TI; Residents, R; Tourists, T) associated with the questions of "How much do you feel fear for place where you live and the world respectively when you think about the climate change risks?" Only significant differences are reported with statistics of mean difference (M.D.), standard error (S.E.), *p* value and lower and upper bound of 95% confidence interval (L.B. and U.B.)

Questions	Groups	M.D.	S.E.	*p*	L.B.	U.B.
Fear for place where I live	DM vs. R	0.81	0.31	0.01	-1.42	-0.21
	DM vs. T	−1.00	0.31	0.00	−1.60	−0.40
	E vs. R	−0.87	0.44	0.05	−1.74	−0.01
	E vs. T	−1.06	0.44	0.02	−1.93	−0.20
	TI vs. R	−0.54	0.19	0.01	−0.91	−0.16
	TI vs. T	−0.73	0.19	0.00	−1.10	−0.36
Fear for the world	DM vs. T	−0.79	0.33	0.02	−1.44	−0.13
	E vs. T	−1.25	0.48	0.01	−2.19	−0.30
	TI vs. T	−0.74	0.21	0.00	−1.14	−0.34
	R vs. T	−0.64	0.11	0.00	−0.86	−0.42

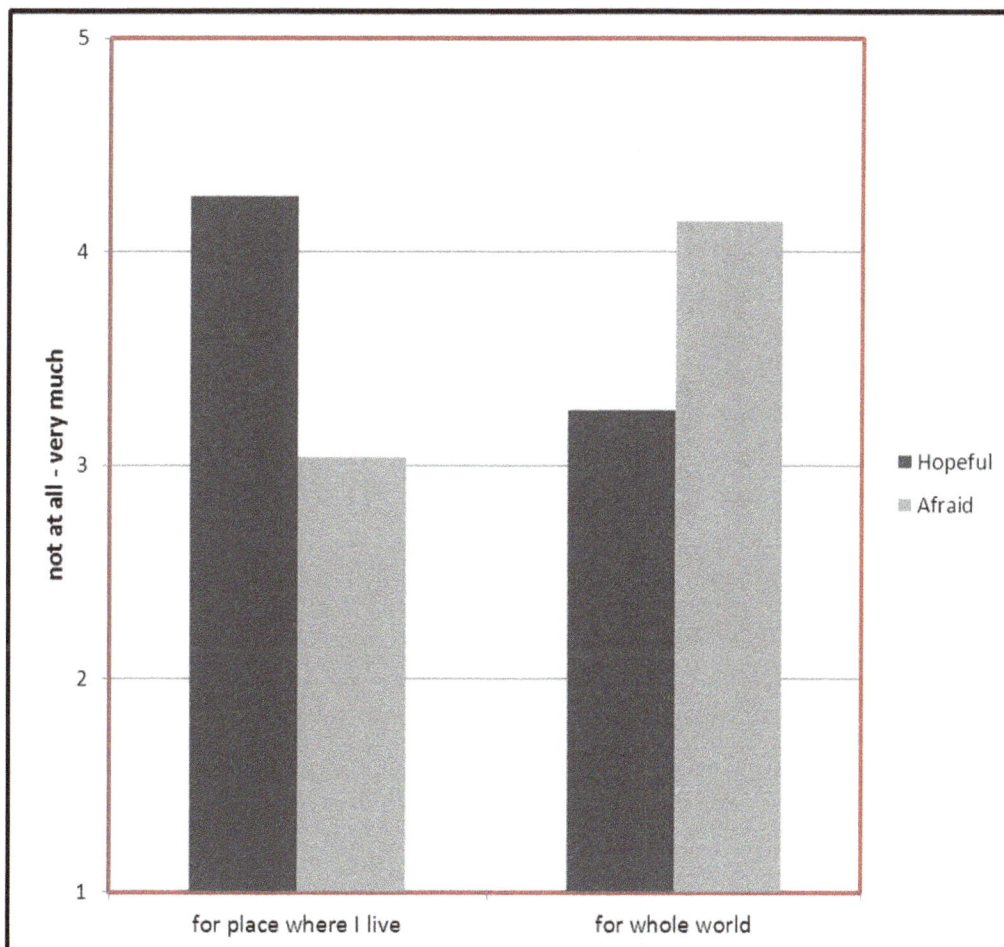

Figure 4. Mean hope (Hopeful) and fear (Afraid) for two types of environments (neighborhood vs. whole world) across five groups of participants.

As can be seen in **Table 4**, the youngest participants (25 years) were most, and the oldest (65+ years) participants least, concerned about their future as related to climate change consequences.

Women were shown to be generally (all 12 statements) more concerned for the effects of climate change (see **Figure 5**). Furthermore, and as can be seen in **Table 5**, the least educated participants were mostly concerned

Table 4. Post hoc, multiple comparisons (LSD, least significant difference) between the Age groups (−25, 26 - 35, 36 - 45, 46 - 55, 56 - 65, 66+) and concern for my future. Only significant differences are reported with statistics of mean difference (M.D.), standard error (S.E.), p value and lower and upper bound of 95% confidence interval (L.B. and U.B.).

Concern for	Age group	M.D.	S.E.	p	L.B.	U.B.
My future	−25 vs. 26 - 35	0.43	0.22	0.05	0.00	0.86
	−25 vs. 36 - 45	0.65	0.21	0.00	0.24	1.07
	−25 vs. 46 - 55	0.83	0.22	0.00	0.41	1.26
	−25 vs. 56 - 65	0.88	0.23	0.00	0.43	1.34
	−25 vs. 66+	1.21	0.26	0.00	0.71	1.71
	26 - 35 vs. 46 - 55	0.40	0.19	0.04	0.01	0.79
	26 - 35 vs. 56 - 65	0.45	0.20	0.03	0.04	0.87
	26 - 35 vs. 66+	0.77	0.21	0.00	0.31	1.24
	36 - 45 vs. 66+	0.55	0.23	0.02	0.09	1.01

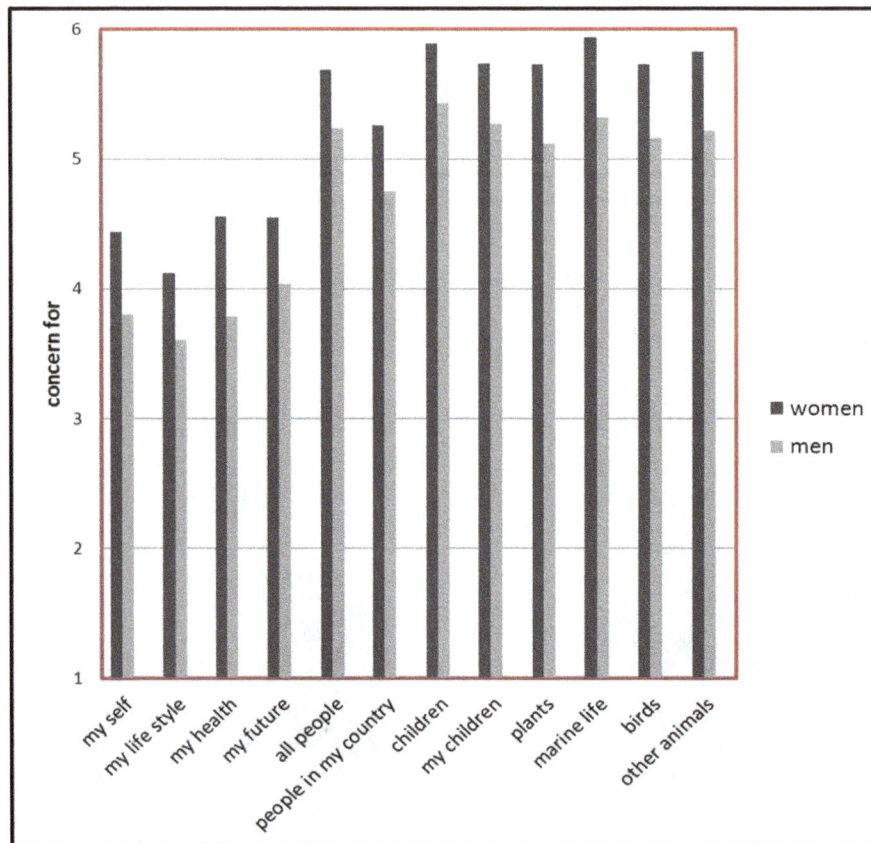

Figure 5. Mean concern in women and men for type of concern.

Table 5. Post hoc, multiple comparisons (LSD, least significant difference) between the Education Level groups (Primary Education, PE, High School, HS, Bachelor Degree, BD, Master Degree, MD, Doctoral Degree, DD) and concerns for my life-style and health. Only significant differences are reported with statistics of mean difference (M.D.), standard error (S.E.), p value and lower and upper bound of 95% confidence interval (L.B. and U.B.).

Concern for	Education level	M.D.	S.E.	p	L.B.	U.B.
My lifestyle	PE vs. HS	0.90	0.23	0.00	0.44	1.35
	PE vs. BD	1.18	0.25	0.00	0.69	1.66
	PE vs. MD	1.05	0.26	0.00	0.55	1.56
	PE vs. DD	0.74	0.35	0.04	0.05	1.43
My health	PE vs. HS	0.96	0.24	0.00	0.49	1.43
	PE vs. BD	1.06	0.25	0.00	0.56	1.55
	PE vs. MD	1.11	0.26	0.00	0.62	1.66
	PE vs. DD	1.72	0.36	0.00	1.01	2.43
	HS vs. DD	0.75	0.30	0.01	0.16	1.34
	BD vs. DD	0.66	0.31	0.04	0.05	1.27

about their *lifestyle* and *health*.

Beliefs. Main significant effects of Age, Wilk's Lambda = 0.95, $F(30, 3946) = 1.88$, $p < 0.01$, Gender, Wilk's Lambda = 0.96, $F(6, 986) = 4.26$, $p < 0.01$, and Education, Wilk's Lambda = 0.94, $F(24, 3440) = 2.47$, $p < 0.01$, were shown.

The first effect was associated with the statements: 1) Pollution generated in one country harms people all over the world ($p = 0.01$); and 2) Claims that there is a climate change is exaggerated ($p = 0.01$), showing that participants aged 35 - 55 agreed mostly with statement 1) and the oldest ones (aged 66+) agreed mostly with statement 2).

See **Table 6** for comparisons between the Age groups.

As can be seen in **Figure 6**, the second effect was associated with the statements of: 1) Laws that protect the environment limit my choices and personal freedom ($p < 0.01$); 2) Protecting the environment will threaten jobs for people like me ($p < 0.01$); 3) Effects of climate change on public health are worse than people realize ($p < 0.05$). That is, compared to women, men were shown to estimate their freedom and jobs to be more threatened by environmental laws and protection. In addition, women estimated the effects of climate change on health as worse.

Table 6. Post hoc, multiple comparisons (LSD, least significant difference) between the Age groups (−25, 26 - 35, 36 - 45, 46 - 55, 56 - 65, 66+) associated with the statements of: (1) Pollution generated in one country harms people all over the world; and (2) Claims that there is a climate change are exaggerated. Only significant differences are reported with statistics of mean difference (M.D.), standard error (S.E.), p value and lower and upper bound of 95% confidence interval (L.B. and U.B.).

Statements	Age group	M.D.	S.E.	p	L.B.	U.B.
Pollution...	36 - 45 vs. 56 - 65	0.40	0.14	0.01	0.12	0.67
	36 - 45 vs. 66+	0.40	0.15	0.03	0.11	0.69
	46 - 55 vs. 56 - 65	0.32	0.14	0.03	0.04	0.60
	46 - 55 vs. 66+	0.32	0.15	0.04	0.02	0.62
Claims...	−25 vs. 66+	−1.03	0.21	0.00	−1.45	−0.62
	26 - 35 vs. 46 - 55	−1.07	0.20	0.04	−1.45	−0.69
	26 - 35 vs. 56 - 65	−1.08	0.19	0.03	−1.45	−0.71
	26 - 35 vs. 66+	−0.75	0.19	0.00	−1.45	−0.70
	36 - 45 vs. 66+	0.75	0.20	0.02	−1.14	−0.35

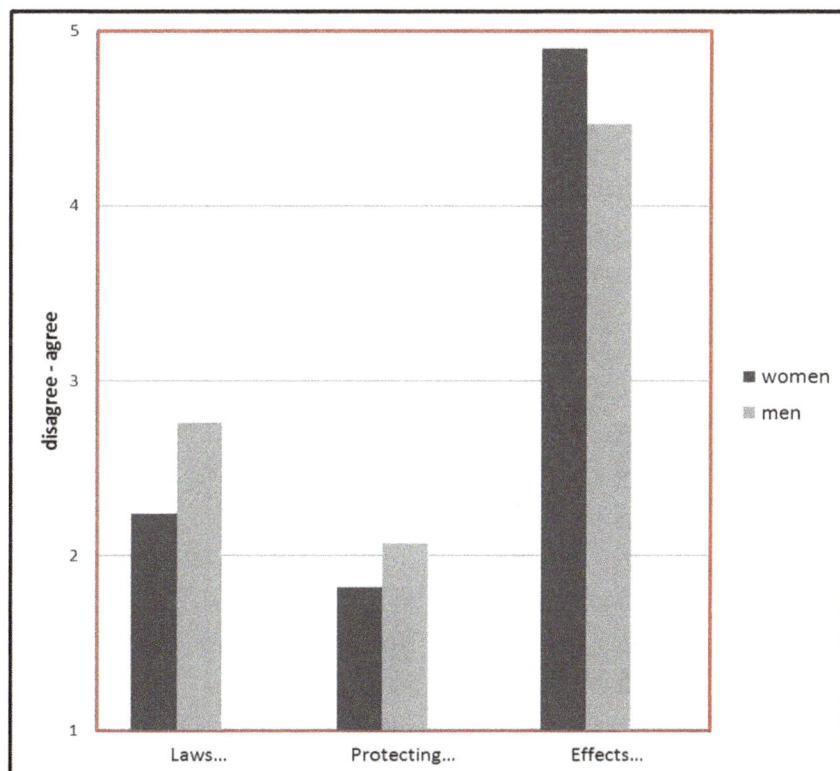

Figure 6. Mean disagreement-agreement in women and men associated with statements of (1) Laws that protect the environment limit my choices and personal freedom; (2) Protecting the environment will threat jobs for people like me; (3) Effects of climate change on public health are worse than people realize.

The third effect was associated with the statements: 1) Protecting the environment will threaten jobs for people like me ($p < 0.01$); 2) Claims that there is climate change are exaggerated ($p < 0.05$). Compared to the other Education groups, the least educated participants considered their jobs to be more threatened by environmental protection and thought that the claims about climate change were exaggerated (see **Table 7**). In addition, and as can be seen in **Figure 7**, a tendency of decreasing agreement with educational level was indicated.

Table 7. Post hoc, multiple comparisons (LSD, least significant difference) between the Education Level groups (Primary Education, PE, High School, HS, Bachelor Degree, BD, Master Degree, MD, Doctoral Degree, DD) associated with the statements of: (1) Pollution generated in one country harms people all over the world; and (2) Claims that there is a climate change are exaggerated. Only significant differences are reported with statistics of mean difference (M.D.), standard error (S.E.), p value and lower and upper bound of 95% confidence interval (L.B. and U.B.)

Statements	Education level	M.D.	S.E.	p	L.B.	U.B.
Protecting...	PE vs. HS	0.39	0.17	0.02	0.07	0.72
	PE vs. BD	0.86	0.18	0.00	0.51	1.21
	PE vs. MD	0.99	0.19	0.00	0.62	1.36
	PE vs. DD	1.08	0.26	0.00	0.56	1.60
	HS vs. BD	0.47	0.12	0.00	0.24	0.70
	HS vs. MD	0.60	0.13	0.00	0.34	0.86
	HS vs. DD	0.69	0.23	0.00	0.24	1.13
Claims...	PE vs. HS	0.70	0.19	0.00	0.32	1.07
	PE vs. BD	1.03	0.21	0.00	0.63	1.43
	PE vs. MD	0.96	0.22	0.00	0.54	1.38
	PE vs. DD	1.33	0.30	0.00	0.74	1.93
	HS vs. BD	0.34	0.14	0.01	0.07	0.60
	HS vs. DD	0.64	0.26	0.02	0.12	1.15

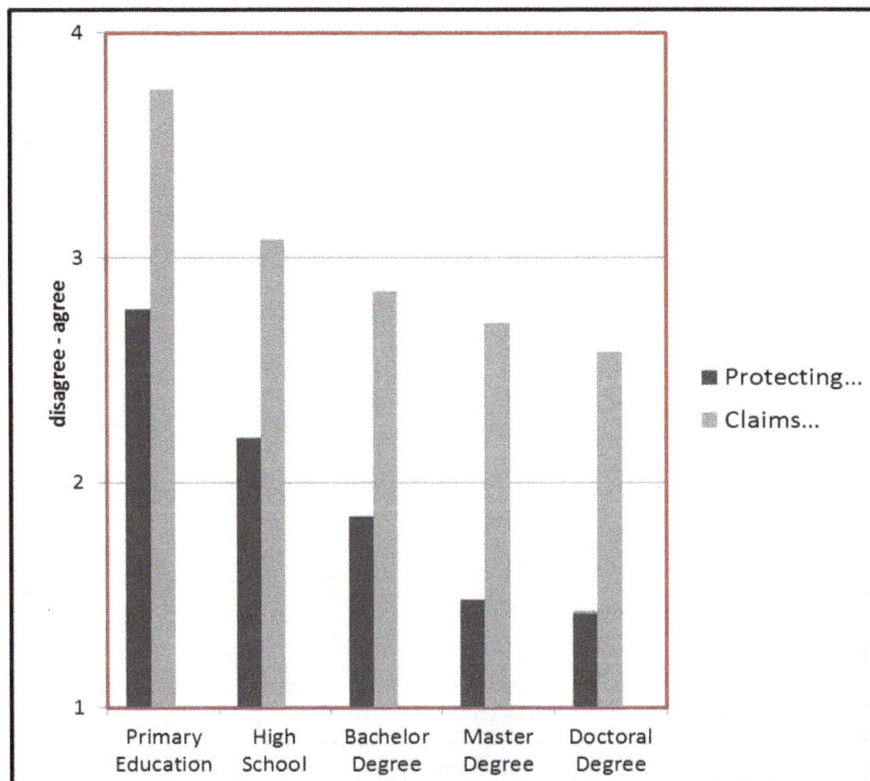

Figure 7. Mean disagreement-agreement in Education groups (Primary Education, High School, Bachelor Degree, Master Degree, Doctoral Degree) associated with statements of: (1) Protecting the environment will threat jobs for people like me; (2) Claims that there is climate change are exaggerated.

Emotions. Main significant effects of Age, Wilk's Lambda = 0.96, $F(20, 3281) = 2.09$, $p < 0.01$, Gender, Wilk's Lambda = 0.99, $F(4, 989) = 2.36$, $p < 0.05$, and Education, Wilk's Lambda = 0.97, $F(16, 3950) = 1.89$, $p < 0.05$, were shown.

As can be seen in **Table 8**, oldest and youngest participants were least hopeful about their neighbouring environment ($p < 0.05$), and youngest participants were most afraid for the whole world ($p < 0.05$) as related to climate change consequences.

Women were shown to be more afraid of climate change impact as related to both neighbouring environment ($p < 0.05$) and the whole world ($p < 0.01$) than men (see **Figure 8**). Concerning the Education effect, no significant association with any particular question was shown, but a general tendency across all four questions indicated more fear than hope about the consequences of climate change. However, and as can be seen in **Figure 9**, participants with the lowest educational level were shown to be the least hopeful and the most afraid.

Table 8. Post hoc, multiple comparisons (LSD, least significant difference) between the Age groups (−25, 26 - 35, 36 - 45, 46 - 55, 56 - 65, 66+) associated with the questions of: (1) How much do you feel afraid when you think about the climate change risks for the whole world? (2) How much do you feel afraid when you think about the climate change risks for the whole world?" Only significant differences are reported) with statistics of mean difference (M.D.), standard error (S.E.), p value and lower and upper bound of 95% confidence interval (L.B. and U.B.).

Questions	Age group	M.D.	S.E.	p	L.B.	U.B.
Hope for place where I live	−25 vs. 46 - 55	0.34	0.16	0.03	−0.66	−0.03
	26 - 35 vs. 36 - 45	0.30	0.14	0.03	−0.58	−0.03
	26 - 35 vs. 46 - 55	0.39	0.14	0.01	−0.67	−0.01
	36 - 45 vs. 66+	0.36	0.16	0.02	0.06	0.67
	46 - 55 vs. 66+	0.45	0.16	0.00	0.14	0.76
	56 - 65 vs. 66+	0.33	0.17	0.05	0.00	0.65
Fear for the world	−25 vs. 26 - 35	0.54	0.20	0.01	0.15	0.93
	−25 vs. 46 - 55	0.63	0.20	0.00	0.24	1.02
	−25 vs. 56 - 65	0.71	0.21	0.00	0.31	1.12
	−25 vs. 66+	0.60	0.21	0.01	0.18	1.02
	36 - 45 vs. 56 - 65	0.38	0.18	0.04	0.02	0.74

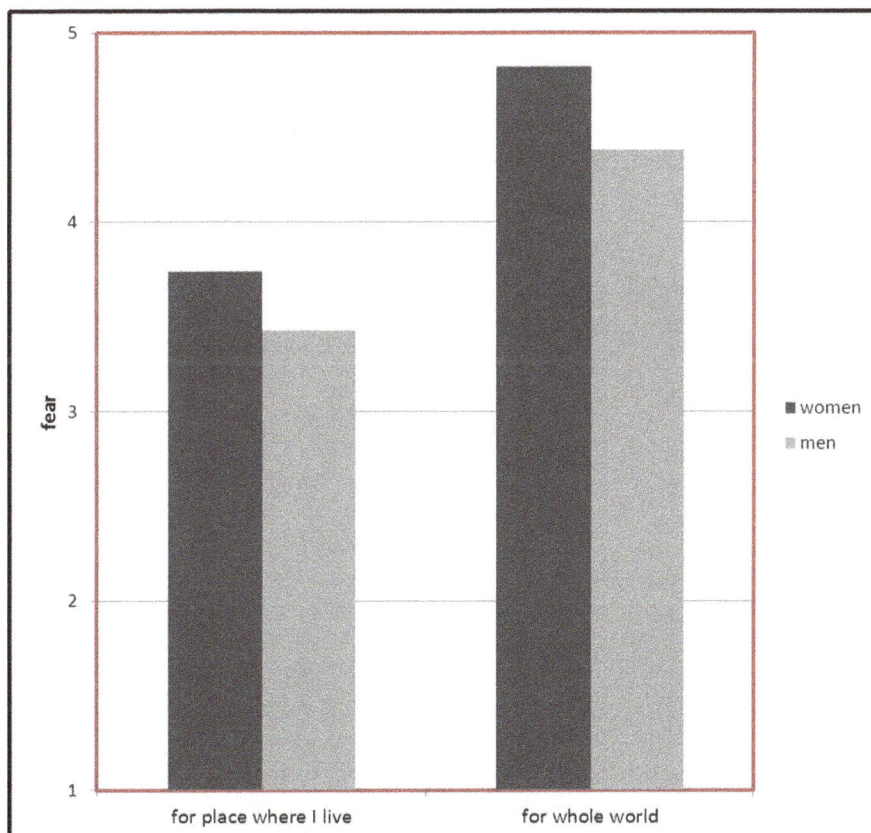

Figure 8. Mean fear in women and men for the place where I live and for the whole world.

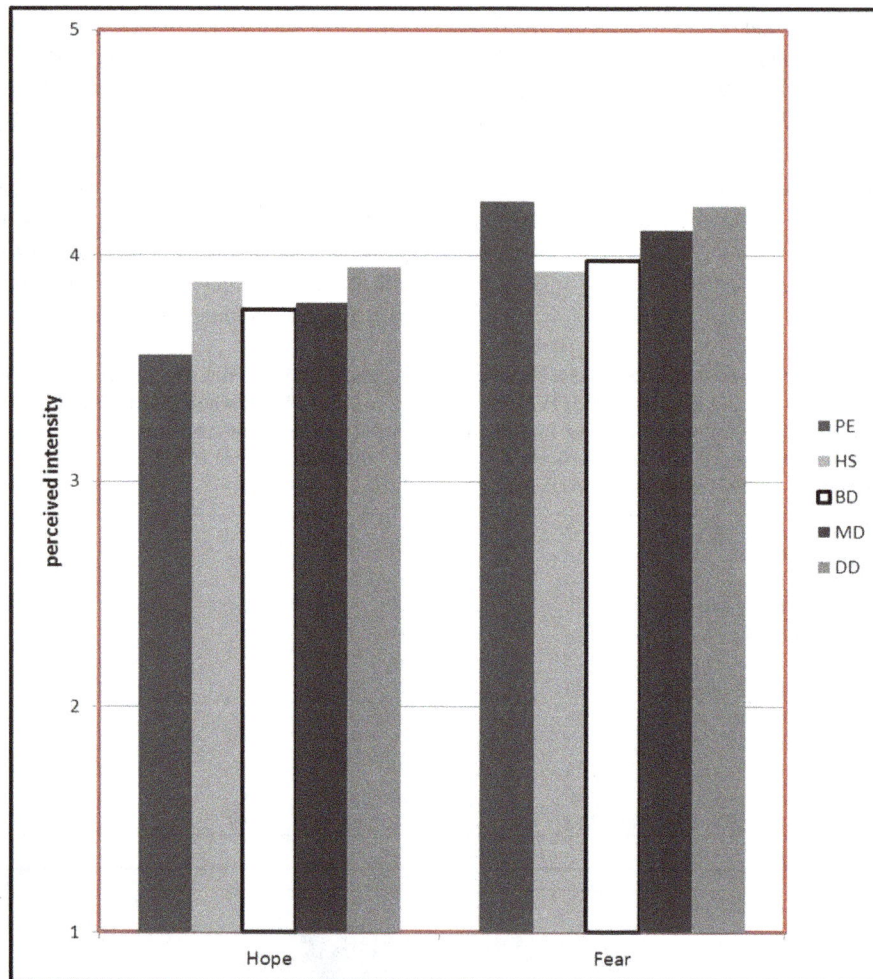

Figure 9. Mean hope and fear in Education groups (Primary Education, PE; High School, HS; Bachelor Degree, BD; Master Degree, MD; Doctoral Degree, DD).

4. Discussion

The aim of this article was to investigate effects of different groups of individuals (residents, tourists, experts, decision makers and members of the tourist industry) and demographic variables (gender, age, education across different groups of individuals) on participants' climate change-related concerns, beliefs and emotions.

Concerning the *type of group* results (Hypothesis 1, see Introduction), Experts (scientists working with climate and environment issues) were shown to be least concerned about and afraid of the global climate change impact. Tourists were shown to be the most concerned and afraid. In line with Sundblad *et al.* [11] this may indicate that scientists, due to their confidence in their own knowledge, are relatively less emotionally involved and concerned compared to "laypersons"; who are more prone to misjudge the "scientifically uncertain" information [9,10]. On the local environmental level, however, Decision Makers (politicians, stakeholders) and the Tourist Industry were shown to react in a similar way to Experts.

The reactions of Residents were similar to those of Tourists. It may be that this highlights a demarcation line between those that know "better" and those that know "poorer"; alluding to [11]. If the Tourist Industry knows "better" or is just lacking in commitment as previously indicated [50], is a question for a future research to investigate.

All participants were shown to be most concerned for children and least for their lifestyle. They also mostly believed that pollution generated in one country harms people all over the world and least that protecting the environment will threaten jobs for people like them. This indicates an environmental value [23,24,26] of social-altruistic (concern for others) and biospheric (concern for world) orientations in participants, grounding their environmental-related beliefs and concerns [17,60]; because beliefs ("awareness" of the climate change impact) and concerns are supposed to causally relate with value orientations. In the words of Hansla *et al.* [29, p. 3]: "...a value orientation biases individuals to select and believe in information that is congruent with their value orientation and deny value-incongruent information".

However, and in contrast to the above, Residents considered the claims about climate change impact to be more exaggerated than the other four groups did, indicating an egoistic (environmental concerns at a personal level; [20,21]) value orientation in this group of individuals. This is, generally speaking, in contrast with some of the previous findings indicating a positive association between urban residents and environmental issues [31].

Finally, all participants were shown to be most hopeful about neighbouring compared to global environment indicating a type of psychological distance, "escaping affect" [61], in calculating long-term risks, or showing compassion, for the environmental facet. That is, a degree of disengagement from involvement in global compared to local milieu issues was indicated, meaning that neighbouring compared to global threats may be mentally represented differently in our mind.

Concerning the *demographic variable* results (Hypothesis 2, see Introduction) and as hypothesised [30,31], youngest participants were indicated to be most, and oldest least, concerned about their future as related to climate change impact. Matching their concerns, youngest participants were the most afraid for both types of milieus. The oldest believed that claims that there is a climate change are exaggerated, suggesting an egoistic value orientation in this group of individuals. They were also shown to be most afraid for their neighbouring milieu compared to the whole world, indicating a type of psychological distance ("escaping affect"; [61]).

In line with some previous findings [30,53] women were shown to be generally more concerned about the consequences of climate change compared to men, and they estimated the effects of climate change on health as worse. Men believed that their freedom and jobs would be more threatened by environmental laws and protection than did women, indicating an egoistic and a social-altruistic value orientation [17,60] in men and women respectively. Women were also shown to be more afraid for both local and global arenas, as related to climate change impact. This is in line with some previous research indicating women to be more expressive than men [62,63] and/or embracing environmental-related altruistic values to a higher degree [35]. In sum, and in line with socialization theory [64,65], this type of result indicates that females compared to males are more: "...interdependent, compassionate, nurturing, cooperative, and helpful in caregiving roles" [53, p. 445].

Compared to participants with higher educational levels, the least educated ones were shown to believe that their jobs would be more threatened by environmental laws and protection, and that the claims about climate change are exaggerated. They were also mostly con-

cerned about their lifestyle and health. All this indicates an egoistic (environmental concerns at a personal level; [17,60]) value orientation in this group of individuals. This is in contrast with some recent findings suggesting that lower class individuals (in this study the lower educated ones; see [66]): "...will act in a more prosocial fashion and do so because of an increased orientation to the needs of others." [67, p. 772; also 68]. In addition, the least educated participants were the least hopeful and the most afraid as related to climate change impact. This may indicate that they are more inclined to misjudge the "scientifically uncertain" information [9-11] and therefore experience fear more intensely than participants with a higher level of education.

5. Conclusions

Taken together and in line with the predictions, we have reported significant influences of type of group (residents, tourists, experts, decision makers and members of tourist industry) and demographic variables (gender, age, education across different groups of individuals) on participants' climate-change-related concerns, beliefs and emotions. Alluding to the words of Weber [69, p. 103]: "It should come as no surprise that the governments and citizens of many countries show little concern about climate change and its consequences." we have shown that some individuals do, and some do not, bother about environmental long-term risks such as climate change. This means that these socio-psychological constructs vary with 1) type of social group we belong to; and 2) our gender, age and education; and in turn ground different types of value orientation towards local and global milieus.

Accordingly, when fostering sustainable behaviour, policy support and commitment [70,71] to protect the environment, we have to take into account this diversity in value orientation and environmental risk-related emotion and awareness found in different groups of individuals. As recently pointed out by Patchen [72]: "Programs to combat climate change should be structured so that individuals see their actions as part of a shared social effort".

REFERENCES

[1] W. Kempton, J. Boster and J. Hartley, "Environmental Values in American Culture," MIT Press, Cambridge, 1995.

[2] L. Saad, "American Sharply Divided on Seriousness of Global Warming," *Gallup Poll Monthly*, Gallup, Princeton, 2002, pp. 43-48.

[3] P. Schmuck and C. Vlek, "Psychologists Can Do Much to Support Sustainable Development," *European Psychologist*, Vol. 8, No. 2, 2003, pp. 66-76.

[4]   S. C. Moser and L. Dilling, "Creating a Climate for Change," Cambridge University Press, Cambridge, 2007.

[5]   M. Bonnes and M. Bonaiuto, "Environmental Psychology: From Spatial-Physical Environment to Sustainable Development," In: R. B. Bechtel and A. Churchman, Eds., *Handbook of Environmental Psychology*, John Wiley & Sons, New York, 2002, pp. 28-54.

[6]   J. Curry, "Reasoning about Climate Uncertainty," *Climatic Change*, Vol. 108, No. 4, 2011, pp. 723-732.

[7]   M. van Vugt, "Central, Individual, or Collective Control?" *American Behavioral Scientist*, Vol. 45, No. 5, 2002, pp. 783-800.

[8]   K. M. Wilson, "Drought, Debate, and Uncertainty: Measuring Reports Knowledge and Ignorance about Climate Change," *Public Understanding of Science*, Vol. 9, No. 1, 2000, pp. 1-13.

[9]   N. Stern, "The Economics of Climate Change: The Stern Review," Cambridge University Press, Cambridge, 2006.

[10] G. Böhm and H.-R. Pfister, "Mental Representations of Global Environmental Risks," *Research in Social Problems and Public Policy*, Vol. 9, 2001, pp. 1-30.

[11] E.-L. Sundblad, A. Biel and T. Gärling, "Knowledge and Confidence in Knowledge about Climate Change among Experts, Journalists, Politicians, and Laypersons," *Environment and Behavior*, Vol. 41, No. 2, 2008, pp. 281-302.

[12] S. Brechin, "Comparative Public Opinion and Knowledge on Global Climate Change and the Kyoto Protocol," *Social Policy*, Vol. 23, No. 10, 2003, pp. 106-134.

[13] M. Oppenheimer and A. Todorov, "The Psychology of Long-Term Risk," *Climatic Change*, Vol. 77, No. 1-2, 2006, pp. 1-6.

[14] W. Viscusi and R. Zeckhauser, "The Perception and Valuation of the Risks of Climate Change," *Climatic Change*, Vol. 77, No. 1-2, 2006, pp. 151-177.

[15] S. H. Schwartz, "Universals in the Content and Structure of Values: Theoretical Advances and Empirical Tests in 20 Countries," In: M. Zanna, Ed., *Advances in Experimental Psychology*, Academic Press, Orlando, 1992, pp. 1-65.

[16] S. H. Schwartz, "Are There Universal Aspects in the Structure and Contents of Human Values?" *Journal of Social Issues*, Vol. 50, No. 4, 1994, pp. 19-45.

[17] P. C. Stern, "Toward a Coherent Theory of Environmentally Significant Behavior," *Journal of Social Issues*, Vol. 56, No. 3, 2000, pp. 407-424.

[18] P. C. Stern and T. Dietz, "The Value Basis of Environmental Concern," *Journal of Social Issues*, Vol. 50, No. 3, 1994, pp. 65-84.

[19] P. C.Stern, T. Dietz and J. S. Black, "Support for Environmental Protection: The Role of Moral Norms," *Population and Environment: Behavioral and Social Issues*, Vol. 8, No. 3-4, 1986, pp. 204-222.

[20] P. C. Stern, T. Dietz and L. Kalof, "Value Orientations, Gender, and Environmental Concern," *Environment and Behavior*, Vol. 25, No. 5, 1993, pp. 322-348.

[21] P. C. Stern, T. Dietz, G. A. Guagnano and L. Kalof, "A Value-Belief-Norm Theory of Support for Social Movements: The Case of Environmentalism," *Human Ecology Review*, Vol. 6, No. 2, 1999, pp. 81-97.

[22] P. W. Schultz, "Empathizing with Nature: The Effects of Perspective Taking on Concern for Environmental Issues," *Journal of Social Issues*, Vol. 56, No. 3, 2000, pp. 391-406.

[23] P. W. Schultz, "The Structure of Environmental Concern: Concern for Self, Other People, and the Biosphere," *Journal of Environmental Psychology*, Vol. 21, No. 4, 2001, pp. 1-13.

[24] P. W. Schultz and L. C. Zelezny, "Values and Predictors of Environmental Attitudes: Evidence for Constancy across 14 Countries," *Journal of Environmental Psychology*, Vol. 19, No. 3, 1999, pp. 255-265.

[25] P. W. Schultz and L. C. Zelezny, "Reframing Environmental Messages to Be Congruent with American Values," *Human Ecology Review*, Vol. 10, 2003, pp. 126-136.

[26] P. W. Schultz, V. V. Gouveis, L. D. Cameron, G. Tanhka, P. Schmuck and M. Franek, "Values and Their Relationship to Environmental Concern and Conservation Behavior," *Environment and Behavior*, Vol. 36, No. 4, 2005, pp. 457-475.

[27] L. Steg, L. Dreijerink and W. Abrahamse, "Factors Influencing the Acceptability of Energy Policies: A Test of VBN Theory," *Journal of Environmental Psychology*, Vol. 25, No. 4, 2005, pp. 415-425.

[28] S. Oreg and T. K. Gerro, "Predicting Proenvioronmental Behavior Cross-Nationally: Values, the Theory of Planned Behavior, and Value-Belief-Norm Theory," *Environment and Behavior*, Vol. 38, No. 4, 2006, pp. 462-483.

[29] A. Hansla, A. Gamble, A. Juliusson and T. Gärling, "The Relationships between Awareness of Consequences, Environmental Concern, and Value Orientations," *Journal of Environmental Psychology*, Vol. 28, No. 1, 2008, pp. 1-9.

[30] A. Olofsson and S. Öhman, "General Beliefs and Environmental Concern: Transatlantic Comparisons," *Environment and Behavior*, Vol. 38, No. 6, 2006, pp. 768-790.

[31] K. D. Van Liere and R. E. Dunlap, "The Social Bases of Environmental Concerns: A Review of Hypotheses, Ex-

planations and Empirical Evidence," *Public Opinion Quarterly*, Vol. 44, No. 2, 1980, pp. 181-197.

[32] T. Dietz, P. C. Stern and G. A. Guagnano, "Social Structural and Social Psychological Bases of Environmental Concern," *Environment and Behavior*, Vol. 30, No. 4, 1998, pp. 450-471.

[33] K. Boehnke, D. Fuss and M. Rupf, "Values and Well-Being: The Mediating Roles of Worries," In: P. Schmuck and K. M. Sheldon, Eds., *Life-Goals and Wellbeing: Towards a Positive Psychology of Human Striving*, Hogrefe & Huber Publishers, Seattle, 2001, pp. 85-101.

[34] R. Garcia-Mira, J. E. Real and J. Romay, "Temporal and Spatial Dimensions in the Perception of Environmental Problems: An Investigation of the Concept of Environmental Hyperopia," *International Journal of Psychology*, Vol. 40, No. 1, 2005, pp. 5-10.

[35] M. Ojala, "Hope and Worry: Exploring Young People's Values, Emotions, and Behaviour Regarding Global Environmental Problems," Doctoral Dissertation, Örebro University, Örebro, 2007.

[36] S. Gössling and C. M. Hall, "Uncertainties in Predicting Tourist Flows Under Scenarios of Climate Change," *Climatic Change*, Vol. 79, No. 3-4, 2006, pp. 163-173.

[37] L. Hein, M. J. Metzeger and A. Moreno, "Potential Impacts of Climate Change on Tourism: A Case Study for Spain," *Current Opinion in Environmental Sustainability*, Vol. 1, No. 2, 2009, pp. 170-178.

[38] D. Weaver, "Can Sustainable Tourism Survive Climate Change?" *Journal of Sustainable Tourism*, Vol. 19, No. 1, 2012, pp. 5-15.

[39] J. M. Hamilton, D. J. Maddison and R. S. J. Tol, "Climate Change and International Tourism: A Simulation Study," *Global Environmental Change*, Vol. 15, No. 3, 2005, pp. 253-266.

[40] D. Viner, "Tourism and Its Interactions with Climate Change," *Journal of Sustainable Tourism*, Vol. 14, No. 4, 2006, pp. 317-322.

[41] S. Gössling and C. Hall, "Tourism and Global Environmental Change: Ecological, Social, Economic and Political Interrelationships," Routledge, London, 2006.

[42] UNWTO "Climate Change and Tourism: Responding to Global Challenges," World Tourism Organixzation and United Nations Environment. World Tourism Organization, Madrid, 2008.

[43] D. Scott, "Why Sustainable Tourism must Address Climate Change," *Journal of Sustainable Tourism*, Vol. 19, No. 1, 2011, pp. 17-34.

[44] B. Amelung and D. Viner, "Mediterranean Tourism: Exploring the Future with the Tourism Climatic Index," *Journal of Sustainable Tourism*, Vol. 14, No. 4, 2006, pp.

349-366.

[45] C. E. Hanson, J. P. Palutikof, A. Dlugolecki and C. Giannakopoulos, "Bridging the Gap between Science and the Stakeholder: The Case of Climate Change Research," *Climate Research*, Vol. 31, No. 1, 2006, pp. 121-133.

[46] D. Maddison, "In Search of Warmer Climates. The Impact of Climate Change on Flows of British Tourists," *Climate Change*, Vol. 49, No. 1-2, 2001, pp. 193-208.

[47] H. M. Wang, F. F. Feng and J. H. Wen, "A Study on Interactions and Impact between Climate Change and Tourism," *Proceedings of the 1st International Conference on Sustainable Construction and Risk Management: I and II.* Chongqing, 2010, pp. 969-973.

[48] I. Eliasson, G. Olshammar, S. Thorsson, I. Knez, A. Eraydin, B. Gedikli, Ö. Edizel, H. Andrade, E. B. Henriques and R. Machete, "Urban Tourism and Climate Change," Urban-Net Research Anthology, Formas, Stockholm, 2010, pp. 41-47.

[49] D. Scott, C. R. de Freitas and A. Matzarakis, "Adaptation in the Tourism and Recreation Sector," In: G. R. McGregor, I. Burton and K. Ebi, Eds., *Biometeorology for Adaptation to Climate Variability and Change*, Springer Netherlands, Houten, 2009, pp. 171-194.

[50] D. Scott, P. Peeters and S. Gössling, "Can Tourism Deliver Its 'Aspirational' Greenhouse Gas Emission Reduction Targets?" *Journal of Sustainable Tourism*, Vol. 18, No. 3, 2010, pp. 393-408.

[51] J. Willms, "Climate Change = Tourism Change? The likely Impacts of Climate Change on Tourism in Germany's North Sea Coast Destinations," In: A. Matzarakis, C. R. De Freitas and D. Scott, Eds., *Developments in Tourism Climatology*, Freiburg University Press, Freiburg, 2007, pp. 246-253.

[52] P. C. Stern, T. Dietz, L. Kalof and G. A. Guagnano, "Values, Beliefs, and Proenvironmental Action: Attitude Formation toward Emergent Attitude Objects," *Journal of Applied Social Psychology*, Vol. 25, No. 18, 1995, pp. 1611-1636.

[53] L. C. Zelezny, P. P. Chua and C. Aldrich, "New Ways of Thinking about Environmentalism: Elaborating Gender Differences in Environmentalism," *Journal of Social Issues*, Vol. 56, No. 3, 2000, pp. 443-457.

[54] B. Gatersleban, L. Steg and C. Vlek, "Measurement and Determinants of Environmentally Significant Consumer Behavior," *Environment and Behavior*, Vol. 34, No. 3, 2002, pp. 335-362.

[55] A. Leiserowitz, "Climate Change Risk Perception and Policy Preferences: The Role of Affect, Imagery, and Values," *Climatic Change*, Vol. 77, No. 1-2, 2006, pp. 45-72.

[56] B. McKercher, S. F. H. Pang and B. Prideaux, "Do Gender and Nationality Affect Attitudes towards Tourism and Environment?" *International Journal of Tourism Research*,

Vol. 13, No. 3, 2011, pp. 266-300

[57] M. Ojala, "Adolescents' Worries ABOUT Environmental Risks: Subjective Well-Being, Values, and Existential Dimensions," *Journal of Youth Studies*, Vol. 8, No. 3, 2005, pp. 331-347.

[58] F. J. McGuigan, "Experimental Psychology: Methods of Research," Prentice Hall, Inc., Upper Saddle River, 1983.

[59] R. M. Liebert and L. L. Liebert, "Science and Behavior: An Introduction to Methods of Psychological Research," Prentice Hall, Upper Saddle River, 1995.

[60] S. H. Schwartz and J. A. Howard, "A Normative Decision-Making Model of Altruism," In: P. J. Rushton and R. M. Sorrentino, Eds., *Altruism and Helping Behavior: Social, Personality, and Developmental Perspectives*, Lawrence Erlbaum, Hillsdale, 1981, pp. 189-211.

[61] C. D. Cameron and B. K. Payne, "Escaping Affect: How Motivated Emotion Regulation Creates Insensitivity to Mass Suffering," *Journal of Personality and Social Psychology*, Vol. 100, No. 1, 2011, pp. 1-15.

[62] F. Fujita, E. Diener and E. Sandvik, "Gender Differences in Negative Affect and Well-Being: The Case for Emotional Intensity," *Journal of Personality and Social Psychology*, Vol. 61, No. 3, 1991, pp. 427-434.

[63] L. R. Brody, "On Understanding Gender Differences in the Expression of Emotion," In: S. L. Ablon, D. Brown, E. J. Khantzian and J. E. Mack, Eds., *Human Feelings: Explorations in Affect Development and Meaning*, Analytic Press, Hillsdale, 1993, pp. 87-121.

[64] A. Eagly, "Sex Differences in Social Behavior: A Social Role Interpretation," Lawrence Erlbaum, Hillsdale, 1987.

[65] A. Beutel and M. Marini, "Gender and Values," *American Sociological Review*, Vol. 60, No. 3, 1995, pp. 436-448.

[66] A. C. Snibbe and H. R. Markus, "You Can't Always Get What You Want: Educational Attainment, Agency, and Choice," *Journal of Personality and Social Psychology*, Vol. 88, No. 4, 2005, pp. 703-720.

[67] P. K. Piff, M. W. Kraus, S. Cote and B. H. Cheng, "Having Less, Giving More: The Influence of Social Class on Prosocial Behavior," *Journal of Personality and Social Psychology*, Vol. 99, No. 5, 2010, pp. 771-784.

[68] Y. Ma, C. Wang and S. Han, "Neural Responses to Perceived Pain in Others Predict Real-Life Monetary Donations in Different Socioeconomic Contexts," *NeuroImage*, Vol. 57, No. 3, 2011, pp. 1273-1280.

[69] E. U. Weber, "Experience-Based and Description-Based Perceptions of Long-Term Risk: Why Global Warming Does Not Scare Us," *Climatic Change*, Vol. 77, No. 1-2, 2006, pp. 103-120.

[70] UNCED, "Convention on Biological Diversity," *United Nations Conference on Environment and Development* Rio de Janeiro, 5 June 1992, p. 79.

[71] S. Menzel and S. Bogeholz, "Values, Beliefs and Norms that Foster Chilean and German Pupils' Commitment to Protect Biodiversity," *International Journal of Environmental & Science Education*, Vol. 5, No. 1, 2010, pp. 31-49.

[72] M. Patchen, "What Shapes Public Reactions to Climate Change? Overview of Research and Policy Implications," *Analyses of Social Issues and Public Policy*, Vol. 10, No. 1, 2010, pp. 47-68.

14

Biome Q_{10} and Dryness

Chuixiang Yi[1*], Daniel Ricciuto[2], George Hendrey[1]

[1]School of Earth and Environmental Sciences, Queens College, City University of New York, New York, USA
[2]Environmental Sciences Division, Oak Ridge National Laboratory, Oak Ridge, USA

ABSTRACT

Temperature sensitivity of soil respiration (Q_{10}) is a critical parameter in carbon cycle models with important implications for climate-carbon feedbacks in the 21st century. The common assumption of a constant Q_{10}, usually with a value of 2.0, was shown to be invalid by a previous model-data fusion study that reported biome-specific values of this parameter. We extend the previous analysis by demonstrating that these biome-level values of Q_{10} also are a function of dryness ($R^2 = 0.54$). When tundra and cultivated lands are excluded, the correlation is much stronger ($R^2 = 0.92$). Therefore dryness is the primary driver for variability in respiration-temperature sensitivity in forest and grassland ecosystems. This finding has important implications for the response of the terrestrial carbon cycle to climate change, as it implies that the increasing dryness would potentially accelerate the respiration temperature sensitivity feedback.

Keywords: Climate Change; Carbon Cycle; Soil Respiration; Dryness

1. Introduction

Globally, soil respiration releases CO_2 annually at a rate that is over an order of magnitude larger than anthropogenic releases [1]. Although soil heterotrophic respiration is currently balanced or slightly exceeded by terrestrial net primary productivity (NPP), relatively small changes in this large flux could have large impacts on the global net carbon balance. The most important climate driver of soil respiration is temperature, and increasing temperature is likely to induce a positive feedback between climate and the carbon cycle. Uncertainty about the strength of this feedback is a primary source of uncertainty for predicted behaviour of terrestrial carbon sinks in the latter half of this century [2,3]. Additionally, this sensitivity of respiration to temperature was shown to vary as functions of temperature, substrate, soil moisture and/or biome [4,5]. Despite this, many global carbon cycle models assume constant temperature sensitivity, or they assume a sensitivity that depends on only a limited subset of these factors.

Here we extend the analysis of Zhou *et al.* [4], who used an inversion approach to assimilate worldwide soil respiration measurements and measured soil organic carbon into a widely used carbon cycle model. The authors

of that study concluded that Q_{10} is a function of biome, and that the assumption of a constant Q_{10} results in an underestimation of the respiration-temperature feedback intensity by 25%. Upon further analysis of this unique dataset, we find that the respiration-temperature sensitivity is also a strong function of dryness at the biome level. This finding has important implications for the behaviour of the carbon cycle in a changing climate: increasing dryness, which is likely in a warming climate [6], may increase the respiration-temperature sensitivity, accelerating decomposition and providing a stronger feedback to the climate system.

2. Methods

This analysis focuses on Q_{10}, the parameter controlling the temperature-dependence of soil respiration in the following way:

$$R(T) = R_{ref} Q_{10}^{(T-T_{ref})/10}, \qquad (1)$$

where $R(T)$ and R_{ref} are soil respiration at measured temperature (T) and reference temperature (T_{ref}), respectively, and Q_{10} is a factor by which respiration is multiplied when temperature is increased by 10°C. At $Q_{10} = 1$ respiration would be independent of temperature, while larger Q_{10} values indicates a stronger temperature dependence. In many ecosystem models, Q_{10} is treated as a

*Corresponding author.

constant, the most common value being 2. However, considerable variation (1.3 to ~10) in Q_{10} values have been reported by numerous investigations [7-9]. Since van't Hoff introduced Q_{10} in 1898 it has been debated weather Q_{10} is a universal constant and what controls Q_{10} [1-9]. Since soil emissions of CO_2 are expected to have a positive feedback on global warming, modelling of climate change, its consequences and control strategies requires a clearer understanding of the Q_{10} of soils.

We examined climate control of Q_{10} at the level of biomes globally. Spatially resolved Q_{10} values were estimated at a resolution of 1° by 1° using a model-data fusion technique to assimilate worldwide soil respiration measurements and measured soil organic carbon [4]. The model used in that analysis was the Carnegie-Ames-Stanford Approach (CASA) model [10,11], which includes a CENTURY-based soil carbon module that simulates soil organic carbon processes using two carbon storage pools, in addition to litter and microbial pools. Following Zhou et al. [4], we averaged the estimated Q_{10} values for each biome except Desert and Shrub & Bare Ground, both of which are subject to prolonged periods of desiccation. Average temperature, precipitation and net radiation were estimated for each 1° by 1° grid cell and these climate variables were used to estimate biome dryness [12]:

$$I = \frac{\overline{R}_n}{L\overline{P}}, \quad (2)$$

where $\overline{R}_n$ ($MJ \cdot m^{-2} \cdot yr^{-1}$) is an annual sum of net radiation, L ($2.5\ MJ \cdot kg^{-1}$) is a latent heat coefficient and $\overline{P}$ ($mm \cdot yr^{-1}$) is the total annual precipitation. We then linked the biome-level Q_{10} values to biome-level climate and dryness data (Table 1).

3. Results and Discussion

We found that biome-level Q_{10} is significantly correlated to dryness ($R^2 = 0.54$, Figure 1) or to precipitation ($R^2 = 0.45$, Figure omitted) and is independent of net radiation and temperature. We expect that the correlation of Q_{10} with dryness is stronger than with precipitation because the soil moisture content is determined not only by precipitation (input), but also by energy available for evaporation (output). In our analysis, both tundra and cultivated soils appear to be outliers. When these systems are excluded, the correlation of biome-level Q_{10} with dryness is much higher ($R^2 = 0.92$).

It is not surprising that the Q_{10} value for agricultural soil is an outlier (Figure 1) because cultivation (including tillage, fertilization, irrigation and drainage) accelerates soil respiration in ways that are not adequately captured by CASA [7]. The high Q_{10} value for tundra may be a consequence of the non-linearity of respiration with respect to temperature, particularly as T approaches 0°C. In systems with permafrost, which are poorly represented by CASA, the high Q_{10} value is also likely related to depth of the active layer. As the soil column warms, the depth of unfrozen soil increases and exposes more soil organic matter to decomposition [13]. In actuality this increases the base respiration rate, but the strong correlation of this effect with temperature causes it to be interpreted by the model-data fusion technique as a higher Q_{10} value. Thus both cultivated soil and tundra Q_{10} values are as high as grasslands even though the dryness of these systems is similar to deciduous forests.

Our results demonstrate that the temperature sensitivity of soil respiration-aggregated at the biome level, is controlled by dryness, and that soils in different biomes respond differently to dryness. As can be seen in Figure 1, the temperature sensitivity of forest soil respiration to dryness is much less than that of grasslands, which may also related to substrate quality (woody and non-woody components). This implies that conversion of forest to pasture and agriculture would increase temperature sensitivity of soil respiration, accelerating CO_2 emissions

Table 1. Climate characteristics of biomes (ten-year average, 1986-1995) and Q_{10} values estimated by an inversion approach developed by Zhou et al. [4]. The vegetation is coded according to the IGBP classification. The sources and calculation method of biome-climate data in the table can be found in Zhou et al. [17].

Code	Biome	P	T	R_n	Dryness	Q_{10}	Area
		($mm \cdot a^{-1}$)	(°C)	($MJ \cdot M^{-2} \cdot a^{-1}$)			($10^4\ km^{-2}$)
1	Broadleaf evergreen forest	2171	25.1	4662	0.86	1.50	13.3
2	Broadleaf deciduous forest and woodland	913	15.2	3650	1.60	1.75	3.3
3	Mixed coniferous and broadleaf deciduous forest and woodland	883	8.6	2694	1.22	1.61	6.5
4	Coniferous forest and woodland	517	−2.5	1944	1.50	1.71	12.9
5	High latitude deciduous forest and woodland	438	−5.6	1889	1.73	1.63	5.8
6	C3 wooded grassland	1097	14.0	3446	1.26	1.67	4.5
7	C4 wooded grassland	1304	23.0	4413	1.35	1.59	17.1
8	C3 grassland	433	7.0	2953	2.73	1.97	11.3
9	C4 grassland	566	23.4	4066	2.87	2.02	8.9
10	Tundra	316	−10.8	1287	1.63	2.03	7.0
11	Cultivation	799	13.6	3262	1.63	1.99	13.3

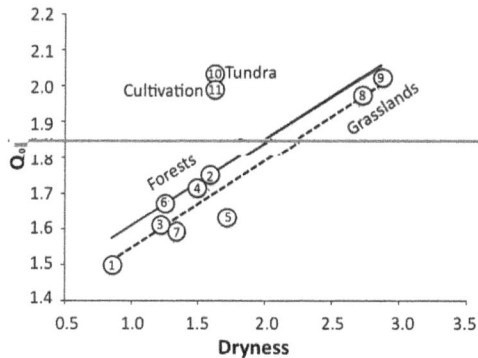

Figure 1. Biome-level Q_{10} versus dryness. The solid regression line with $R^2 = 0.54$ includes all data points, while the dashed regression line with $R^2 = 0.92$ excludes two outliers (tundra and cultivation). Dryness is defined as $R_n/(LP)$, where R_n (MJ·m^{-2}·a^{-1}) and P (mm·a^{-1}) are global annual mean net radiation and precipitation for a biome respectively, and L = 2.5 MJ·kg^{-1} is the enthalpy of vaporization. The number on each data point indicates vegetation type that can be found in Table 1.

and global warming [14,15]. Furthermore, we reiterate the findings of Zhou *et al.* [4] that models of soil responses to climate change, at least at the biome level, should not assume $Q_{10} = 2$, but need to accommodate the sensitivity of soil respiration in different soil types to dryness. Thus, the feedback to terrestrial ecosystems from respiration temperature sensitivity may accelerate more than previously predicted if climate change causes an increase in dryness.

Uncertainties may result from the methods of Zhou *et al.* [4], in which the International Geosphere-Biosphere Programme (IGBP) soil organic carbon (SOC) dataset was used to optimize the Q_{10} parameter in the CASA model. A significant part of the total SOC is relatively recalcitrant and resistant to decomposition, while soil respiration is mostly produced from the newly shed plant litter, and surface SOC. This may lead to an underestimate of the optimized Q_{10} because the decomposition of the recalcitrant SOC is slower than assumed by the model [16]. However, at the global scale, this underestimation of the optimized Q_{10} is mitigated because Zhou *et al.* [4] constrained the optimization process such that the global mean optimized Q_{10} is equal to the global mean value of Q_{10} (1.72) of soil respiration measurements in major ecosystems of the world as reported by Raich *et al.* [1]. While local biases in Q_{10} may result if the modeled SOC quality is not correct, comparisons against site-level soil respiration data in Zhou *et al.* [4] show consistent improvement when optimized values are used compared to a globally constant value. Theoretically, the biome Q_{10} reflects apparent temperature sensitivity that is controlled by environmental constraints [16]. Our results demonstrate that dryness is the most important control on Q_{10} among environmental constraints at biome level.

4. Acknowledgements

This work was financially supported by the National Science Foundation (NSF-DEB-0949637). Authors are grateful for useful discussion with Tao Zhou and for valuable comments from an anonymous reviewer.

REFERENCES

[1] J. W. Raich, C. S. Potter and D. Bhagawati, "Interannual Variability in Global Soil Respiration, 1980-1994," *Global Change Biology*, Vol. 8, No. 8, 2002, pp. 800-812.

[2] P. Friedlingstein, *et al.*, "How Positive Is the Feedback between Climate Change and the Carbon Cycle?" *Tellus*, Vol. 55, No. 2, 2003, pp. 692-700.

[3] P. Friedlingstein, *et al.*, "Climate-Carbon Cycle Feedback Analysis: Results from the (CMIP)-M-4 Model Intercomparison," *Journal of Climate*, Vol. 19, No. 14, 2006, pp. 3337-3353.

[4] T. Zhou, P. Shi, D. Hui and Y. Luo, "Global Pattern of Temperature Sensitivity of Soil Heterotrophic Respiration (Q(10)) and Its Implications for Carbon-Climate Feedback," *Journal of Geophysical Research-Biogeosciences*, Vol. 114, No. G4, 2009, pp. 2156-2202.

[5] E. A. Davidson, I. A. Janssens and Y. Q. Luo, "On the Variability of Respiration in Terrestrial Ecosystems: Moving beyond Q(10)," *Global Change Biology*, Vol. 12, No. 2, 2006, pp. 154-164.

[6] G. A. Meehl, *et al.*, "Climate Change Projections for the Twenty-First Century and Climate Change Commitment in the CCSM3," *Journal of Climate*, Vol. 19, No. 11, 2006, pp. 2597-2616.

[7] J. W. Raich and W. H. Schlesinger, "The Global Carbon-Dioxide Flux in Soil Respiration and Its Relationship to Vegetation and Climate," *Tellus*, Vol. 44, No. 2, 1992, pp. 81-99.

[8] M. U. F. Kirschbaum, "The Temperature-Dependence of Soil Organic-Matter Decomposition, and the Effect of Global Warming on Soil Organic-C Storage," *Soil Biology & Biochemistry*, Vol. 27, No. 6, 1995, pp. 753-760.

[9] P. Ciais, *et al.*, "Horizontal Displacement of Carbon Associated with Agriculture and Its Impacts on Atmospheric CO_2," *Global Biogeochemical Cycles*, Vol. 21, No. 2, 2007, pp. 1944-1951.

[10] C. S. Potter, *et al.*, "Terrestrial Ecosystem Production—A Process Model-Based on Global Satellite and Surface Data," *Global Biogeochemical Cycles*, Vol. 7, No. 4, 1993, pp. 811-841.

[11] C. B. Field, J. T. Randerson and C. M. Malmstrom, "Global Net Primary Production—Combining Ecology and Remote-Sensing," *Remote Sensing of Environment*,

Vol. 51, No. 1, 1995, pp. 74-88.

[12] M. I. Budyko, "Climate and Life," Academic, New York, 1974.

[13] E. A. G. Schuur, *et al.*, "Vulnerability of Permafrost Carbon to Climate Change: Implications for the Global Carbon Cycle," *Bioscience*, Vol. 58, No. 8, 2008, pp. 701-714.

[14] C. Yi, *et al.*, "Climate Control of Terrestrial Carbon Exchange across Biomes and Continents," *Environmental Research Letters*, Vol. 5, No. 3, 2010, pp. 1-10.

[15] R. Valentini, *et al.*, "Respiration as the Main Determinant of Carbon Balance in European Forests," *Nature*, Vol. 404, 2000, pp. 861-865.

[16] E. A. Davidson and I. Janssens, "Temperature Sensitivity of Soil Carbon Decomposition and Feedbacks to Climate Change," *Nature*, Vol. 440, 2006, pp. 165-173.

[17] T. Zhou, C. Yi, P. S. Bakwin and Li Zhu, "Links between Global CO_2 Variability and Climate Anomalies of Biomes," *Science in China Series D: Earth Sciences*, Vol. 51, No. 5, 2008, pp. 740-747.

15

Cyclones and Societies in the Mascarene Islands 17th-20th Centuries

Emmanuel Garnier[1,2,3], Jérémy Desarthe[3,4]
[1]Churchill College, University of Cambridge, Cambridge, UK
[2]Institut Universitaire de France, Paris, France
[3]Centre de Recherche d'Histoire Quantitative (UMR CNRS), University of Caen, Caen, France
[4]Institute for Sustainable Development and International Relations, Sciences Po, Paris, France

ABSTRACT

The recent IPCC-SREX report focuses on the impact of extreme weather events on societies and underlines the absence of reliable data to assert a solid link between them and the current global climate change. Thanks to the unpublished materials that are contained in historic archives, this article suggests studying the cyclones which affected the Mascarene islands between 1654 and 2007 and which supply us with a catalog of hitherto unpublished events. Inspired by the Simpson-Saffir hurricanes Wind Scale, the research proposes a relative evaluation of the extremes of the region. It underlines the big fluctuations in the last three centuries and partially answers the current debate on the reliability of the data in relation to hurricanes and their link with the contemporary climate. The available archives show that this type of meteorological event has occurred frequently during the relevant historical period and that for that reason, has given rise to original strategies of adaptation on the part of the societies affected. The results presented here constitute new and reliable data which could make an important contribution to the decision-makers and to climatologists trying to design strategies which the populations of small islands facing the climatic hazards of the future will have to adopt.

Keywords: History; Archives; Cyclones; Mascarene Islands; Society; Risk; Vulnerability; Adaptation; Resilience

1. Introduction

The very recent SREX [1] report points out that the major part of available historical data relating to hurricanes is limited to a very recent and precise period which in the case of the tropical cyclones generally includes from the middle of the 20th century to the present. In addition to the limited period of records, the uncertainties in the historical tropical data and the extent of tropical variability don't presently allow for the detection of any clear trends in tropical hurricane activity that can be attributed to Global Climate Change. In the face of these uncertainties, the SREX report thus prefers to remind us that it is very difficult and careless to predict too rapid an increase in the frequency of the strongest storms in the coming decades. It therefore concludes that there is no clear signal between these extreme events and SRES A1B Warming scenario.

In response to this assessment a recent piece of research tried to remedy these gaps by suggesting reconstructing an older historic series concerning the cyclonic activity in the southwest of the United States [2]. From the data supplied by six tide gauges of the region since 1923, in other words for a period longer than that covered by satellites, this work tries to produce a series based on the intensity of the storm surge. Very new, this research shows clearly, while not proving the link between global climate change and the cyclonic activity, that there is indeed a difference in the frequency of these events between cold years and the warm years. More interesting still, it reveals that there is also a sustainable increase of the number of cyclones during the 20th century, at least since 1923.

Nevertheless, it is possible to widen considerably the chronology of storms and surge storm to propose a much more historic approach of the coastal extreme events [3] [4]. Nevertheless, such an objective requires opening the subject up to the involvement of the social sciences and more particularly of history. Indeed, this enables one to offer to the climatologists and to the climate modelers new original and unpublished data materials extracted from rather homogeneous and at the same time plentiful archives. Better still, the archives of the historian offer the climate scientists the opportunity to reconstruct very long series of events about 300 or 500 years according to the regions of the world. In this paper, we suggest studying the historic example of the Mascarene Islands between the middle of the 17th a century and 2007 according to two generally new approaches. They study successively the frequency and the severity of cyclones in this part of Indian Ocean as well as the impact which they had on the island societies.

More than the mere reconstruction of series of events to the exclusive service of a more reliable modeling, the major stake in the interdisciplinary approach lies from now on in a better knowledge of the impacts to which the populations were exposed and the strength of the strategies which they were able to elaborate to survive. In this respect, the SREX report insists without ambiguity on the strategic character of this aspect of the climate change from now on. The systematic use of the terms of risk, exposure, vulnerability and sustainability is an obvious proof. It shows that the social dimension has become henceforth a strategic objective of the IPCC which concluded that we could not confine themselves only to mathematical projections of the climate which are often badly received by public opinion. Yet, the history of the climate can exactly enlighten the human reality of these extreme climate events of the past and serve as a tool of mediation between the scientific community and the citizens.

2. The Contexts of the Research

2.1. Study Area and Risk of Hurricane

The archipelago of Mascarenes is situated in the southwest of Indian Ocean, east of Madagascar, between the 19th south parallel and the tropic of Capricorn (**Figure 1**). It groups the Reunion Island, Mauritius, island Rogriguez, islands Agalega and Cargados. These rather distant islands have however common points: they are of volcanic origin and they regularly undergo hurricanes of strong intensity which increase considerably their vulnerability during the season of cyclones between December and March.

Paradoxically, the bibliography dedicated to this risk in the region is very poor. Generally, the north part of Indian Ocean held the interest of the researchers more. The variation of the activity of the tropical cyclones in the South of Indian Ocean is today difficult to understand and only Chang-Hoi et al. tried to make it from a comparison with the El-Nino-Southern Oscillation and the oscillation of Madden-Julian Oscillation [5]. Hoarau et al. were very interested recently in the variability of the most severe tropical cyclones (category 3 - 5) for period 1980-2009 [6]. They showed that the ten-year distribution of cyclones did not reveal clear trend of an increase of the cyclones of category 3 - 5 during the last 30 years, in spite of a doubling of the cyclonic activity with regard to 1980s and 2000. In conclusion, the authors of the study underline the importance of a more historic ap-

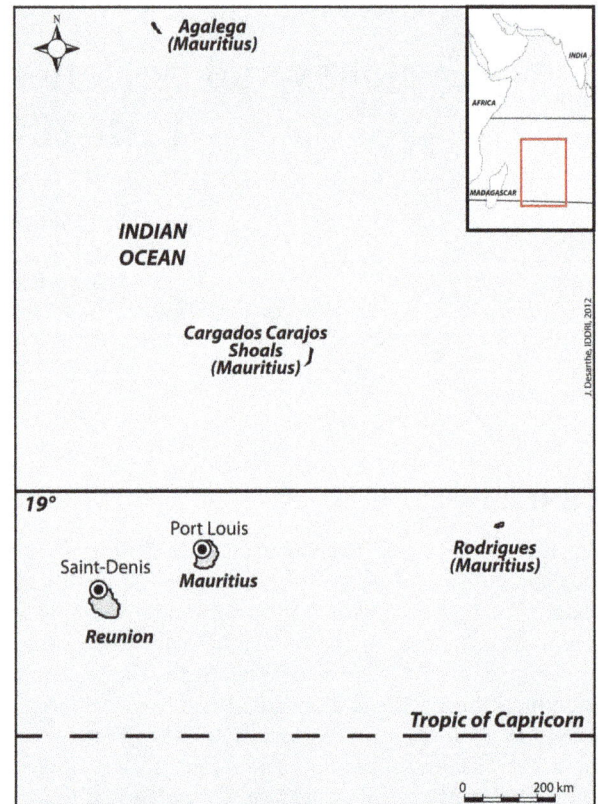

Figure 1. Map of the Mascarenes.

proach. They insist on the weakness of their results because of the absence of older data which would have allowed them to discern natural cycles in the decadal variations of intense cyclone activity.

The historic available data for the researchers interested in the tropical cyclones are extremely limited and very scattered [7]. NOAA's National Climatic Data Center (NCDC), in concert with the World Data Center (NCDC) has developed a comprehensive dataset through the collection of global tropical cyclone best-track: the International Best Track Archive for Climate Stewardship (IBTrACS). The major goal of the IBTrACS dataset was to provide the first publicly available centralized repository of global tropical cyclone best-track data from different agencies. The inventory of cyclones and storms per decade so obtained indicates clearly that the majority of the available data only begins in the 1950s. For the South of Indian Ocean, the most reliable and plentiful historic information appears only by 1880.

2.2. The Historical Material

The volume and the quality of the available archives for the South of Indian Ocean result directly from an original political history based on the major strategic interest that the Mascarene Islands represented for the British and French powers.

2.2.1. A Colonial and Slave History

The archipelago of Mascareignes was known by Arabs from the 10th century but it is the Portuguese sailor Pedro de Mascarenhas who made it known in Europe in the first half of the 16th century. The first real colonization of the island begins in 1638 with the installation of the Dutch people. Struck by a very violent hurricane in February 1695, the colony was not able to recover and finally, the Governor decided to leave in 1706 by evacuating about 200 people. In 1715, the French replaced the Dutch until 1810, the year of the British conquest whose dominion did not end until 1968 [8]. Throughout the French period, this island was called "Ile de France". The nearby island of Reunion was occupied by the French in 1665 and took the name "Ile Bourbon" until 1806. It remained under French rule until today except for a very short period of British occupation between 1810 and 1815, during the Napoleonic Wars.

Now under the exclusive dominion of the King of France, the archipelago was entrusted to the French East India Company to develop trade between the Kingdom and colonies overseas. From 1720, Reunion and Mauritius were organized settlements designed to produce mainly coffee and sugar for the metropolis. The economic system thus created was based on the use of slave labor from the Senegal Coastline, East Africa and Madagascar. Almost independent, the Company created a business empire consisting of counters in India (Chandernagore, Pondicherry, Mahe) in the Mascarene Islands, on the Persian Gulf, Burma and Indonesia. To ensure its dominance, the company got a government, an army and a navy controlled by a governor [9]. This bureaucracy finally changed little after the French Revolution of 1789, and explains the richness and quality of records available for the study of hurricanes.

2.2.2. Archives and Their Content

For convenience, we distinguish the pre-revolutionary period (before 1789) the contemporary period between 1789 and today. It is important to remember that the period after 1789 saw a significant political alternation between the monarchy (1815-1848) and the French Republic (1848-present).

Before 1789, the archives for studying hurricanes mainly concern the correspondence of Directors of the East India Company between 1700 and 1789. Of course, these sets of archives do not contain files specifically devoted to hurricanes and must therefore make very long and random analysis in this voluminous stock of archives. Primarily devoted to the management of maritime trade and the plantation economy, these records regularly mention extreme weather events and the main threat is hurricanes. In fact, the winds usually accompanied by heavy rainfall often have an impact on the fleet at anchor in the

ports and on the plantations where they can cause the death of many slaves, destruction of crops and sugar refineries. Such damage had an immediate impact on the fragile economy of the islands and it is not uncommon that famine and disease followed after a hurricane. The examples below give an idea of the extent and nature of damage caused by these events.

Letter of the governor of the Ile-de-France (Mauritius) of the April 1st, 1718:

The hurricane was particularly violent. The harvest of coffee was almost lost. The hurricane provoked the filling of a river with pebbles pulled by rains and torrents. This filling transformed lands near the "river of Pebbles" into a desert.

Letter of the governor of the Ile-de-France (Mauritius) of March 8th, 1743:

We had a hurricane on March 8th. The big rashness of the wind lasted only from ten o'clock in the evening till two o'clock at night. Several vessels ran aground in the port because of very high waves which reached the store of the port. The harvest was almost completely destroyed, in particular the corn, the potatoes and the sugar canes. On the other hand, the rice and the manioc were protected. As soon as our port (Port Louis) will be repaired, I shall send to you by boat of the peas of the Cape (South Africa) and the beans which you can distribute in the poorest and to the blacks.

After 1789, the available material becomes even more precise about hurricanes and about their effects. On the British side, the archives of Colonial Office and of Secretary of State for the Colonies situated today in the national archives (Kew) supply extremely detailed reports on hurricanes having struck Mauritius fin the 19th and 20th centuries. In a rather systematic way, these documents give climatic information resulting from Meteorological Offices of the island. They concern the chronology and the meteorological parameters (temperature, pressure, winds) of the event. Then, the authors describe exactly the nature of the damage caused by the extreme event by enumerating buildings, trees, infrastructures as well as crops. At the same time, the cost of the damage is estimated because it justifies the measures of assistance taken by the government in London. These English archives focus on the consequences that cyclones can have on the Mauritius's big source of wealth: the sugar production. The economic stakes explain the publication of numerous governmental reports and planters' labor unions in the Mauritius's agricultural Bulletin. The administrative archives can be completed by the colonial press of period, in particular The Mercury, The Brisbane Courrier and The Straits Times Weekly Issue. Besides meteorological and social data (deaths, destruction), newspapers get the first photos during the big hurricane of 1892.

For La Réunion, archives also become stronger with

the development of the republican centralist model which is translated by the increase of the administrations. The services of the governor, the Conseil Général and the Farmers' association guarantee the historian an excellent quality of the information. So from 1820 the archives of the government of the colony of La Réunion contain files specially dedicated to hurricanes. We find in particular the very precise statements of the colonial gendarmes (French military policemen) whose barracks were distributed throughout the whole island. They supply from now on a state of the disaster for every part of La Réunion, which facilitates the evaluation of the severity of the hurricane. From the point of view of the vulnerability of territories, the geographical precision of these reportsauthorizes a cartography of zones destroyed on the scale of a street or field. Finally, let us add inquiries made by the state employees of the forest administration about trees broken or uprooted by winds. This silvicultural information is a very precious resource which allows us to evaluate the severity of these historic events today even without information from meteorological instruments or also where events may have been totally underestimated by the various Meteorological Offices of Mascareignes (Météo-France and Mauritius Meteorological Services) whose catalogs of events begin in the first half of the 20th century.

3. Reconstruction of Cyclones 17th-20th Centuries

3.1. Method

The archives can therefore establish a longer time series than those currently offered by the meteorological services of the Indian Ocean that can't consider exploring the historical archives. Historical material involves work that only professional climate historians can achieve due to their mastery of the archives, the historical method and technical knowledge in fields such as palaeography. However, the results obtained from the archives can only be useful to climatologists and meteorologists if the historian is able to translate his "social" data in more quantitative data. Indeed, archives supply information on the extreme events only according to the damage which they cause to societies and it is necessary to wait for 1850s to find finally data of instrumental nature (strength and orientation of winds, barometric pressure). And yet, nowadays, the meteorologists estimate the severity of hurricanes in order to compare them more easily according to their destructive power. To achieve this, they use the Simpson-Saffir hurricanes Wind Scale (SSHWS) which gives an estimation of the damage engendered by hurricanes. Set up in 1969 by Herbert Saffir, a consulting engineer, and Bob Simpson, Director of the National Hurricane Center, this scale provides examples of the types

of damage caused by winds [10] and classifies hurricanes into five categories [11]. Subsequently, it has been improved by adding other additional parameters such as storm surge and floods. If the scale does not give full satisfaction to assess a hurricane today, however, it is particularly well suited to the material historical sources mentioned above. We therefore decided to resume headings modifying to adapt to historical descriptions (Appendix 1). The criteria to estimate the hurricanes of the past can be divided into 12 criteria:

- Meteorological details;
- People, livestock;
- Infrastructure;
- Buildings (slave huts, colonial houses, barracks, stone churchs, stone stores and Governor's Palace);
- Plantations;
- State of the sea;
- Ships;
- Landscape;
- Trees;
- Surge;
- Power and water;
- Social and economic consequences.

Thus organized, the "historic" SSHWS assigns to each event an index between 1 and 5. Nevertheless, the method has a certain limit when it comes to record events of lesser intensity between categories 1 and 3 of the scale. Indeed, the historical data available for the 17th and 18th centuries mainly provide social and economic information, that is to say information that provide information on the damage suffered by the island societies. The traces left by hurricanes in the archives thus mainly interested the most severe events that can be classified among categories 3 - 5 scale. Consequently, the approach taken in this work struggles to register extreme events located in the low or medium frequency severity before 1850. After this date, the instrumental quality archives (weather reports and instrumental data) significantly refines the interpretation and allows to better take into account the low-intensity hurricanes (category 1 - 3). In addition, sometimes a hurricane archive relate to a specific date without providing sufficient information to place on the scale. In this case, we have chosen to assign it an index −1, so it appears on the chronological chart but without estimation of its severity.

3.2. Frequency and Severity

We have found 89 cyclones in the archives between 1656 and 2007 for the Mascarene archipelago. They show very strong fluctuations according to the century (**Table 1** and **Figure 2**). The distribution of hurricanes over the last three centuries displays strong contrasts. Weakness during the 17th century is very likely due to incomplete re-

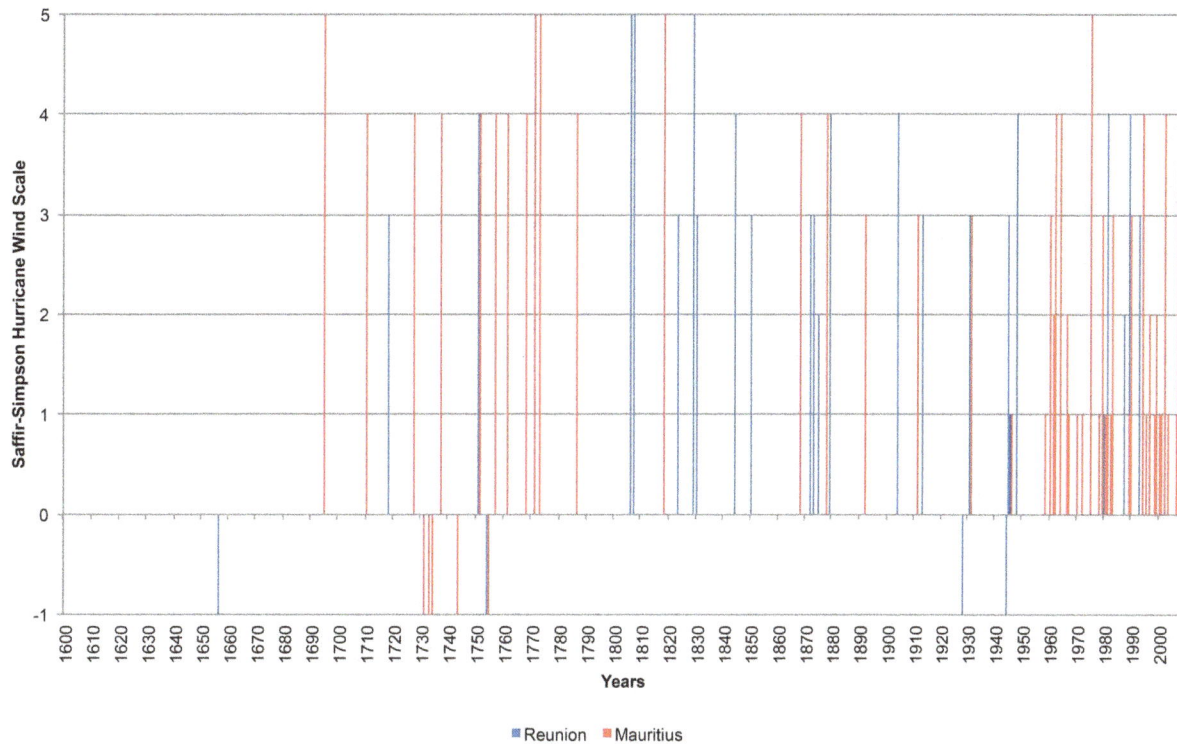

Figure 2. Chronology and severity (according to the SSHWS) cyclones in Mascarene islands between 1656 and 2007. The blue columns correspond to the events of La Réunion and the red to those of Mauritius. Source: see manuscript sources.

Table 1. Distribution by century and by category of cyclones in the Mascarene Islands.

Category	−1	1	2	3	4	5	Total
1600-1699	1					1	2
1700-1799	7			1	9	3	20
1800-1899			1	6	5	5	17
1900-1999	1	16	7	9	9	1	43
2000-2007		4		1	2		7
Total	9	20	8	17	25	10	89

cords and the beginning of the European colonization of the region. For the subsequent period, the results show that the 18th century was more frequently affected by the extremes than the 19th century. However, the observation of the entire series clearly shows a rising trend during the 20th century, which displays a score of 43 cyclones, the little less than half of the total. The clear domination of the 20th century cannot be explained by a better recording of extremes in the 20th century than in the archives from the period 1700-1900.

Besides the fact of confirming the increase in the number of cyclones in the 20th century, **Figure 3** specifies the chronology of this development. Years 1900-1959 seem rather sparse whereas a turning point occurs in 1960 with an almost continuous increase of the number

of cyclones till the end of the millennium and a peak of frequency during decade 1980. However, the increase in the number of the events does not seem to be mirrored by an escalation of their severity estimated according to the SSHWS.

On the contrary, the **Figure 4** which presents the cycles of 50 years shows clearly that the period 1950-1999 is almost exempt from hazards of category 5 (one event) while the category 1 offers near half of the number. We can thus conclude that the increase of cyclones in Mascarene islands is not mirrored by an escalation of their severity. On the other hand, we can think that this greater frequency, of low or average intensity, will probably increase the vulnerability of the island societies of the region in the next decade.

4. Vulnerability and Adaptation of Society of the Mascarene in the Past

IPCC-SREX-report identifies small islands in Pacific, Indian and Atlantic as the areas most at risk of climate extremes. It considers that the main climate models give the same signal: the future risk increased. The report addresses the issue from a strategic perspective, since it speaks of the challenge of adapting and lists the vulnerabilities of these islands whose future development is directly threatened. However, it does not seem able to truly assess the particular risk to the Indian Ocean, for

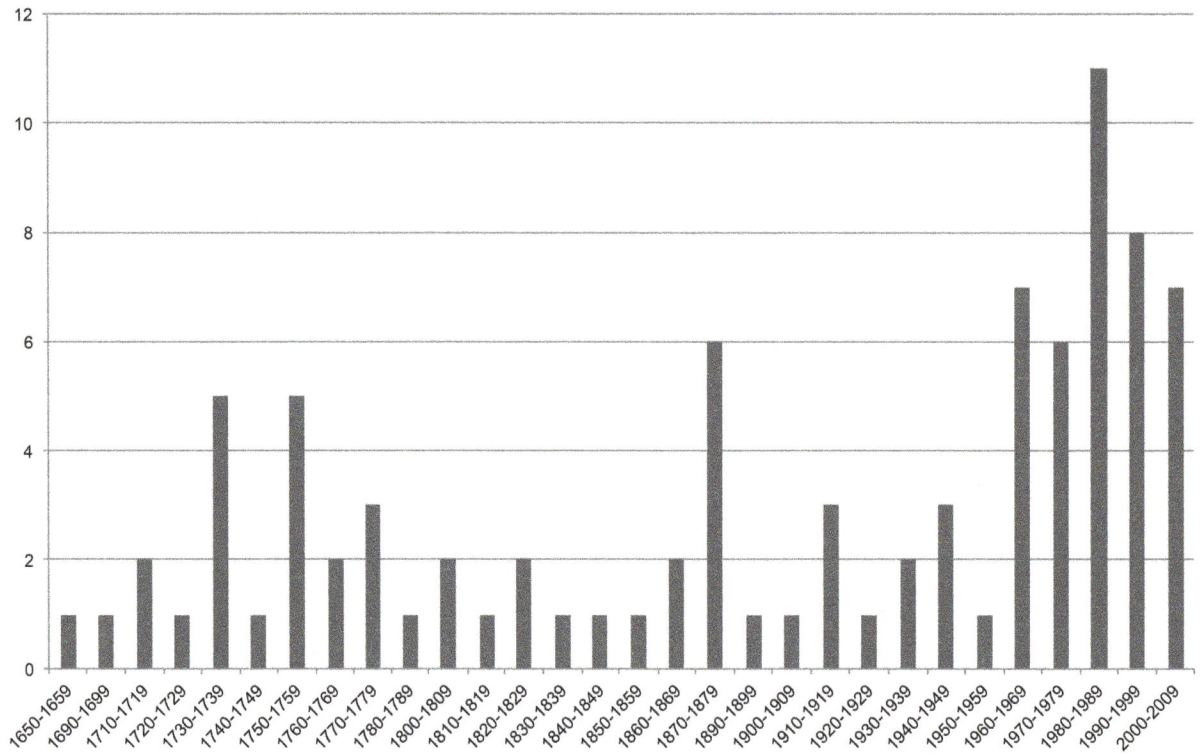

Figure 3. Distribution of cyclones by ten-year period.

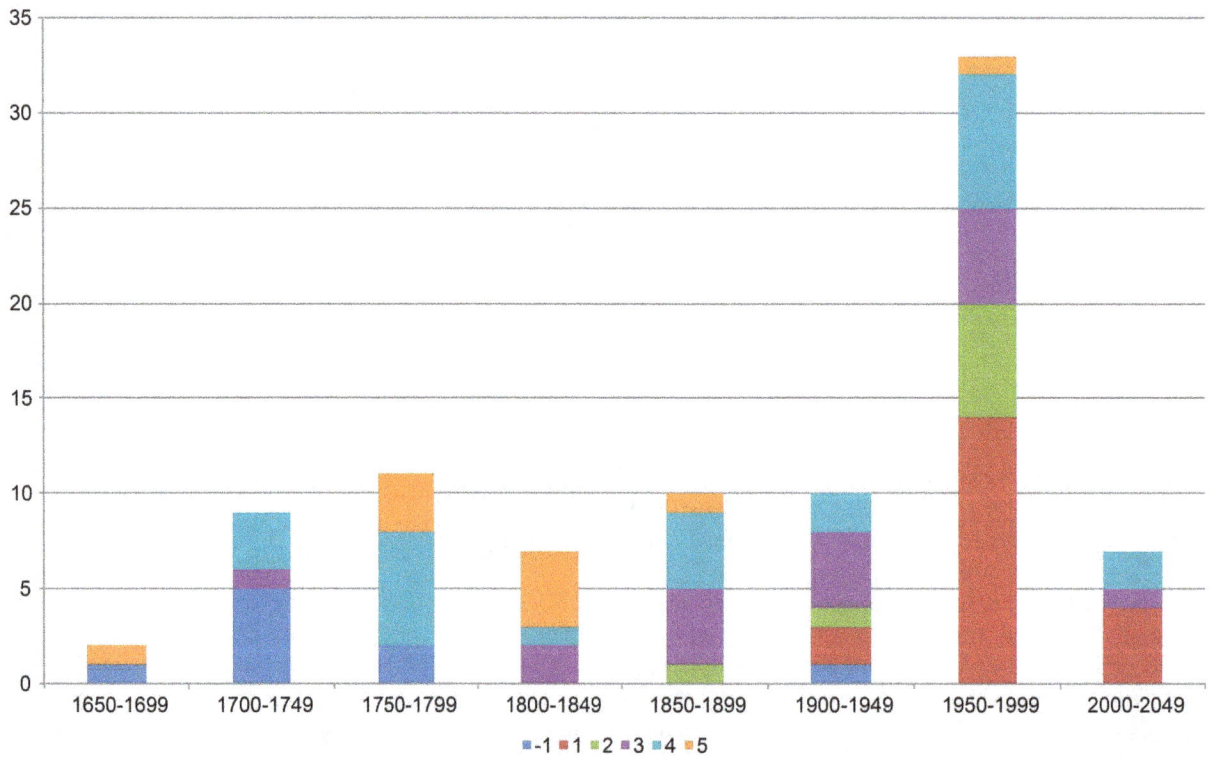

Figure 4. Distribution by period of 50 years and by category (from 1 to 5) of the SSHWS in Mascarene islands. The category-1 (blue color) indicates a historic event mentioned in archives but insufficiently informed to be estimated by means of the SSHWS.

lack of sufficient research. With the exception of the Maldives, the chapter "Case Studies" focuses instead on the Pacific and the Caribbean. Mark Pelling and Juha I. Uitto follow a more dynamic approach by questioning the factors that make these islands so vulnerable to natural disasters [12]. Following their investigation, they conclude that their vulnerability results essentially from limited mitigation capabilities and their limited integration into the global economy. More recently, other studies recommend reducing the impact of climate change and extremes of these small islands by a better transfer of technologies for adaptation and mitigation [13]. Facing these issues and recommendations present a big challenge to the global economy which is currently already struggling, so interrogating the archives is a relevant new resource as they deliver multiple historical experiences of adaptation which may be helpful for the present and the future.

4.1. The Semantic Precondition

4.1.1. A Transdisciplinary Stake

In recent years, under the influence of the social sciences the scientific community has been introduced to the key terms of vulnerability, exposure and resilience. The IPPC appropriated these terms and now employs them regularly. Their social significance is so strategic in the debate on global change it was thought necessary to define them in the IPCC-SREX-report. Fairly systematically used in environmental discourse, these terms are not necessarily well understood and defined by scientists and citizens. Indeed, the meaning of these words is different according to whether they are used by a researcher or climatologist, geographer, ecologist or economist. "Vulnerability" identified in the IPCC-SREX-report is thus presented as "the propensity or predisposition to be affected by a threat", without any more precision than that. The definition is so vague as to make its use not necessarily valid based on the data available. However, the definition of adaptation is more accurate as it relates to environmental realities paradoxically, social and cultural might be corrupted. Given these uncertainties and the lexical content of particular historical archives, it is necessary to first undertake a semantic exercise in order to best match the objects mentioned in historical records (hurricane, storm, vulnerability) with the expectations of the current scientific community. This methodological precaution is a condition of dialogue between the disciplines involved in the debate on climate change and the necessary step to a lasting collaboration between the exact sciences and the social sciences.

4.1.2. The Words of the Past and Their Contemporary Significance

In his book on the transition from a disaster in a risk so-

ciety, the sociologist Ulrich Beck contrasts a pre-modern society described as "traditional", to a "modern" society [14]. In the first case, the perception of risk is presented as non-existent, replaced by a social conviction: that threats result from natural and unpredictable disasters. To this collective fatalism, he contrasts the industrialized society, whose birth is placed as around 1840. For Beck, generating risk, industrialization finally allowed its definition and quantification by instrumental rationalization and scientific progress.

However, the reality apprehended in the archives about hurricanes strongly contradicts the idea of the total vulnerability of ancient societies. It reveals that the roots of the concept of risk can be found much further upstream. Dictionaries of the 17th and 18th centuries deliver a point of view much more pragmatic when they define risk as a "danger", a "great peril" [15], they immediately put two notions related to the fact that "exposure to risk" [16]. It appears in the historical record as a social uncertainty which should be prepared for by developing strategies of resilience. Historical definitions are relevant today as they introduce an approach based not on a variable (risk) but on a social reality that is vulnerability already defined centuries ago in terms of what "may be injured" (populations) and "destroyed" (infrastructure and housing). Historical depth allows us to understand the roots of vulnerability [17] in the Mascarene Islands and facilitates the understanding of the trajectories of vulnerability [18], an assumption according to vulnerability results from a double evolution of society and the natural environment over the centuries [19]. Use of analytical methods helps us apply historical experience to the present and the future.

A research dedicated to the history of cyclones in the Mascarene Islands also helps to solve the question of the old meteorological terms. They illustrate remarkably well the evolution and dissemination of weather knowledge in the Indian Ocean and the assessment of hurricane risk by the authorities. From 1666 until about the 1850s the British and French governments routinely used the term "hurricane" to denote a very high winds causing significant human and material losses. The French regularly used the word "houragan", spelled with an H. It disappears permanently in the archives after the First World War The first mention of French "cyclones" appears in 1868 in a report by the Governor who speaks about the center of a cyclone, a term borrowed from the British terminology. This loan can be explained by the proximity of Mauritius, which was British island since 1810. Indeed, the word cyclone was invented by the sea captain Anglo-Indian Henry Piddington (1797-1858) following an observation made on a ship offshore of Mauritius in February 1845. Noticing that the ship caught up in a meteorological event made a trajectory in a circle, he compared the phenomenon with a snake winding in circle and is inspired by the Greek word kyklos (circle) to call it "cyclone" [20]. A new lexical evolution occurred in the 1930s with the rise of the term "atmospheric depression" in administrative and weather reports. This was a milestone which reflected the scientific progress of Meteorological Sciences because it designated more exactly the circulation around a closed center of low pressure in tropical regions.

4.2. The Historical Vulnerabilities

4.2.1. People

Contrary to what one might think, it seems that hurricanes did not cause exceptional loss of life. In this regard, the archives remain unclear for the pre-revolutionary period (before 1800), while the following period provides some statistics. Generally, human losses from cyclones did not exceed fifty dead. However, this mortality rate had racial and social disparities. Unsurprisingly, the main victims were slaves working in the plantations or ports, while those located in cities were less vulnerable. After the abolition of slavery in Mauritius (British colony) in 1835, then in Reunion (French colony) in 1848, the inequality of the death rate became social. In addition to the black people, new Indian migrants (coolies) and Malagasy came to replace slaves in agriculture and "small white" (poor white people) were now the main victims of cyclones due to their precarious living conditions. Thus, the majority of them perished in the collapse of their homes or drowned because their neighborhoods were located in or near rivers and ravines that suddenly became disastrous torrents.

More than the cyclone itself it was its side effects that affected most people. In fact, by destroying or damaging crops and infrastructure, cyclones affected the operation of the island's economy. In addition, the destruction of warehouses in ports or in the villages, threatened crop supplies and thus the food security of the population. The 1770s were particularly difficult for the people who underwent a series of cyclones which were level 5 on the SSHWS. However, in respect of La Réunion or Mauritius, the turning point was really in the years 1806-1808 when they suffered repeated and extremely violent hurricanes which destroyed practically all the coffee trees, clove trees and corn. During these three years, the price of corn rose sixfold because of the shortage which reigned in the Mascarene Islands. After the hurricane of 1832, the supply difficulties gave rise to a social protest against the French and British authorities in Reunion and Mauritius which were accused of incompetence and passivity. Reunion protesters stormed "out of the woods like wolves and subjected those to a colonial gale violent enough to send them join their destroyed cassava and bananas" [21], while a riot broke out in Port Louis during

which the British Governor was insulted and threatened [22].

Another side effect of cyclones, the health risk was a major concern of the authorities from the second half of the 19th century, as they feared the problems of water pollution because of the destruction of pipelines and wells. In 1913, outbreaks of typhoid occurred in several towns on the island of Reunion in the isolated mountains after the destruction of roads and bridges. In an attempt to contain the disease, military doctors were sent as a matter of urgency to these villages on horses and donkeys. On the spot, they treated and distributed some bleach to the populations.

4.2.2. Housing

The destruction caused by cyclones in the housing varied widely depending on the quality of the building. That is the reason why huts lived in by the slaves and the poor people, constructed of wood and roofed with straw or palm leaves, represented the majority of the damaged houses. Mostly, they are uncovered, overturned or sucked up by the cyclone and thrown far off. A traditional component of the architecture of islands even today, the more substantial colonial houses constructed of wood and covered with shingles were less vulnerable even if they regularly suffered damage.

Graver for the resilience of the island societies, the destruction or damage of public buildings were always felt as a profound trauma, in particular when it concerned churches, city halls, schools and high schools which are places of power and sociability. Often built stone from 1740, these buildings were the first to be reconstructed after a disaster with public money and donations.

4.2.3. Agriculture

Storm surges regularly caused floods and accumulation of pebbles and sand. In 1718, such an event led to the sealing of the estuary of the "River of Pebbles" (Réunion) and the agricultural devastation of part of the nearby coast. More systematically, cyclones disrupt agricultural production brutally and massively. Winds and floods which accompany them hit foodcrop production vital to the population such as corn, manioc, rice and potatoes. In fact, it is especially floodwaters which ravage fields by ripping up the crops or by causing them to rot. Nevertheless, these small family-run farms show themselves more resistant than the other more speculative commercial plantations of coffee and clove trees which, because of their size are more exposed to the wind. They decline more and more after the cyclone of 1844 for the benefit of the sugar cane, considered more resilient.

4.2.4. Navigation

The violent winds of cyclones engender grave conesquences for the ships which navigate or which drop anchor in the ports of Port Louis and Saint-Denis. The strategic position of these islands for the British and the French until the 1850s added to the vulnerability because the total or partial destruction of the fleet could arouse an attack of the enemy. In 1771, the English council of Bengal indicates in a report that the French fleet of Port Louis underwent very heavy losses after the passage of a cyclone of category 5 on the SSHWS. Ships were effectively very exposed and it could take less than one hour for them to be wrecked in the port which welcomes them. In 1823, the officers of the French navy looked on powerless at the wreck of nine ships on the beach of Saint-Denis. The loss of ships from the fishing and trade fleets was also supplemented by the loss of rowboats and other small boats which were sent to the Rodrigue islands to gather the tortoises consumed on Mauritius and Reunion.

4.2.5. Infrastructure

The major part of the damage caused to infrastructure was generally connected to the floods and to the torrents caused by the rains which accompany cyclones. Water cut roads and destroyed bridges provoking a break of the land communications and the isolation of populations for days, sometimes weeks or months. Roads and railroads were mostly taken by the flows of water and mud as in 1875 on Reunion where all the rivers of the island overflowed at the same time and interrupted the traffic between the capital Saint-Denis and the South of the colony. In the 19th century, the frequency of cyclones led to a deterioration in the situation. Roads barely had time to be repaired before they suffered new damage. The sugar industry was also very exposed to the floods and the wind. From its development in the 1820s, the mentions in the archives of the destruction of sugar refineries and their steam pumps are more and more frequent, and this served to weaken an economy more and more dependent on its sugar exports. Tidal waves also weakened infrastructure by damaging port facilities. Waves repeatedly destroyed the dikes and piers of the port of Barrachois on La Reunion. Finally, new factors of vulnerability appear from 1870 with the technical progress of the industrial revolution. It quickly becomes apparent that the new telegraphic and electric networks have little resilience. The official reports about cyclones speak from now on about overturned posts or about flooded power stations. This progress, which also gave rise to new weaknesses, contributed to aggravate the concern of the population which has become used to a new quality of life. In 1921, the storm caused the break of the submarine cables and interrupted the connections between Mauritius, Mozambique and France.

4.3. Adaptation and Forms of Resilience

The terms of adaptation and resilience are anachronistic with regard to this historical research because they first appeared in the second half of the 20th century in the disciplines of the physics of materials and child psychiatry. In the 1970s, they were assimilated by the ecologists then by the climatologists only one decade ago. The IPCC-SREX-Report defines the latter as the ability of a system and its component parts to anticipate, absorb, accommodate, or recover from the effects of a hazardous event in a timely and efficient manner. Whereas adaptation corresponds in the glossary to the process of adjustment to moderate the negative effects of the climate or to take advantage of it. In these conditions, we can wonder about the relevance of such terms from a historical point of view. In reality, ancient societies were not content with passively suffering the negative effects of the climate and in the Mascarenes Islands, the populations learnt to live with cyclones very early on and actively tried to limit their dangers.

4.3.1. Individual Behavior

The historical testimonies evoke original behavior which is always usual by the inhabitants. In 1666, inhabitants take refuge in the nearby forests to shelter from the wind but also from the risk of surge. Others leave their houses on the coast and find shelter in the hills and mountains where they wait for the end of the extreme event. In 1771, the fishermen identify the risk of cyclone by observing the foam of the sea, the sky. They pull their dugouts and their boats far from the sea. Meanwhile, the owners strengthen their houses by attaching ropes to them and by nailing boards on doors and windows.

At the end of the 18th century, the inhabitants noticed that the colonial houses were particularly exposed to the damage and the widespread use of wooden shingles to roof their houses instead of palm leaves and straw began. The municipal authorities even published regulations from 1786 which made the use of this construction material compulsory, which also helped to limit fire risk in the city.

4.3.2. The Agricultural Revolution of the Sugar Cane

The series of disastrous cyclones of 1806 and 1807 plays a major role in the agricultural mutation of the Mauritius and La Réunion islands. After the cyclones of February and March, 1806 then those of 1807, planters and farmers gave up reconstituting the plantations of coffee and spices which were very badly damaged by winds and floods. They were largely replaced by the cultivation of sugar cane which was considered more resistant to the winds. The incidence of cyclones on the cultivation of the cane depends essentially on the rhythms of the plant. As a agronomist reminds us in 1884, the plant bears the storm well when it has only 6 to 8 months growth [23]. On the other hand, if the cyclone occurs a few months away from the harvest, the cane is eradicated or broken. Indeed, reports drafted after the passage of cyclones always insist on the biggest resilience of sugar canes compared with other crops. Besides, Great Britain and France were then at war and it become vital to produce sugar in colonies. So, the outbreak of cyclones and their escalation contributed with economic interests to amplify a profound agricultural transformation which even today still characterizes the Mauritius and La Réunion islands to a lesser extent.

Figure 5 shows that despite of six cyclones of categories 3 - 4, the production of sugar and the land area dedicated to it are little impacted by cyclones. The relative and short decline after the cyclone of 1863 until 1864 is attributable in the appearance of the Borer insect (Chilo sacchariphagus), which lives as a parasite on some sugar cane and destroyed part of the harvest in Mascarenes.

4.3.3. The Collective Alarm Systems
From the 18th century, archives reveal the existence of rudimentary but relatively effective alarm systems. In ports, the authorities launched the alert thanks to a code known by captains. It consisted of raising a flag and firing two cannon-shot in case of bad visibility. Immediately, ships cut their ropes and sail. The efficiency of the plan implies a good knowledge of the meteorological phenomenon in a period when measuring instruments

were very rare or non-existent. For the inhabitants, the alert was given by the churches which rang their bells at the approach of the cyclone.

From 1820, the political will to manage the risk better by an alarm system improved considerably with the widespread use of meteorological instruments. In La Réunion, the captain in charge of the port of Saint-Denis had a major role. From 1829 he used barometers which allowed him to plan the arrival of an atmospheric depression better and to estimate the severity so as to know which plan alerts he had to organize. In Mauritius, the British were very sensitive to the meteorological risk. They created the Royal Alfred Observatory whose first instrumental data were published in 1862. The task of forecasting the tracks of these tropical storms was one of the most important of the duties of the Royal Alfred Observatory; in the eyes of the layman it constituted its raison d'être: and indeed it was of great local importance, especially to the Railway and Harbour Departments [24].

This work was greatly facilitated by the receipt of daily weather telegrams from Rodrigues and Réunion during the cyclone season, while in other seasons it only served to exemplify an oft-quoted adage that "a little knowledge is a dangerous thing". The development of the means of communication facilitated the prevention cyclone damage. To strengthen the relations between the various islands, a project of submarine cable was voted

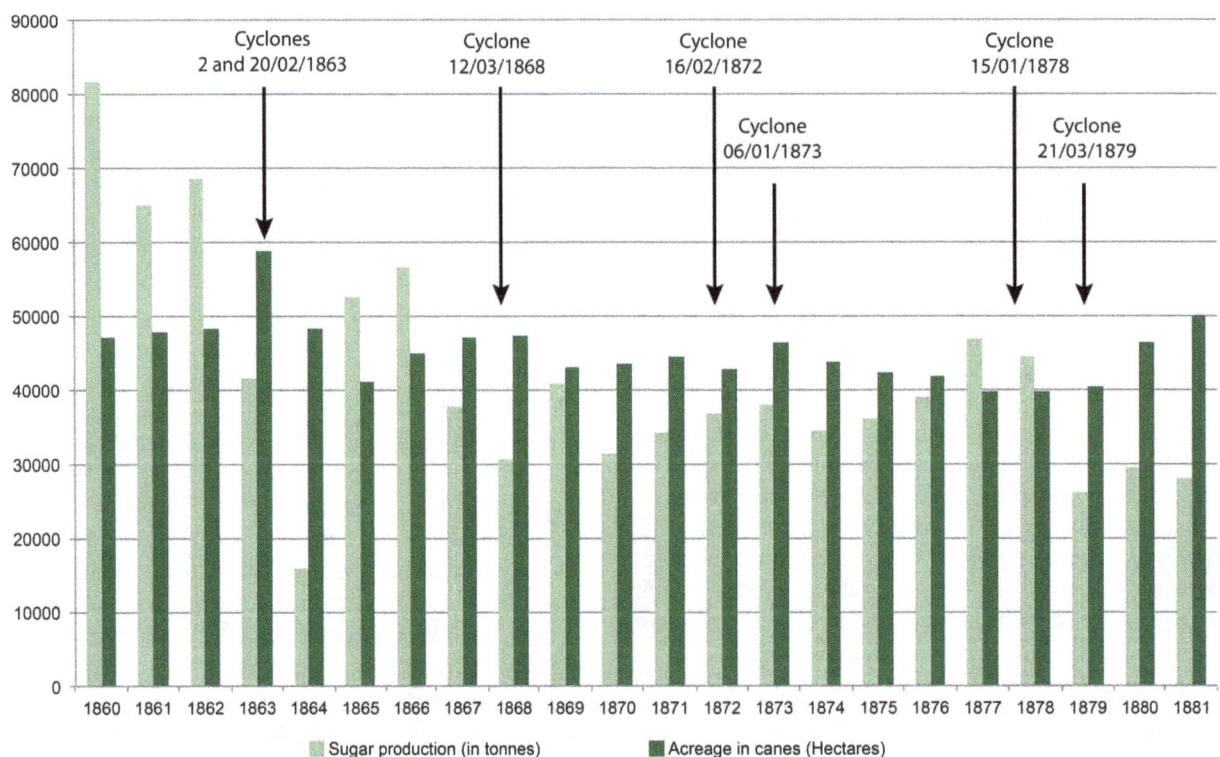

Figure 5. Production and surface of the sugar cane to La Réunion (1860-1881). Bars green-darkened indicate the production of sugar in ton and bars green-clear surfaces in hectares.

by the French National Assembly in 1903. It consisted of connecting La Réunion to Madagascar and to Mauritius. Three years later, an agreement between Britain and France ratified this project of a submarine connection between La Réunion and Mauritius [25]. So, from the end of the 19th century a regional alarm system was built which connected the various islands of the Indian Ocean (**Figure 6**). The main observations were given by Tananarive and Port Louis. These two cities broadcast various bulletins mainly to Rodrigue, the Seychelles and La Réunion. The various ships present in the Indian Ocean also supplied numerous instrumental data to help plot the route of the cyclones.

Such cyclone warning systems were continuously developed and fine-tuned. Today, the governments of the region assert that the regional alarm system is a recent creation which is based on the recent collaboration between international organizations such as World Meteorological Organization (WMO) and Regional Specialized Meteorological Centre (Météo-France, La Réunion) [26]. The information exchange between them guarantees the announcement and the prevention of cyclones. It's the same for the three-tier warning system which according to them, was established in the 1950s, in particular by including police forces. In reality, the British and especially French authorities used the network of gendarmeries and police stations from the beginning of the 19th century with great effectiveness to warn then help the victims.

5. Conclusions

The historical method enriches the current debate concerning two strategic questions about global climate change. The first one is to know if the extreme events can be connected with climate change or not. The results presented in this historical research demonstrate that since the end of the 17th century, the chronology of cyclones in the Southern Indian Ocean underwent important fluctuations which were probably attributable to the natural cycles of the climate. On the other hand, archives of the time which were numerous and can be considered reliable prove that the trend in the number of cyclones is sharply upwards from 1800 and that from 1960 that trend

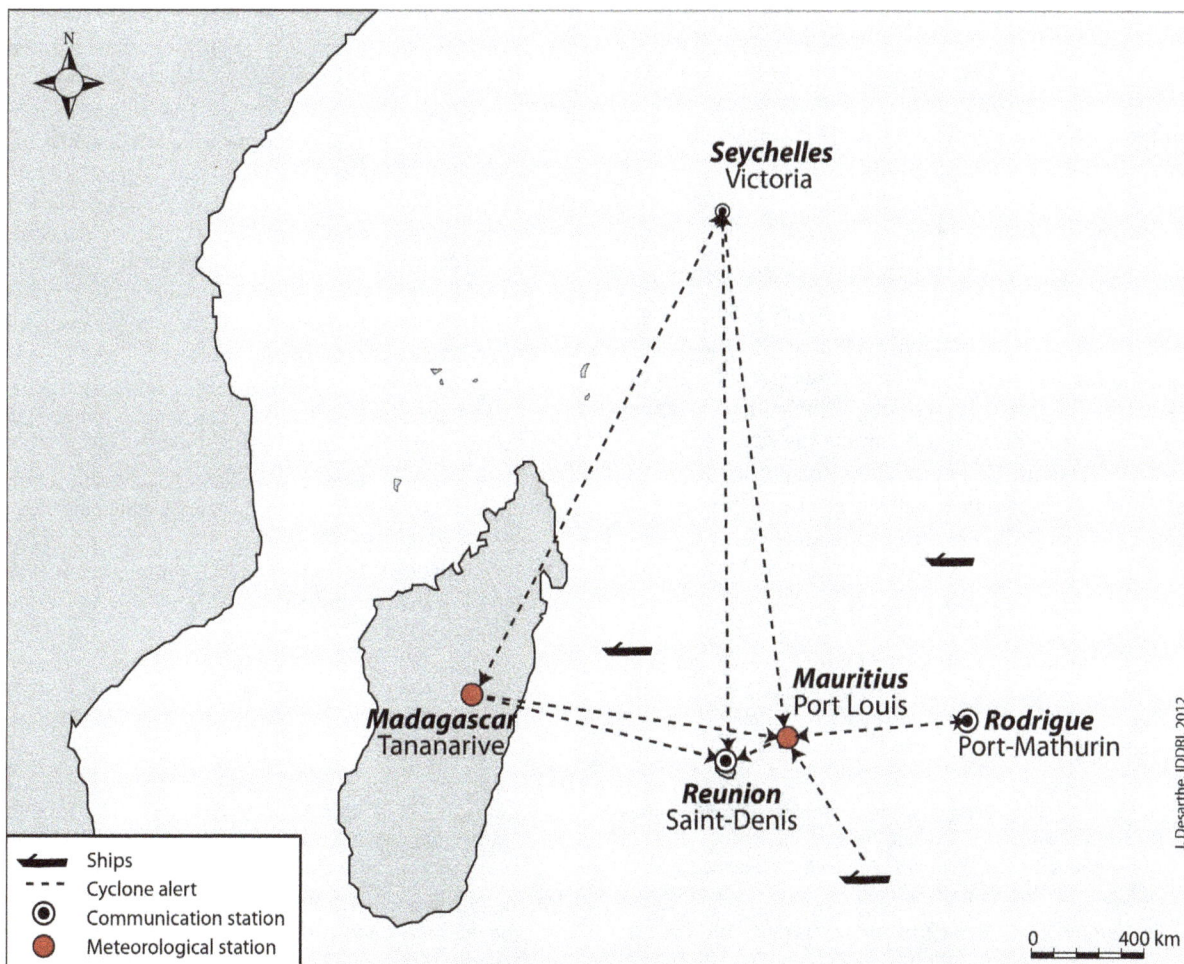

Figure 6. Regional network of alert and forecast against cyclones in the South of Indian Ocean by 1900.

has increased dramatically. This observation thus strengthens, for the Indian Ocean at least, the thesis of a strong link between climate change and extreme meteorology. The other major stake in the current climatic debate, is the question of the resilience and the possibilities of adaptation offered to the increasingly exposed populations. Once again, historical experience enlightens the debate thanks to new and original data offering a strong social dimension. Yet, the awareness of the climatic risk by international public opinion, in particular on the particularly exposed Small Islands and the Least developed countries, depends very widely on the message delivered by the scientific community and the decision-makers.

In this respect, the historical information collected about the vulnerabilities and strategies developed by the old societies can be regarded as tools of dialogue allowing a more effective mediation. By showing the historical character of the climatic risk and by explaining how the populations tried to face the extreme events, history can also participate in the scientific debate about global climate change and contribute to our knowledge of what current societies can expect from it. The wealth of available material in archives worldwide indicates the possibility of multiple worldwide approaches but unfortunately, it will require long painstaking work to collect the data from archives which may in some cases be endangered for lack of interest on the part of the states or institutions which are the owners of the archives.

6. Acknowledgements

The article is partly result of three projects. The HEALTH project (Historic Extremes and heaLTH) is supported by the French Embassy of London, Churchill College, Cambridge and the Universities of Cambridge (Department of Geography) and Caen. The project "Climate and extreme climatic events in the last millennium" is funded by the Institut Universitaire de France. The VulneraRe project is financed by the Fondation Nationale de France. We are indebted to George Turner for proofreading our text.

REFERENCES

[1] IPCC, "Managing the Risks of Extreme Events and Disasters to Advance Climate Change Adaptation," Special Report of Intergovernmental Panel on Climate Change, Cambridge University Press, Cambridge, 2012, pp. 161-163.

[2] A. Grinsted, J. C. Moore and S. Jevrejeva, "Homogenous Record of Atlantic Hurricane Surge Threat since 1923," Proceedings of the National Academy of Sciences, Vol. 109, No. 48, 2012, pp. 19513-19514.

[3] E. Garnier, "Les Dérangements du Temps. 500 ans de Chaud et de Froid en Europe," Plon, Paris, 2010.

[4] E. Garnier, N. Henry and J. Desarthe, "Visions Croisées

de l'Historien et du Courtier en Reassurance sur les Submersions. Recrudescence de l'aléa ou vulnérabilisation croissante?" Quae, Paris, 2012, pp. 105-128.

[5] H. Chang-Hoi, K. Joo-Hong, J. Jee-Hoon, K. Hyeong-Seog and C. Deliang, "Variation of Tropical Cyclone Activity in the South Indian Ocean: El Nino-Southern Oscillation and Madden-Julian Oscillation Effects," Journal of Geophysical Research, Vol. 111, No. D22, 2006, p. 9.

[6] K. Hoarau, J. Bernard and L. Chalonge, "Review Intense Tropical Cyclone Activities in the Northern Indian Ocean," International Journal of Climatology, Vol. 32, No. 13, 2012, pp. 1935-1945.

[7] K. R. Knapp, M. C. Kruk, D. H. Levinson, H. J. Diamond and C. J. Neumann, "The International Best Track Archive for Climate Stewardship (IBTrrACS)," Bulletin of the American Meteorological Society, Vol. 91, No. 3, 2010, pp. 363-376.

[8] H. Dundas, "The History of Mauritius, or the Isle of France, and the Neighbouring Islands from Their First Discovery to the Present Time," Bulmer, London, 1801.

[9] H. Maurin and J. Lentge, "Le Mémorial de la Réunion," Vol. 1, Australes Éditions, Saint-Denis, 1979.

[10] H. S. Saffir, "Hurricane Wind and Storm Surge," The Military Engineer, Vol. 423, 1973, pp.4-5.

[11] R. H. Simpson, "The Hurricane Disaster Potential Scale," Weatherwise, Vol. 27, No. 169, 1974, pp. 169-186.

[12] M. Pelling and J. I. Uitto, "Small Island Developing States: Natural Disaster Vulnerability and Global Change," Environmental Hazards, Vol. 3, No. 2, 2001, pp. 49-62.

[13] De Comarmond and R. Payet, "Small Island Developing States: Incubators of Innovative Adaptation and Sustainable Technologies?" In: M. David and A. Pandya, Eds., Coastal Zones and Climate Change, 2010, pp. 51-68.

[14] U. Beck, "Risk Society: Towards a New Modernity," Sage, Londres, 1992.

[15] Furetière, "Dictionnaire Universel Contenant Générale-ment Tous les Mots Français," Arnout et Reiner, Leers, La Haye and Rotterdam, 1694.

[16] "Dictionnaire Universel Français et Latin Vulgairement Appelé Dictionnaire de Trévoux," Compagnie des Libraires Associés, Paris, 1704.

[17] P. T. Blaikie, I. Cannon and B. D. Wisner, "At Risk: Natural Hazards, People's Vulnerability and Disaster," Routledge, London, 1994.

[18] Magnan, "La Vulnérabilité des Territoires Littoraux au Changement Climatique: Mise au Point Conceptuelle et Facteurs d'Influence," Vol. 1, Analyse Iddri SciencesPo, 2009.

[19] Magnan, V. Duvat and E. Garnier, "Reconstituer les Trajectoires de Vulnérabilité Pour Penser Différemment l'Adaptation au Changement Climatique," Natures Sciences Sociétés, Vol. 20, No. 20, 2012, pp. 82-91.

[20] H. Piddington, "The Horn-Book for the Law of Storms for the Indian and China Seas," Bishop's College Press, Calcutta, 1844.

[21] N. Dodille, "Journal d'Un Colon de l'Île Bourbon,"

L'Harmattan, Paris, 1990.

[22] J. Jeremie and J. Reddie, "A Pamphlet. Recent Events at Mauritius," Hatchard and Son, London, 1835.

[23] A. Delteil, "La Canne à Sucre," Chalamel, Paris, 1884.

[24] A. Walter, "The Sugar Industry of Mauritius. A Study in Correlation. Including a Scheme of Insurance of the Cane Crop against Damage Caused by Cyclones," Arthur L.

Humphreys, London, 1910.

[25] "Bulletin des lois de la République Française," Vol. 68 and 72, Imprimerie Nationale, Paris, 1904 and 1906.

[26] "Managing Tropical Cyclones: Mauritius," Mauritius Meteorological Services, Vol. 12, United Nations Office for South-South Cooperation, 2011.

Manuscript Sources

Archives Départementales de la Réunion, Séries C, 11 C, 20 C et 22 C.

Archives Départementales de la Réunion, Série 2 C

Archives Départementales de la Réunion, Série M

(Administration Coloniale), 1 M 4076-4087

The National Archives, IOR/F/4/588/14256

The National Archives, PRO 30/43

The National Archives, CO 1069/746

16

Possible Trajectories of Agricultural Cropping Systems in China from 2011 to 2050[*]

Junfang Zhao, Jianping Guo
Chinese Academy of Meteorological Sciences, Beijing, China

ABSTRACT

Predicting the possible impacts of future climate change on cropping systems can provide important theoretical support for reforming cropping system and adjusting the distribution of agricultural production in the future. The study was based on the daily data of future B2 climate scenario (2011-2050) and baseline climate condition (1961-1990) from high resolution regional climate model PRECIS (~50 km grid interval). According to climatic divisions of cropping systems in China, the active accumulated temperature stably passing the daily average temperature of 0°C, the extreme minimum temperature and the termination date passing the daily average temperature of 20°C which were justified by dominance as a limitation of different cropping systems in zero-grade zone were investigated. In addition, the possible trajectories of different cropping systems in China from 2011 to 2050 were also analyzed and assessed. Under the projected future B2 climate scenario, from 2011 to 2050, the northern boundaries of double cropping area and triple cropping area would move northward markedly. The most of the present double cropping area would be replaced by the different triple cropping patterns, while current double cropping area would shift towards areas presently dominated by single cropping systems. Thus the shift of multiple cropping areas would lead to a significant decrease of single cropping area. Compared with China's land area, the percentage cover of single cropping area and double cropping area would decrease slowly, while percentage cover of triple cropping area would gradually increase.

Keywords: Climate Change; Agriculture in China; Northern Boundary of Cropping System

1. Introduction

Climate change, which is largely a result of burning fossil fuels, has already affected the Earth's temperature, precipitation, and hydrological cycles. Continuous changes in the frequency and intensity of precipitation, heat waves, and other extreme events are likely, and will put increasing stresses on agricultural production [1]. Furthermore, compounded climate changes can decrease plant productivity, resulting in price increases for many important agricultural crops, especially in the developing world [1]. In China, there is a growing contradiction between rapid and continuous population growth and the associated increasing demands on agricultural products associated with continuous population growth and the limited water and land resources that will likely be intensified as consequences climate change. How the climate change affects the agriculture and cropping system has attracted increasingly great attentions of scholars home and abroad [2-6].

Climate change and agriculture are interrelated processes, both of which take place on a global scale. Global warming is projected to have significant impacts on conditions affecting agriculture, including temperature, carbon dioxide, glacial run-off, precipitation and the interaction of these elements [7-9]. In general, higher CO_2 concentration increases plant productivity due to higher rates of photosynthesis and increased water use efficiency [10,11]. Increased temperatures can reduce plant productivity through heat stress and increased water demand in some areas [3-12]. Changes in temperatures and rainfall can also change the phenological requirements of future crops [12,13]. The response of cropping systems to future climate change also depends strongly on management practices, such as the type and levels of water and nutrient applications. It is well-known that water limitation tends to enhance the positive crop response to elevated CO_2, compared to well-watered conditions [14, 15]. The contrary is true for nitrogen limitation: well-

[*]Supported by the China Meteorological Administration Special Climate Change Research Fund (CCSF201346), and China Meteorological Administration Special Public Welfare Research Fund (GYHY2011-06020).

fertilised crops respond more positively to CO_2 than less fertilised ones [16,17]. Therefore, a wide range adaptation of cropping systems to climate change may exist in order to maintain or even increase crop yields under future climate change compared to current conditions.

Studies on the responses of cropping systems to climate change in China have been highlighted and made great progress [6,8,18-20]. Wang [18] simulated the potential impacts of climate change on cropping systems in China based on the regional climate change scenario for China estimated by composite GCM and found that the northern boundary of triple cropping area would shift from its current border at the Changjiang River to the Huanghe River, a shift of more than 5 degrees of latitude in the year 2050, while the current double cropping area would shift towards the central part of the present single cropping area. Zhang [19] analyzed the response of cropping systems to climate change, and pointed out the northern boundary of triple cropping area and double cropping area would move northwardly, and the cropping range, output and quality of crops would be changed in future. Yang *et al.* [6] analyzed the possible effects of climate warming on the countrywide northern limits of cropping systems, the northern limits of winter wheat and double rice, and the stable-yield northern limits of rainfed winter wheat-summer maize rotation in China from 1981 to 2007, and drew a conclusion that during the past 50 years, the climate warming caused the northwards movement of the northern limits of cropping system, and the northern limits of winter wheat and double rice. The changes might increase the unit grain yield in the changing area. However, the stable-yield northern limits of rainfed winter wheat-summer maize rotation moved southeastwards due to the decreasing rainfall. These results will contribute to better planning of resources, productivity and environmental sustainability. However, because of the complicated Chinese farming patterns, the complex social and economic environment of agricultural development and, especially, great scientific uncertainties in the prediction of climate change [21], and differences in the climate scenarios produced by different global climate models [22-24], accurately predicting the impacts of climate change on cropping systems in China is still an critical and unsolved issue in scientific community, either quantitatively or qualitatively.

The objectives of this work were: 1) to investigate the active accumulated temperature stably passing the daily average temperature of 0°C, the extreme minimum temperature and the termination date passing the daily average temperature of 20°C which are justified by dominance as a limitation of different cropping systems; 2) to find out the possible changes of different cropping systems in zero-grade zone of China from 2011 to 2050; 3) to get better understanding on the potential implications

of climate change for the agricultural sustainability in China.

2. Materials and Methods

2.1. Data

Daily climate variables (maximal and minimal air temperature, average air temperature, precipitation, solar radiation, relative humidity, and wind speed) from regional climate model PRECIS were provided by the Chinese Academy of Agricultural Sciences for time series of 40 and 30 years from 2011 to 2050 and from 1961 to 1990, respectively.

2.2. Climate Change Scenarios Selection for China

In order to estimate the impact of greenhouse gases on global climate and socio-economy, the Intergovernmental Panel on Climate Change (IPCC) developed the emission scenarios of global greenhouse gas (SRES scenarios) in the next 100 years based on analyzing a large number of models [25]. The four SRES scenarios represent different world futures, which describe the ways in which global population, economies and non-climate policies may evolve over the coming decades. Even though the scenarios include a multidimensional space and no simple metric can be used to classify them it has been useful to describe the scenarios using two dimensional spaces. The first dimension designates a more economic (A) or a more environmental (B) orientation and the second dimension a more global (1) or a more regional (2) orientation. Accordingly, the scenarios are termed as A1, A2, B1, and B2.

In this study, regional climate scenarios were generated using a relatively high resolution (~50 km grid size) atmospheric regional model (PRECIS—Providing Regional Climates for Impacts Studies, [26-28]). The data of future B2 scenario from PRECIS model outputs, which fitted broadly with China's national social and economic development plans over the medium to long term, were selected in this study. According to PRECIS scenario output, air temperature and precipitation in 2050s would increase by 2.4°C and 6%, respectively, compared to a baseline period (1961-1990) under future B2 climate scenario.

2.3. Indices for Agricultural Cropping System

Indices for agricultural cropping system were based on existing climatic divisions of cropping systems in China (**Table 1**). According to climatic divisions of cropping systems in China, the cropping systems in zero-grade zone were divided by the heat, while the cropping systems in first zone and secondary zone were divided by

Table 1. Indices for the zero-grade zone in climatic divisions of cropping systems in China.

Cropping patterns	Accumulated temperature stably passing the daily average temperature of 0°C (°C·d)	Extreme minimum temperature (0°C)	Termination date stably passing the daily average temperature of 20°C
Single cropping area	>4000 - 4200	<−20	Early August-Early September
Double cropping area	>4000 - 4200	<−20	Early September-Beginning in late September
Triple cropping area	>5900 - 6100	>−20	Beginning in late September-Early November

heat, moisture, topography and plant [6]. This study was mainly focused on the changes of different cropping systems in zero-grade zone caused by climate change. The indices largely were determined by dominance of the active accumulated temperature stably passing the daily average temperature of 0°C, the extreme minimum temperature and the termination date passing the daily average temperature of 20°C as a limitation of different cropping patterns.

The 5-day gliding average method was used to determinate the initial and final dates of stably passing certain threshold temperature [29].

3. Results

3.1. Northern Boundaries of Cropping Systems under Baseline Period

The main cropping patterns were defined by the number of different crops that grew successively and successfully during a single growing season. In order to illustrate present distribution and possible change of northern boundary of cropping system in future, here we used cropping system rather than individual crop variety as an integrated unit. According to climatic divisions of cropping systems in China, the major Chinese cropping patterns in zero-grade zone were determined with active accumulated air temperature stably passing the daily average temperature of 0°C, the extreme minimal air temperature and the termination date passing the daily mean air temperature of 20°C (see **Table 1**).

The results illustrated that the distributions of northern boundaries of double and triple cropping systems in China under baseline period (**Figure 1**).

The active accumulated temperature values of 4000°C and 5900°C stably passing the daily average temperature of 0°C were significant in distinguishing between the single-double and double-triple cropping systems respectively. The vast western area was dominated by single cropping pattern owing to lower temperature and earlier

termination date stably passing the daily average temperature of 20°C. Generally speaking, this simulated map was well fitted with the actual distribution of cropping patterns in China and relevant findings from other researches [18].

3.2. Potential Changes of Northern Boundaries of Cropping Pattern

There were potential changes of northern boundaries of double and triple cropping areas projected by the regional climate model PRECIS every 10 years from 2011 to 2050, respectively, compared to the baseline period (1961-1990) (**Figures 2** and **3**).

Great changes might occur almost everywhere in China except for the high altitude Qinghai-Xizang Plateau in southwestern China and the northern most part of Northeast China. The northern boundaries of double and triple cropping areas might shift northward in different degrees with the increased accumulated temperature from 2011 to 2050. For the northern boundary of double cropping areas, the larger spatial displacement might occur over provinces of Shannxi, Shanxi, Hebei, Beijing, Liaoning, Jilin, Gansu and Sichuan (**Figure 2**).

Especially, the average moving distance of northern boundary in Shaanxi, Shanxi and Hebei Province would be about 220 km during the 2041-2050, and the northern boundary in south of Liaoning Province might shift from the 38°43'N to 46°8'N in Jilin Province. For the northern boundary of triple cropping areas, the larger spatial displacement might occur in provinces of Sichuan, Yunnan, Hunan, Hubei, Anhui, Guizhou, Zhejiang, Chongqing, and Jiangsu Province (**Figure 3**). For example, the northern boundary might shift from 27°24'N in Zhejiang Province (1961-1990) to 33°26'N in Jiangsu Province (2041-2050), and the average moving distance of northern boundary in Hubei Province would be 229 km. The results showed that the fractional cover of multiple cropping areas to China's land area might change significantly under B2 climate scenarios without changing the variety and level of future production (**Figure 4**). Compared with China's land area, the percentage cover of single cropping area and double cropping area to China's land area would decrease slowly, while the percentage cover of triple cropping area would gradually increase. The percentage cover of single cropping area was 71.7% in 1961-1990, but would likely change to 69.8% in 2011-2020, 69.6% in 2021-2030, 68.4% in 2031-2040, and 66.6% in 2041-2050, respectively. The percentage covers of double cropping area were 17.6% (1961-1990), 16.3% (2011-2020), 15.4% (2021-2030), 12.8% (2031-2040), and 13.3% (2041-2050), respectively. The percentage covers of triple cropping area were 10.6% (1961-1990), 13.9% (2011-2020), 15.0% (2021-2030), 18.8% (2031-2040), and 20.1% (2041-2050), respec-

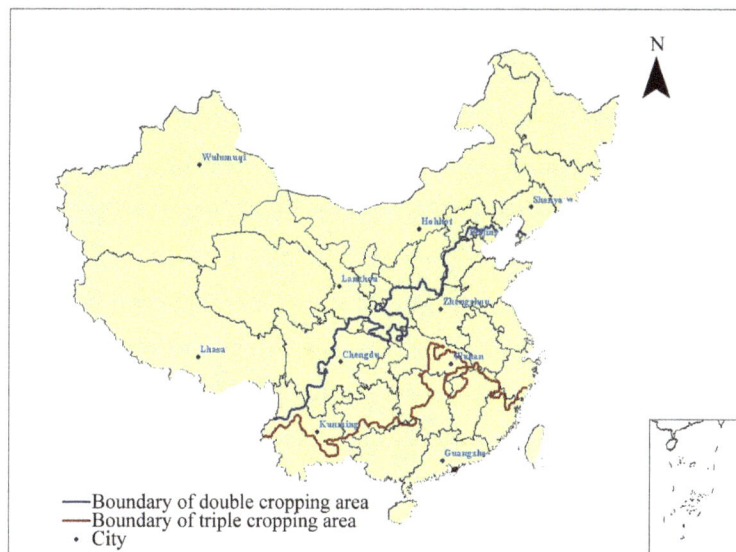

Figure 1. The average northern boundaries of double and triple cropping areas from 1961 to 1990 under future B2 climate change.

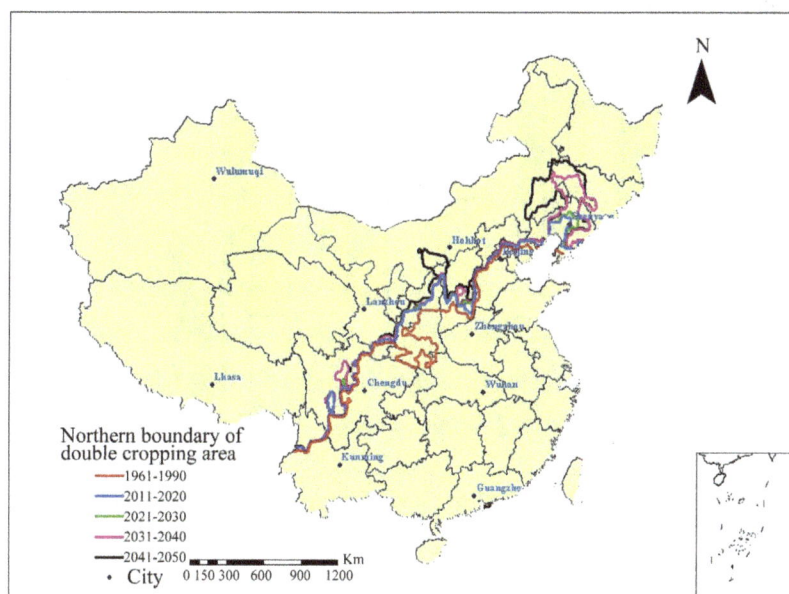

Figure 2. The possible changes of northern boundaries of double cropping areas every ten years from 2011 to 2050 under future B2 climate change compared to baseline period (1961-1990).

tively.

4. Conclusions and Discussion

Possible trajectories of different cropping systems in China from 2011 to 2050 were simulated and assessed based on the daily data of future B2 climate scenario (2011-2050) and baseline period (1961-1990) from regional climate model PRECIS at spatial resolution of 50 km × 50 km. Our study indicated that major changes would occur almost everywhere in China except for the Qinghai-Xizang Plateau in southwestern China and the

northern most part of Northeast China. Under the projected future B2 climate scenario, from 2011 to 2050, the northern boundaries of double cropping area and triple cropping area would move northward markedly. For the northern boundaries of double cropping areas, great spatial displacement might occur in provinces such as Shannxi, Shanxi, Hebei, Beijing, Liaoning, Jilin, Gansu and Sichuan. For the northern boundaries of triple cropping areas, major spatial displacement might occur in provinces of Sichuan, Yunnan, Hunan, Hubei, Anhui, Guizhou, Zhejiang, Chongqing, and Jiangsu. The most parts of the present double cropping area would be re-

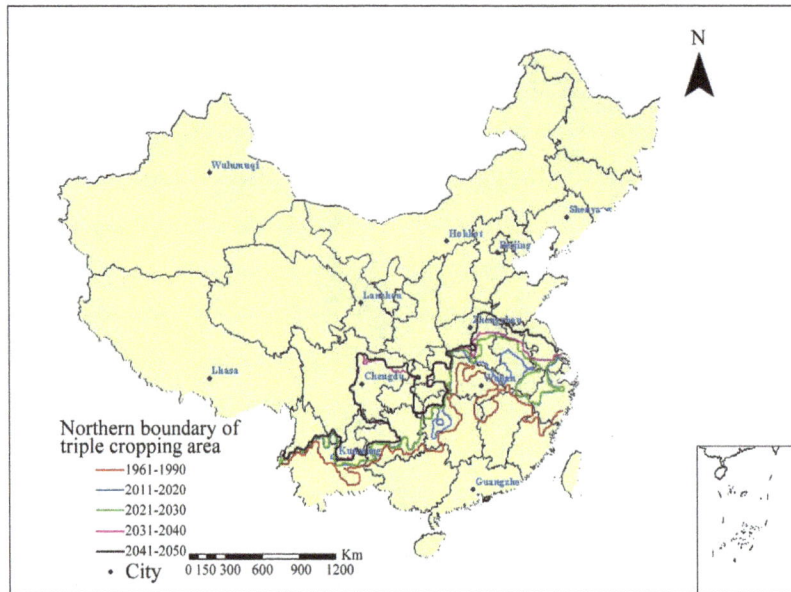

Figure 3. The possible changes of northern boundaries of triple cropping areas every ten years from 2011 to 2050 under future B2 climate change compared to baseline period (1961-1990).

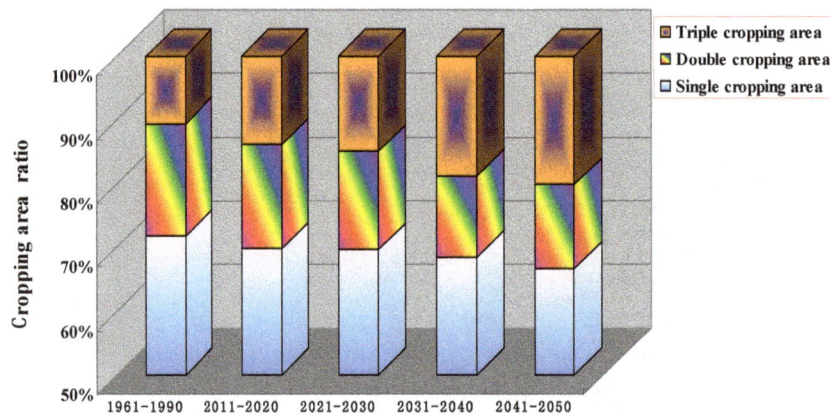

Figure 4. The percentage covers of single, double and triple cropping areas to China's land area under future B2 climate change.

placed by triple cropping systems, while the current double cropping area would shift towards the present single cropping area. Thus the shift of multiple cropping areas would lead to a significant decrease of single cropping area. The percentage covers of single cropping area to China's land area were 71.7% in 1961-1990, but will likely change to 69.8% in 2011-2020, 69.6% in 2021-2030, 68.4% In 2031-2040, and 66.6% in 2041-2050, respectively. The percentage covers of double cropping area to China's land area were 17.6% (1961-1990), 16.3% (2011-2020), 15.4% (2021-2030), 12.8% (2031-2040), and 13.3% (2041-2050), respectively. The percentage covers of triple cropping area to China's land area were 10.6% (1961-1990), 13.9% (2011-2020), 15.0% (2021-2030), 18.8% (2031-2040), and 20.1% (2041-2050), respectively. In summary, this study pro-

vides further evidence for reforming cropping system and making decision of agricultural sustainable development planning in China under climate change. The exact dynamics of these changes require further investigation with both modeling and field-based studies.

This study mainly focuses on the possible impacts of climate change on northern boundaries of cropping systems in zero-grade zone of China, which provides a scientific basis for the cropping system's change. The results of this study are supported by the recent reports [6,18]. However, there is a discrepancy between our result and the conclusion of Zhang [19] that she figured out the ratio of single cropping area would decrease, while the ratios of double and triple cropping areas would gradually increase. The possible reason for this difference is Zhang used the national cultivated area instead of land area. However, with expansion of population and improvement in living standard, there is an increase in demand for grains in Mainland China; in contrast the area of cultivated lands tends to decrease with time [4].

But whether the potential changes in cropping systems induced by climate warming can become a reality is largely determined by factors such as water, economic benefits, and social benefits etc. Take the precipitation which is one of main climatic factors limiting agricultural production in northern China as an example; multiple cropping planting will require more water, which is incompatible between water supply and demand. In theory, the northern boundary of double cropping area will move northward significantly, that is to say the northern boundary of winter wheat will move northward drastically. However, the climate in the northern China is characterized by scarce rainfall in autumn and spring, and vast amount of water that winter wheat required for normal growth must rely on irrigation from groundwater, which will lead to declining of water table, and further affecting water use of harvest crops in autumn. Therefore, moving northward of multiple cropping systems is not simple issue and still should be cautious.

In this paper, we try to use the indices of cropping systems in zero-grade zone of China from previous research results [6] because it's better performance in comparison with other methods. However, new indicators for agricultural cropping system nationwide under climate change are notoriously difficult to get. Uncertainties may be involved in estimating the impacts of future climate change on cropping systems. Therefore, in order to get a more precise estimation, it is necessary to carry out more detailed studies, and to further adjust indices of cropping systems. In addition, our study is mainly focused on the changes of different cropping systems in zero-grade zone divided by the heat under climate change. Restrictions of variety characteristics, water levels, economic conditions etc. are not sufficiently considered. Large uncertainties on the boundaries of the predictions exist. Differences in the daily characteristics of climate variables and differences between ensemble experiments will also affect the overall results. None the less, there is a broad agreement that, in addition to increased temperature, climate change will bring about regionally dependent increase or decrease in rainfall, an increase in cloud cover and increases in sea level. Extreme weather events will also increase in intensity or frequency, such as higher maximum temperatures, more intense precipitation events, increased risk and duration of drought, and increased peak wind intensities of cyclones [30-33]. However, climate models, either global or regional, are unlikely to capture fully the spatial and temporal detail of many extreme events across China. This is due to the imperfect understanding of their physical causes and limitations

related to model structure. Thus the impacts of climate fluctuations and extremes on changes in cropping systems are not taken into account in this paper.

REFERENCES

[1] IPCC, "Climate Change 2007: The Physical Science Basis, Summary for Policy Makers," IPCC WGI 4th Assessment Report, Paris, 2007.

[2] M. Cao, S. Ma and C. Han, "Potential Productivity and Human Carrying Capacity of an Agro-Ecosystem: An Analysis of Food Production Potential of China," *Agricultural Systems*, Vol. 47, No. 4, 1995, pp. 387-414.

[3] F. L. Tao, M. Yokozawa, Y. Hayashi and E. D. Lin, "Changes in Agricultural Water Demands and Soil Moisture in China over the Last Half-Century and Their Effects on Agricultural Production," *Agricultural and Forest Meteorology*, Vol. 118, No. 3-4, 2003, pp. 251-261.

[4] J. K. Zhang, F. R. Zhang, D. Zhang, D. X. He, L. Zhang, C. G. Wu and X. B. Kong, "The Grain Potential of Cultivated Lands in Mainland China in 2004," *Land Use Policy*, Vol. 26, No. 1, 2008, pp. 68-76.

[5] W. Xiong, D. Conway, E. D. Lin, Y. L. Xu, H. Ju, J. H. Jiang, I. Holman and Y. Li, "Future Cereal Production in China: The Interaction of Climate Change, Water Availability and Socio-Economic Scenarios," *IOP Conference Series*: *Earth and Environmental Science*, Vol. 6, 2009, pp. 34-44.

[6] X. G Yang, Z. J. Liu and F. Chen, "The Possible Effects of Global Warming on Cropping Systems in China I. The Possible Effects of Climate Warming on Northern Limits of Cropping Systems and Crop Yields in China," *Scientia Agricultura Sinica*, Vol. 43, No. 2, 2010, pp. 329-336.

[7] C. Rosenzweig and D. Hillel, "Climate Change and the Global Harvest," Oxford University Press, Oxford, 1998.

[8] N. T. Francesco, D. Marcello, C. Rosenzweig and O. S. Claudio, "Effects of Climate Change and Elevated CO_2 on Cropping Systems: Model Predictions at Two Italian Locations," *European Journal of Agronomy*, Vol. 13, No. 2-3, 2000, pp. 179-189.

[9] S. D. Gryze, A. Wolf, S. R. Kaffka, J. Mitchell, D. E. Rolston, S. R. Temple, J. Lee and J. Six, "Simulating Greenhouse Gas Budgets of four California Cropping Systems under Conventional and Alternative Management," *Ecological Applications*, Vol. 20, No. 7, 2010, pp. 1805-1819.

[10] F. Ewert, D. Rodriguez, P. Jamieson, M. A. Semenov, R. A. C. Mitchell, J. Goudriaan, J. R. Porter, B. A. Kimball, P. J. Pinter, R. Manderscheid, H. J. Weigel, A. Fangmeie, E. Fereres and F. Villalobos, "Effects of Elevated CO_2 and Drought on Wheat: Testing Crop Simulation Models for Different Experimental and Climatic Conditions," *Agriculture, Ecosystems & Environment*, Vol. 93, No. 1-3, 2002, pp. 249-266.

[11] P. J. Gregory, S. N. Johnson, A. C. Newton and J. S. I. Ingram, "Integrating Pests and Pathogens into the Climate Change/Food Security Debate," *Journal of Experimental Botany*, Vol. 60, No. 10, 2009, pp. 2827-2838.

[12] D. W. Lawlor and R. A. C. Mitchell, "Crop Ecosystem Responses to Climatic Change: Wheat. Climate Change and Global Crop Productivity," CAB International, Cambridge, 2000, pp. 57-80.

[13] F. N. Tubiello, M. Donatelli, C. Rosenzweig and C. O. Stockle, "Effects of Climate Change and Elevated CO_2 on Cropping Systems: Model Predictions at Two Italian Locations," *European Journal of Agronomy*, Vol. 13, No. 2-3, 2000, pp. 179-189.

[14] U. N. Chaudhuri, M. B. Kirkam and E. T. Kanemasu, "Root Growth of Winter Wheat under Elevated Carbon Dioxide and Drought," *Crop Science*, Vol. 30, No. 4, 1990, pp. 853-857.

[15] B. A. Kimball, P. J. J. Pinter, R. L. Garcia, R. L. LaMorte, G. W. Wall, D. J. Hunsaker, G. Wechsung, F. Wechsung and T. Kartschall, "Productivity and Water Use of Wheat under Free-Air CO_2 Enrichment," *Global Change Biology*, Vol. 1, No. 6, 1995, pp. 429-442.

[16] N. Sionit, D. A. Mortensen, B. R. Strain and H. Hellmers, "Growth Response of Wheat to CO_2 Enrichment and Different Levels of Mineral Nutrition," *Agronomy Journal*, Vol. 73, No. 6, 1981, pp. 1023-1027.

[17] R. A. C. Mitchell, V. J. Mitchell, S. P. Driscoll, J. Franklin and D. W. Lawlor, "Effects of Increased CO_2 Concentration and Temperature on Growth and Yield of Winter Wheat at Two Levels of Nitrogen Application," *Plant, Cell & Environment*, Vol. 16, No. 5, 1993, pp. 521-529.

[18] F. T, Wang, "Impact of Climate Change on Cropping System and Its Implication for Agriculture in China," *Acta Meteorological Sinica*, Vol. 11, No. 4, 1997, pp. 407-415.

[19] H. X. Zhang, "The Response of China's Cropping Systems to Global Climatic Changes I. The Effect of Climatic Changes on Cropping Systems in China," *Chinese Journal of Agrometeorology*, Vol. 21, No. 1, 2000, pp. 9-13.

[20] T. Axel, "Agricultural Irrigation Demand under Present and Future Climate Scenarios in China," *Global and Planetary Change*, Vol. 60, No. 3-4, 2008, pp. 306-326.

[21] J. Gupta, X. Olsthoorn and E. Rotenberg, "The Role of Scientific Uncertainty in Compliance with the Kyoto Protocol to the Climate Change Convention," *Environmental Science & Policy*, Vol. 6, No. 6, 2003, pp. 475-486.

[22] W. E. Easterling, L. O. Mearns, C. J. Hays and D. Marx, "Comparison of Agricultural Impacts of Climate Change Calculated from High and Low Resolution Climate Change Scenarios: Part II. Accounting from Adaptation and CO_2 Direct Effects," *Climate Change*, Vol. 51, No. 2, 2001, pp. 173-197.

[23] F. N. Tubiello, C. Rosenzweig, R. A. Goldberg, S. Jagtap and J. W. Jones, "Effects of Climate Change on US Crop Production: Simulation Results Using Two Different GCM Scenarios. Part I: Wheat, Potato, Maize and Citrus," *Climate Research*, Vol. 20, No. 3, 2002, pp. 259-270.

[24] E. A. Tsvetsinskaya, L. O. Mearns, T. Mavromatis, W. Gao, L. McDaniel and M. W. Downton, "The Effect of Spatial Scale of Climatic Change Scenarios on Simulated Maize, Winter Wheat and Rice Production in the Southern United States," *Climate Change*, Vol. 60, No. 1-2, 2003, pp. 37-72.

[25] N. Nakićenović, J. Alcamo, G. Davis, B. de Vries, J. Fenhann, S. Gaffin, K. Gregory, A. Grubler, T. Y. Jung, T. Kram, E. Emilio la Rovere, L. Michaelis, S. Mori, T. Morita, W. Pepper, H. Pitcher, L. Price, K. Riahi, A. Roehrl, H.-H. Rogner, A. Sankovski, M. E. Schlesinger, P. R. Shukla, S. Smith, R. J. Swart, S. van Rooyen, N. Victor and Z. Dadi, "Special Report on Emissions Scenarios," Cambridge University Press, Cambridge, 2000.

[26] R. G. Jones, M. Noguer, D. C. Hassell, D. Hudson, S. S. Wilson, G. J. Jenkins and J. F. B. Mitchell, "Generating High Resolution Climate Change Scenarios using PRECIS," Met Office Hadley Centre, Exeter, 2004, p. 35.

[27] Y. L. Xu, *Setting up PRECIS over China to Develop Regional SRES Climate Change Scenarios*," *Proceedings of the International Workshop: Prediction of Food Production Variation in East Asia under Global Warming*, Tsukuba, 2004, pp. 17-21.

[28] Y. L. Xu, X. Y. Huang, Y. Zhang, Z. P. Wen and W. B. Li, "Validating PRECIS' Capacity of Simulating Present Climate over South China," *Acta Scientiarum Naturalium Unversitatis Sunyatseni*, Vol. 46, No. 5, 2007, pp. 93-97.

[29] J. F. Zhao, J. P. Guo, Y. P. Ma, Y. H. E, P. J. Wang and D. R. Wu, "Change Trends of China Agricultural Thermal Resources under Climate Change and Related Adaptation Countermeasures," *Chinese Journal of Applied Ecology*, Vol. 21, No. 11, 2010, pp. 2922-2930.

[30] J. A. Dracup, K. S. Lee and E. G. Paulson, "On the Definition of Droughts," *Water Resources Research*, Vol. 16, No. 2, 1980, pp. 297-302.

[31] M. Lal, K. K. Singh, L. S. Rathore, G. Srinivasan and S. A. Saseendran, "Vulnerability of Rice and Wheat Yields in NW India to Future Change in Climate," *Agriculture, Ecosystems and Environment*, Vol. 89, No. 2, 1998, pp. 101-114.

[32] J. Q. Zhang, "Risk Assessment of Drought Disaster in the Maize-Growing Region of Songliao Plain, China," *Agriculture, Ecosystems and Environment*, Vol. 102, No. 2, 2004, pp. 133-153.

[33] L. Y. Zhong, L. M. Liu and Y. B. Liu, "Natural Disaster Risk Assessment of Grain Production in Dongting Lake Area, China," *Agriculture and Agricultural Science Procedia*, Vol. 1, 2010, pp. 24-32.

17

Future Changes in Drought Characteristics over Southern South America Projected by a CMIP5 Multi-Model Ensemble

Olga C. Penalba[1], Juan A. Rivera[1,2]
[1]Department of Atmospheric and Oceanic Sciences, University of Buenos Aires, Buenos Aires, Argentina
[2]National Council of Scientific and Technical Research, Buenos Aires, Argentina

ABSTRACT

The impact of climate change on drought main characteristics was assessed over Southern South America. This was done through the precipitation outputs from a multi-model ensemble of 15 climate models of the Coupled Model Inter-comparison Project Phase 5 (CMIP5). The Standardized Precipitation Index was used as a drought indicator, given its temporal flexibility and simplicity. Changes in drought characteristics were identified by the difference for early (2011-2040) and late (2071-2100) 21st century values with respect to the 1979-2008 baseline. In order to evaluate the multi-model outputs, model biases were identified through a comparison with the drought characteristics from the Global Precipitation Climatology Centre database for the baseline period. Future climate projections under moderate and high-emission scenarios showed that the occurrence of short-term and long-term droughts will be more frequent in the 21st century, with shorter durations and greater severities over much of the study area. These changes in drought characteristics are independent on the scenario considered, since no significant differences were observed on drought changes. The future changes scenario might be even more dramatic, taking into account that in most of the region the multi-model ensemble tends to produce less number of droughts, with higher duration and lower severity. Therefore, drought contingency plans should take these results into account in order to alleviate future water shortages that can have significant economic losses in the agricultural and water resources sectors of Southern South America.

Keywords: CMIP5 Models; Drought; Standardized Precipitation Index; Climate Change; Southern South America

1. Introduction

Among extreme meteorological events, droughts are possibly the most slowly developing ones, that often have the longest duration, and at the moment the least predictability among all atmospheric hazards [1]. In the last 20 years, 1 billion people worldwide were affected by droughts [2]. Southern South America (SSA) was no exception to this hazard, whose impacts were evident in the reduction in crop yields, reduced cattle products, streamflow deficiencies and consequently problems for hydroelectric power generation.

Climate change refers to any change in climate over time, whether due to natural variability or as a result of human activity [3]. These changes can lead to changes in the statistical properties of the distribution of the variable considered, as changes in their mean values and its am-plitude or variability. Changes in precipitation variability can include more frequent and damaging extreme events such as drought [4]. Therefore, climate change is expected to primarily affect the frequency and severity of droughts. However, most of the research in SSA concentrated on changes in the mean state of the climate rather than in changes in its temporal variability, and the impact of climate change on drought characteristics remains partially unknown. Several works have evidenced the occurrence of positive trends in precipitation totals at different time scales during the second half of the 20th century over most of SSA [5-7]. In agreement to these variations, a decrease in the annual number of dry days was observed over a great portion of Argentina [8]. These trends exhibited a high degree of non-linearity, and some regions presented a reversion in their sign after 1990s [8,9]. Hence, climate model outputs are necessary in order to evaluate if modeled precipitation shows these low-frequency changes and if these trends' reversion will continue during the 21st century, given the importance of these trends in the economy of the region. If a return to dry conditions is projected, droughts can become a more frequent hazard, and its severity and duration characteristics can be altered as well. Therefore, the quantitative knowledge of the characteristics of droughts in the region is an important aspect of the planning and management of agricultural practices and water resources.

Taking this into consideration, the aim of this research is to evaluate how well a multi-model ensemble from the Coupled Model Intercomparison Project Phase 5 (CMIP5) Global Climate Models (GCMs) represents the drought characteristics in SSA, and to evaluate future changes in drought frequency, duration and severity. Given the rising demand on water resources, governments and water agencies will face increased planning for drought alleviation. Therefore it is important to identify if climate change will aggravate water issues by changing drought characteristics at a regional level.

2. Data

2.1. Reference Data

Observed monthly rainfall totals were obtained from the Global Precipitation Climatology Centre (GPCC) Full Data Reanalysis v6 gridded at $1° \times 1°$ resolution [10]. This dataset spans the Atmospheric Model Intercomparison Project (AMIP) period (1979-2008). The study area corresponds to the portion of South America south of 19°S, which comprises 617 grid points excluding the southern oceans. The GPCC dataset was selected because its agreement with the spatial and temporal patterns of rain gauge data over the region. This is evident especially over the La Plata Basin, one of the largest basins and producers of hydroelectric power in the world, which is located in the central-eastern portion of SSA. Moreover, several studies used the GPCC dataset in order to characterize precipitation main features in the region [11,12] which also confirm its appropriateness.

2.2. Model Outputs

Assessments of future drought have traditionally used only few climate models to assess possible impacts [13]. Because the outputs of GCMs vary widely within the same scenarios, the use of GCM ensemble means with some acknowledgement of the uncertainty in ensemble outputs has become a standard practice in climate science research [14]. In this work we used an ensemble of monthly modeled precipitation data from 15 GCMs belonging to the CMIP5 [15]. **Table 1** lists the selected models used in this study, with their respective modeling groups. Criteria for the selection of models were based

Table 1. List of the 15 GCMs considered for the multi-model ensemble.

Model	Institute (country)	Resolution (Lat × Lon)
ACCESS1-0	Commonwealth Scientific and Industrial Research Organization (Australia)	1.24° × 1.88°
BCC-CSM1-1	Beijing Climate Center, China Meteorological Administration (China)	2.81° × 2.81°
BNU-ESM	College of Global Change and Earth System Science, Beijing Normal University (China)	2.81° × 2.81°
CCSM4	National Center for Atmospheric Research (USA)	0.94° × 1.25°
CNRM-CM5	Centre National de Recherches Meteorologiques (France)	1.41° × 1.41°
CSIRO-Mk3-6-0	Commonwealth Scientific and Industrial Research Organization (Australia)	1.87° × 1.87°
GISS-E2-R	NASA Goddard Institute for Space Studies (USA)	2.0° × 2.5°
INMCM4	Institute for Numerical Mathematics (Russia)	1.5° × 2.0°
IPSL-CM5A-LR	Institut Pierre-Simon Laplace (France)	1.87° × 3.75°
IPSL-CM5B-LR	Institut Pierre-Simon Laplace (France)	1.87° × 3.75°
MIROC5	Atmosphere and Ocean Research Institute, University of Tokyo (Japan)	1.41° × 1.41°
MPI-ESM-LR	Max Planck Institute for Meteorology (Germany)	1.87° × 1.87°
MPI-ESM-MR	Max Planck Institute for Meteorology (Germany)	1.87° × 1.87°
MRI-CGCM3	Meteorological Research Institute (Japan)	1.13° × 1.13°
NorESM1-M	Norwegian Climate Centre, Norway	1.87° × 2.5°

on the availability of data, primarily for future projections. Given that most of the models have different spatial resolutions (**Table 1**), all the model outputs were regridded to $1° \times 1°$ resolution using bilinear interpolation [16]. The 15 GCM runs considered in this study cover two of the four representative concentration pathways (RCPs) designed as a new set of scenarios for the Fifth Assessment Report of the IPCC: the RCP4.5 and RCP8.5 scenarios [17]. In contrast to the SRES scenarios, RCPs represent pathways of radiative forcing, not detailed socio-economic narratives or scenarios [18], where, for example, RCP4.5 reaches a radiative forcing of 4.5 W/m^2 by the year 2100. RCP4.5 show a stabilizing CO_2 concentration, close to the median range of the existing literature; while RCP8.5 follows the upper range of available literature, with rapidly increasing concentrations [19]. Three periods of 30 years were considered in order to evaluate changes in future climate: the AMIP period (1979-2008); and projections for the early 21st century (2011-2040) and late 21st century (2071-2100).

3. Methodology

Precipitation is the primary factor controlling the formation and persistence of drought conditions, but evapotranspiration is also an important variable [20]. Given the difficulties in obtaining reliable observed and modeled evapotranspiration measures over SSA, a drought estimator based solely on precipitation totals was choosen.

In order to identify drought characteristics we used the Standardized Precipitation Index (SPI), developed by [21] for drought definition and monitoring. The SPI only requires monthly precipitation as input variable, which is a common variable in all GCMs outputs. This is a powerful, flexible index that is simple to calculate, was widely used in SSA proving to be a good estimator of both wet and dry conditions [9,22,23]. A detailed description of the calculation of the SPI can be found in [20]. A brief summary of the main assumptions for its calculation is presented. The SPI calculation for every gridpoint is based on the accumulated precipitation for a fixed time scale of interest. These series of accumulated precipitation were fitted to a gamma probability distribution, which is found to be one of the most suitable to fit the precipitation distribution in the region [24]. This procedure was performed for each time scale of interest and for each month of the year. The cumulative probability is then transformed to the standard normal random variable z with mean zero and variance of one, which is the value of the SPI [25]. In summary, the SPI quantifies the number of standard deviations that the accumulated rainfall in a given time scale deviates from the average value of a location in a particular period.

In this work we consider time scales of 3 (SPI3) and 12 (SPI12) months, which represents short-term and long-term droughts, respectively. Short-term droughts used to affect the agricultural sector, while long-term droughts have impacts on the water resources. Both sectors are extremely important in SSA. The SPI was calculated for the 3 and 12 months accumulated precipitation from the GPCC database and the simulated precipitation from the multi-model ensemble in the same time scales for the three time periods considered. Several thresholds for the SPI values can be found on literature in order to define drought events [21,26,27]. In this work we consider a drought event as the period of time where SPI values are below to -1.0, which means that precipitation departures from average conditions exceed one standard deviation. Three different parameters were used for drought characterization: a) frequency—number of droughts over the period of interest; b) duration—average duration of all drought events; and c) severity—average SPI value of all drought events. In order to compute the difference between baseline and future drought characteristics, we calculated the percentage of change of the early and late 21st century drought parameters with re-

spect to the 1979-2008 period for both RCPs and time scales considered.

In the case where accumulated precipitation time series does not fit to the gamma probability function, those gridpoints were removed from the analysis. It is expected that the spatial distribution of grid points with non-significant fits show differences among the time scales considered for the accumulation and the precipitation data considered.

4. Results

4.1. Drought Climatology and Evaluation of Multi-Model Control Simulation (1979-2008)

The main characteristics of drought events computed from SPI3 based on the CMIP5 multi-model ensemble for the 1979-2008 baseline are shown in **Figure 1**. Regarding drought frequency, most of the study area Registered between 15 and 30 droughts. The regions with higher drought frequency are located over the Argentinean Patagonia, northern Argentina and southern Brazil (**Figure 1(a)**). Those areas where the occurrence of droughts is more (less) common, corresponds with regions where its duration is shorter (longer) (**Figure 1(b)**). In average, droughts lasted between 2 and 3 months in most of the region, although in southern Chile short-term droughts can last almost 4 months. Concerning drought mean severity, higher values are located over the northwest and southwest portions of the domain (**Figure 1(c)**). Taking into account the drought categories defined for the SPI [20], this result indicates that, on average, severe droughts (SPI ≤ -1.5) were recorded in the above mentioned regions, as well as over most of Paraguay.

When this drought climatology is compared to the one obtained through the GPCC database, it appears that the CMIP5 ensemble subestimate drought frequencies over most of the study area (**Figure 2(a)**), mainly in the regions where drought frequency is low. The areas of subestimation in drought frequency are in concordance with areas with an overestimation of drought mean duration (**Figure 2(b)**). These areas are characterized by a subestimation of 20% - 40% for drought frequency and an overestimation of 20% - 60% in drought duration. There is also a subestimation of drought mean severity, of about 5% - 15%, mostly in the areas where mean severities are lower (**Figure 2(c)**). In summary, in most of SSA the CMIP5 ensemble tends to produce less number of short-term droughts, with higher duration and lower severity. There are some specific areas where the opposite occurs, with higher drought frequency followed by a decrease in their mean duration, as the case of southern Argentinean Patagonia, a portion of northern Argentina and a small region over the coast of Brazil.

Analyzing the characteristics of the long term drought,

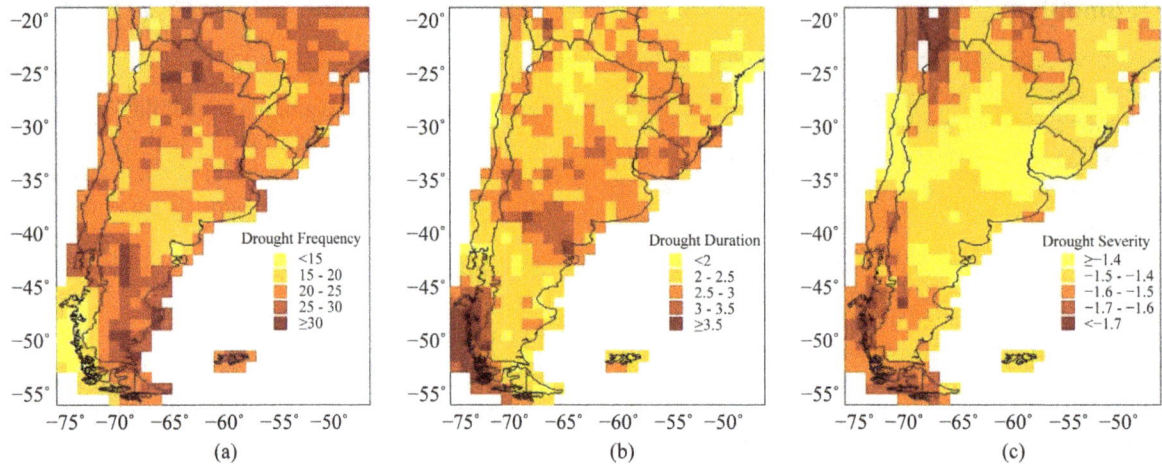

Figure 1. Drought climatology based on the multi-model ensemble for the SPI3. (a) Frequency; (b) mean duration and (c) mean severity of droughts. Period: 1979-2008. The white grid points over the continent correspond to locations where the gamma probability function does not significantly fit to the 3-month accumulated precipitation distribution.

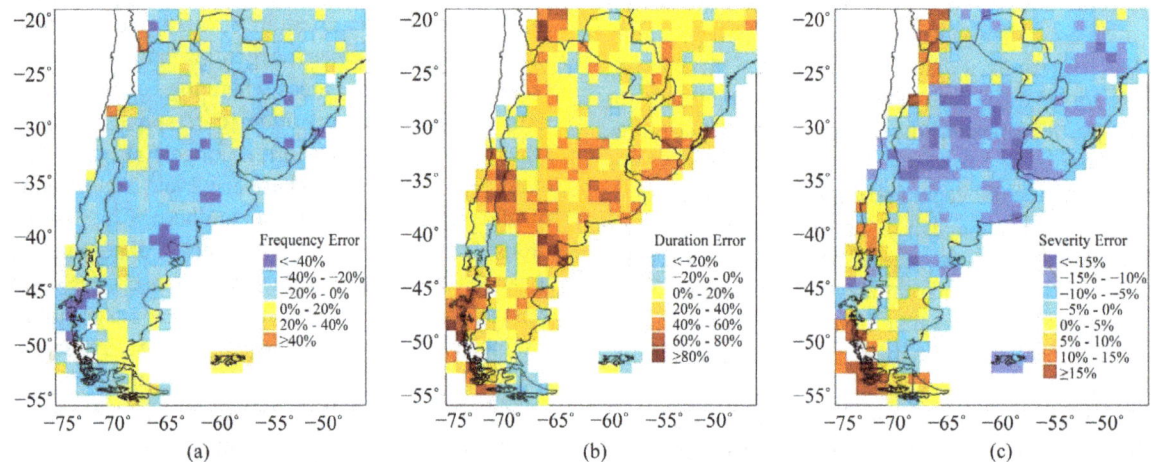

Figure 2. Percentage errors in representation of drought characteristics by the multi-model ensemble when compared with GPCC dataset for the SPI3. Period: 1979-2008.

the multi-model ensemble based on SPI12 approximately shows the same areas of maximum and minimum severities, durations and frequencies than based on SPI3 (**Figure 3**). As the time period is lengthened to 12 months, the SPI responds more slowly to changes in precipitation, therefore, droughts were less frequent (~8 - 14 droughts during 1979-2008) and lasted longer (~4 - 10 months). When these results are compared with the drought climatology obtained through the GPCC database on a time scale of 12 months, the discrepancies seems to be higher than for short term droughts (**Figure 4**). This is indicative that the multi-model ensemble skill varies with the time scale considered in the calculation of the SPI. The northwestern and southwestern portions of SSA show errors towards an increase of drought events, with shorter durations and longer severities. The central portion of the domain shows the opposite pattern, with a heterogeneous spatial structure.

The above results could be indicative that long term droughts are spatially more complex than short term ones, although their temporal pattern is less variable. The discrepancies between the multi-model ensemble and the GPCC database could respond to biases in the simulation of mean precipitation patterns in SSA, which was identified in [28,29]; and also to biases unrelated to errors in the mean patterns, like the representation of seasonal and inter-annual precipitation variabilities [30] that are important for drought development.

4.2. Projected Changes in Drought Characteristics: 2011-2040

The RCPs scenario projections for the period 2011-2040 show that, in average, the occurrence of short-term and long-term droughts will be more frequent with respect of 1979-2008 period, with shorter durations and greater

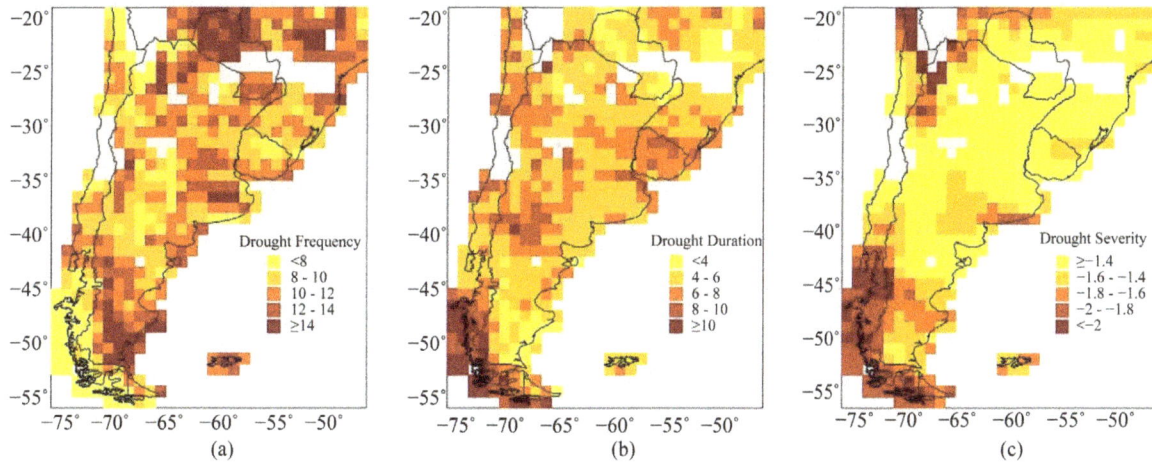

Figure 3. Same as Figure 1 for SPI12.

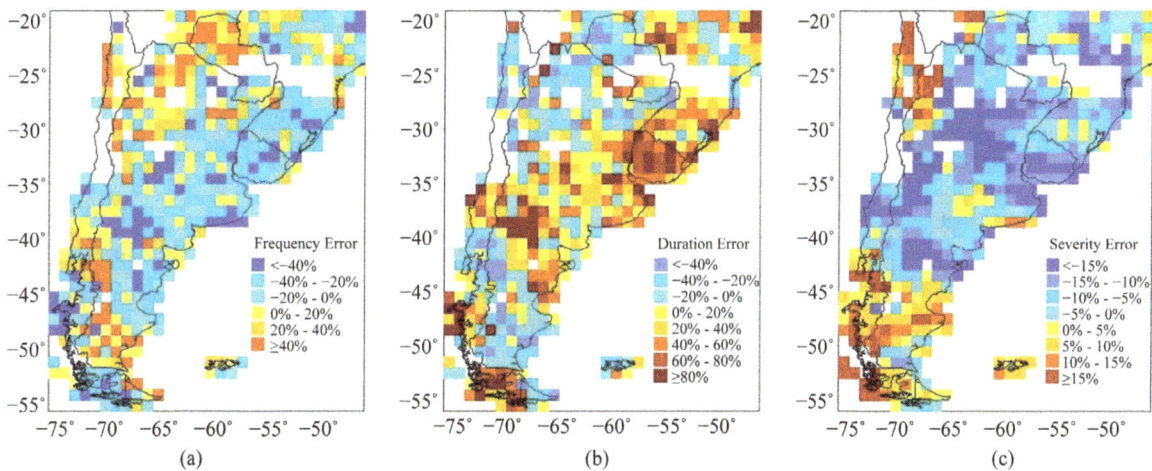

Figure 4. Same as Figure 2 for SPI12.

severities (**Figures 5** and **6**). This is indicative of a future increase in the inter-annual variability of precipitation in SSA, which was already documented for the second half of the 20th century by [7,24]. Therefore, changes in the mean state of climate, and more specifically, precipitation, will have superimposed an increase in the seasonal and inter-annual variabilities that will enhance the occurrence of precipitation extremes, in this case, droughts. Regarding SPI3, both RCP4.5 and RCP8.5 scenarios show approximately the same spatial pattern, which indicates that an increase in greenhouse gasses concentration will have little impact in future changes of short-term droughts during early 21st century (**Figure 5**). A large part of the study area will experience decreases of 10% - 30% in drought duration and increases of 10% - 60% in drought frequency. The major changes in drought severity are located in the central portions of Argentina and Chile, Uruguay and southern Brazil, with a well defined increase. Severity increases will be between 5% and 15% for the central portion of SSA.

In the case of long-term droughts, the regional pattern of changes is not as well defined as for short-term droughts (**Figure 6**). This could be related with the spatial structure of its climatology (**Figure 3**) and also can be associated to model uncertainties in the representation of precipitation accumulated over longer time scales. The expected changes in long-term drought parameters are greater than the changes for short-term droughts. In the case of the RCP4.5 projection, it is expected that La Plata Basin experience more and severe droughts in the 2011-2040 period, with increases of more than 30% in drought frequency and more than 10% in drought severity. This is also verified for the RCP8.5 projection, although with a diminished magnitude in the changes. Variations according to RCP8.5 scenario show an important increase in drought severity in central-west Argentina and central Chile and in northern Patagonia. Southern Patagonia shows a clear dipole pattern, with higher (lower) drought frequencies over the Chilean (Argentinean) portion, accompanied with lower (higher)

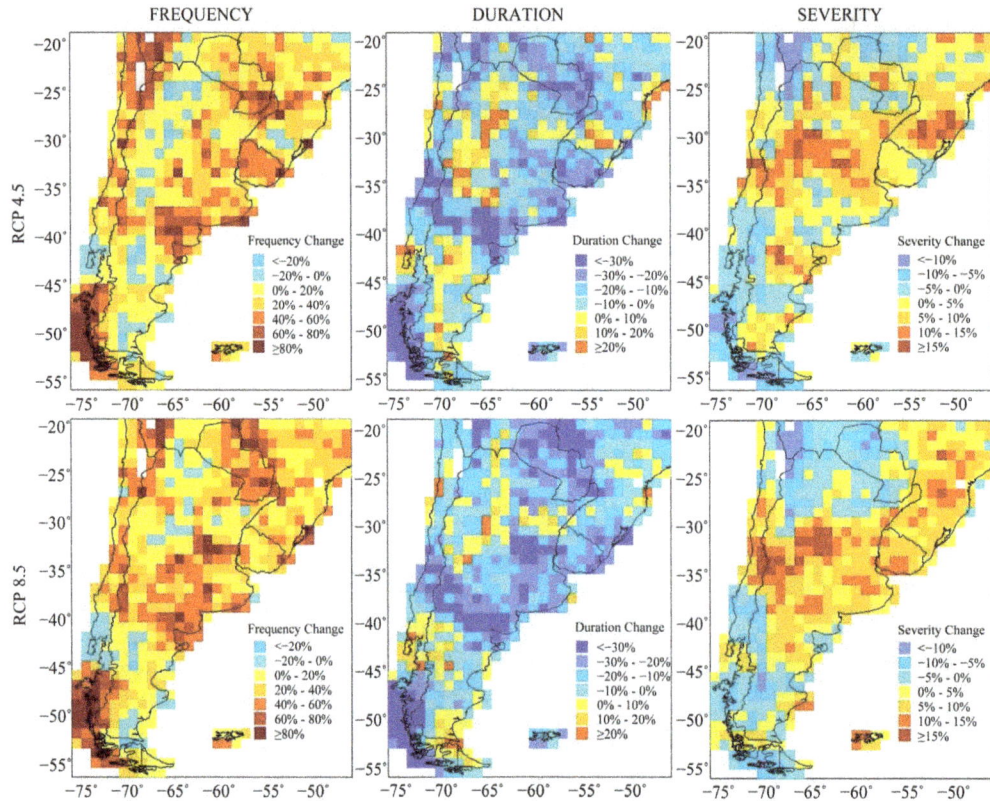

Figure 5. Mean changes in SPI3 drought characteristics projected by the multi-model ensemble for the period 2011-2040 relative to the 1979-2008 baseline along the RCP4.5 (top) and RCP8.5 (bottom) scenarios.

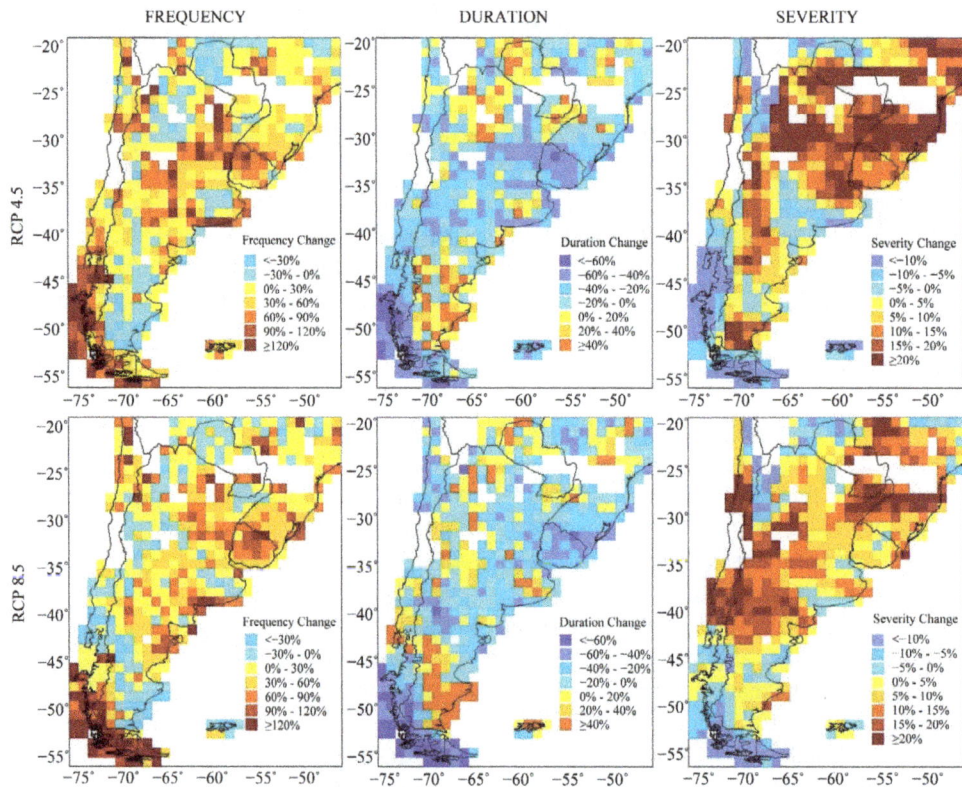

Figure 6. Same as Figure 5 for SPI12.

durations and severities (**Figure 6**).

4.3. Projected Changes in Drought Characteristics: 2071-2100

Projections for the late 21st century also show approximately the same spatial pattern of expected changes (**Figures 7** and **8**). The multi-model ensemble produces more drought events of shorter duration and greater severities over a large portion of SSA. The spatial extension of changes in short-term drought characteristics is greater for 2071-2100 than for 2011-2040 (**Figures 5** and **7**), although the magnitude of the expected changes remains similar. This is not evident in the case of long-term droughts (**Figures 6** and **8**). Major changes will be experienced over central Argentina and Uruguay in all the SPI3 drought parameters (**Figure 7**). Changes in long-term drought severity are important in magnitude and spatial extension, and are located over the central and northeastern portions of SSA (**Figure 8**). In the case of drought duration, higher uncertainty exists at a regional

scale for long-term droughts, with a noisy spatial pattern. As in the case of the projections for 2011-2040, both scenarios show the same changes for long-term drought characteristics, although increases in drought severity are larger under future scenario RCP4.5 than future scenario RCP8.5 (**Figures 7** and **8**).

5. Summary and Conclusions

This study used the SPI as a drought indicator, in order to evaluate the present day simulations of a CMIP5 multi-model ensemble for the AMIP period (1979-2008) against observations using the GPCC $1° × 1°$ dataset; and to establish future changes in drought characteristics over SSA projected by two RCP scenarios for early and late 21st century.

For the 1979-2008 baseline period, the CMIP5 ensemble tends to produce less number of droughts over most of SSA, with higher duration and lower severity, although there are some specific areas where the opposite occurs. The multi-model ensemble skill varies with the

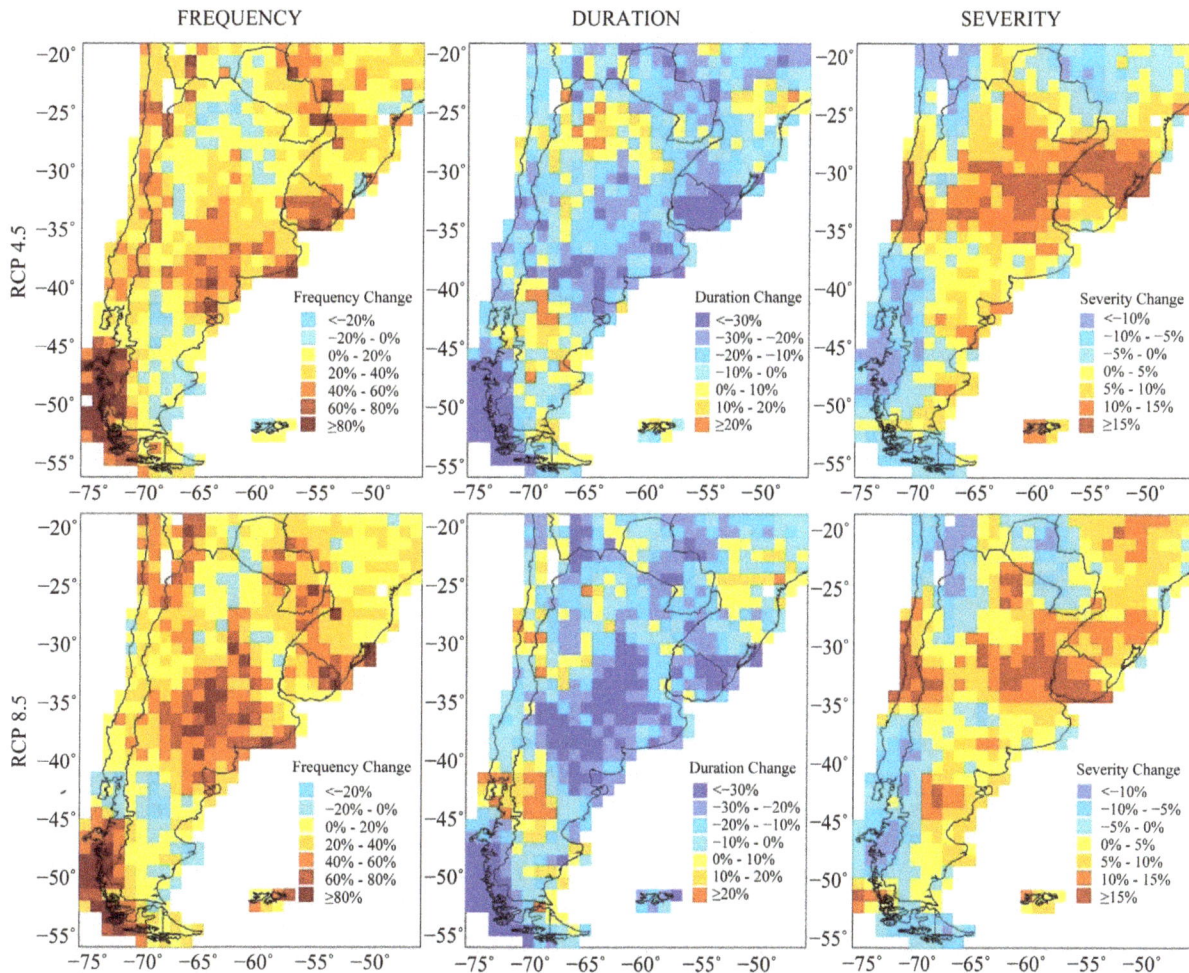

Figure 7. Mean changes in SPI3 drought characteristics projected by the multi-model ensemble for the period 2071-2100 relative to the 1979-2008 baseline along the RCP4.5 (top) and RCP8.5 (bottom) scenarios.

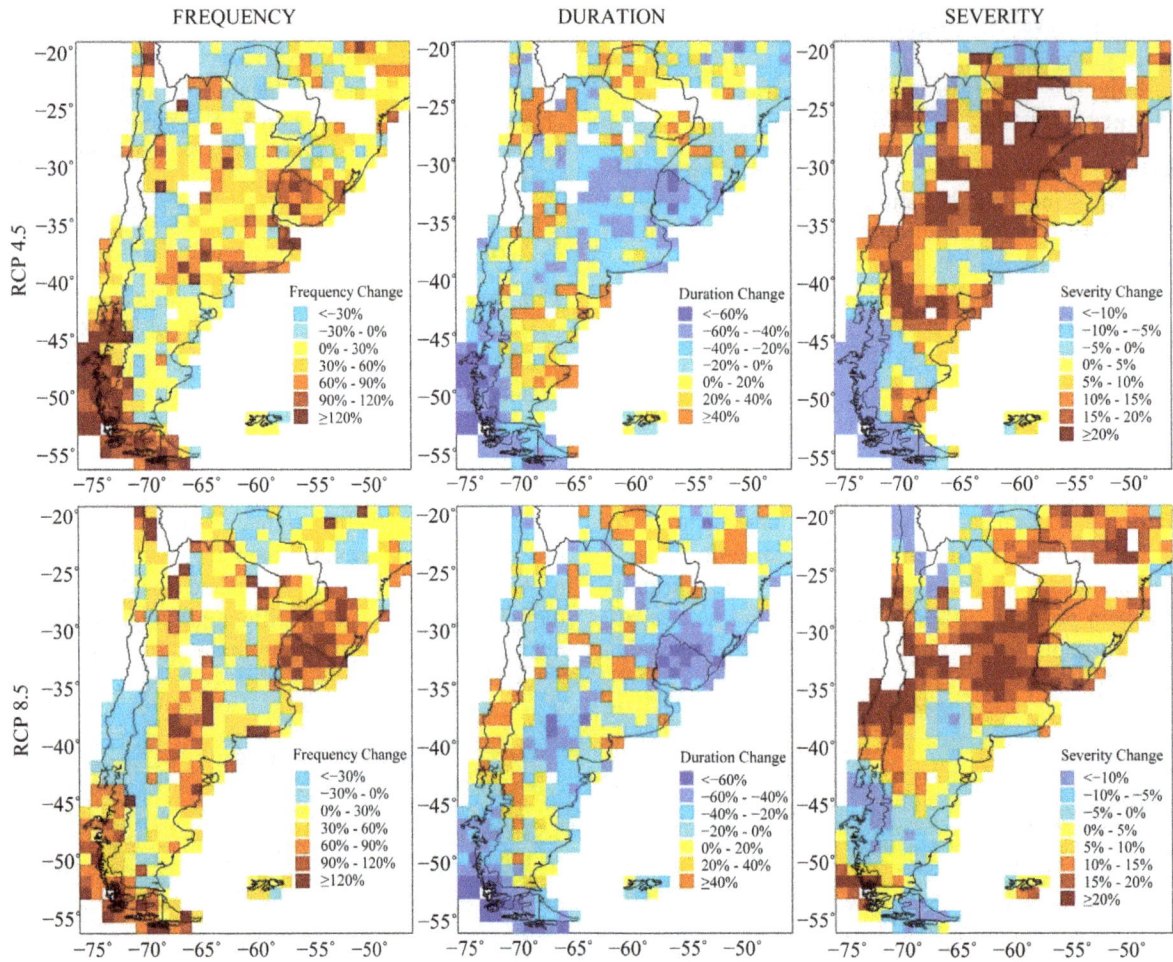

Figure 8. Same as Figure 7 for SPI12.

time scale considered in the calculation of the SPI, with larger discrepancies for the estimation of long-term drought characteristics (SPI12). Taking into account the heterogeneous spatial pattern for the SPI12 climatology, we can conclude that there is a high uncertainty in the estimation of long term droughts through the CMIP5 ensemble; although this kind of droughts could by spatially more complex than short term droughts, something already mentioned in [31]. The discrepancies between the multi-model ensemble and the GPCC database could respond to biases in the simulation of mean precipitation patterns over SSA and biases in the representation of seasonal and inter-annual variabilities, which are important for drought development. A bias correction could be applied to the precipitation data for each of the model outputs, but, as stated in [32], this is a troubling procedure and it is not known whether the same bias correction to model outputs will be valid in a future climate.

Future projections indicate that, under the two RCPs used, climate change will have large effects on drought characteristics over SSA. The occurrence of short-term and long-term droughts will be more frequent in the 21st

century, with shorter durations and greater severities over much of SSA. The increases in drought frequency changes will be about 10% - 30%; accompanied with increases of 5% - 15% in the mean drought severity and a decrease of 10% - 30% in the mean drought duration. Therefore, it is expected a future increase in the seasonal and inter-annual variability of precipitation in SSA that can result in economic losses for the region if proper adaptation measures are not proposed timely. Hence, drought contingency plans should take these results into account in order to alleviate future drought effects in the agricultural and water resources sectors. The RCP4.5 and RCP8.5 scenarios will have approximately the same effect on droughts, since no significant differences are observed on drought frequencies, durations and severities. This result was already found in [33] for the SRES emission pathways at a global scale. The authors stated that the already accumulated greenhouse gases and the thermal inertia of the oceans could contribute to the increase in drought occurrence. If we consider that the multi-model ensemble tends to underestimate drought frequency, if this bias remains for the future, the panorama for the 21st century could be more dramatic.

6. Acknowledgements

This work has been supported by the projects UBA-20020100100789 from the University of Buenos Aires and CONICET PIP 227 from the National Council of Scientific and Technical Research. The authors would like to thank Hernán Bechis for his technical contribution analyzing the model outputs at an early stage.

REFERENCES

[1] A. K. Mishra and V. P. Singh, "A Review of Drought Concepts," *Journal of Hydrology*, Vol. 391, No. 1-2, 2010, pp. 202-216.

[2] UNISDR: The United Nations Office for Disaster Reduction, "Impacts of Disasters Since the 1992 Rio de Janeiro Earth Summit," 2012. http://www.unisdr.org/files/27162_2012no21.pdf

[3] IPCC, "Climate Change 2007: The Physical Science Basis," In: S. Solomon, D. Qin, M. Manning, Z. Chen, M. Marquis, K. B. Averyt, M. Tignor and H. L. Miller, Eds., *Contribution of Working Group I to the Fourth Assessment Report of the Intergovernmental Panel on Climate Change*, Cambridge University Press, Cambridge, New York, 2007, p. 996.

[4] C. A. S. Cohelo and L. Goddard, "El Niño-Induced Tropical Droughts in Climate Change Projections," *Journal of Climate*, Vol. 22, No. 23, 2009, pp. 6456-6476.

[5] J. L. Minetti, W. M. Vargas, A. G. Poblete, L. R. Acuña and G. Casagrande, "Non-linear Trends and Low Frequency Oscillations in Annual Precipitation Over Argentina and Chile," *Atmósfera*, Vol. 16, 2003, pp. 119-135.

[6] O. C. Penalba and W. M. Vargas, "Interdecadal and Interannual Variations of Annual and Extreme Precipitation Over Central-Northeastern Argentina," *International Journal of Climatology*, Vol. 24, No. 12, 2004, pp. 1565-1580.

[7] V. R. Barros, M. E. Doyle and I. A. Camilloni, "Precipitation Trends in Southeastern South America: Relationship with ENSO Phases and with Low-Level Circulation," *Theoretical and Applied Climatology*, Vol. 93, No. 1-2, 2008, pp. 19-33.

[8] J. A. Rivera, O. C. Penalba and M. L. Bettolli, "Inter-Annual and Inter-Decadal Variability of Dry Days in Argentina," *International Journal Of Climatology*, Vol. 33, No. 4, 2012, pp. 834-842.

[9] C. M. Krepper and G. V. Zucarelli, "Climatology of Water Excess and Shortages in the La Plata Basin," *Theoretical and Applied Climatology*, Vol. 102, No. 1-2, 2012, pp. 13-27.

[10] U. Schneider, A. Becker, P. Finger, A. Meyer-Christoffer, B. Rudolf and M. Ziese, "GPCC Full Data Reanalysis Version 6.0 at 1.0°: Monthly Land-Surface Precipitation from Rain-Gauges Built on GTS-based and Historic Data," 2011.

[11] S. C. Chou, J. F. Bustamante and J. L. Gomes, "Evaluation of Eta Model Seasonal Precipitation Forecasts over South America," *Nonlinear Processes in Geophysics*, Vol. 12, No. 4, 2005, pp. 537-555.

[12] D. A. Vila, L. G. G. de Goncalvez, D. L. Toll and J. R. Rozante, "Statistical Evaluation of Combined Daily Gauge Observations and Rainfall Satellite Estimates over Continental South America," *Journal of Hydrometeorology*, Vol. 10, No. 2, 2009, pp. 533-543.

[13] S. Blenkinsop and H. J. Fowler, "Changes in European Drought Characteristics Projected by the PRUDENCE Regional Climate Models," *International Journal of Climatology*, Vol. 27, No. 12, 2007, pp. 1595-1610.

[14] K. Strzepek, G. Yohe, J. Neumann and B. Boehlert, "Characterizing Changes in Drought Risk for the United States from Climate Change," *Environmental Research Letters*, Vol. 5, No. 4, 2010, pp. 1-9.

[15] K. E. Taylor, R. J. Stouffer and G. A. Meehl, "An Overview of CMIP5 and the Experiment Design," *Bulletin of the American Meteorological Society*, Vol. 93, No. 4, 2012, pp. 485-498.

[16] C. Accadia, S. Mariani, M. Casaioli, A. Lavagnini and A. Speranza, "Sensitivity of Precipitation Forecast Skill Scores to Bilinear Interpolation and a Simple Nearest-Neightbor Average Method on High-Resolution Verification Grids," *Weather and Forecasting*, Vol. 18, No. 5, 2003, pp. 918-932.

[17] R. H. Moss, *et al.*, "The Next Generation of Scenarios for Climate Change Research and Assessment," *Nature*, Vol. 463, 2010, pp. 747-756.

[18] R. K. Chaturvedi, J. Joshi, M. Jayaraman, G. Bala and N. H. Ravindranath, "Multi-Model Climate Change Projections for India under Representative Concentration Pathways," *Current Science*, Vol. 103, No. 7, 2012, pp. 791-802.

[19] K. A. Hibbard, D. P. van Vuuren and J. Edmonds, "A Primer on the Representative Concentration Pathways (RCPs) and the Coordination Between the Climate and Integrated Assessment Modeling Communities," *CLIVAR Exchanges*, Vol. 16, No. 2, 2011, pp. 12-15.

[20] B. Lloyd-Hughes and M. A. Saunders, "A Drought Climatology for Europe," *International Journal of Climatology*, Vol. 22, No. 13, 2002, pp. 1571-1592.

[21] 21T. B. McKee, N. J. Doesken and J. Kleist, "The Relationship of Drought Frequency and Duration to Time Scales," *Proceedings of the 8th Conference on Applied Climatology*, California, 17-22 January 1993, pp. 179-184.

[22] R. A. Seiler, M. Hayes and L. Bressan, "Using the Standardized Precipitation Index for Flood Risk Monitoring," *International Journal of Climatology*, Vol. 22, No. 11,

2002, pp. 1365-1376.

[23] J. A. Rivera and O. C. Penalba, "How Temporal Changes in Gamma Distribution Parameters Influence the Standardized Precipitation Index Estimation? Error Analysis in Drought Categorization in Southeastern South America," *Proceedings of the XI Argentinean Meteorology Congress*, Mendoza, 28 May-1 June 2012, CD-ROM.

[24] O. C. Penalba and J. A. Rivera, "Using the Gamma Distribution to Represent Monthly Rainfall in Southeastern South America. Spatio-Temporal Changes in its Parameters," *Proceedings of the XI Argentinean Meteorology Congress*, Mendoza, 28 May-1 June 2012, CD-ROM.

[25] D. C. Edwards and T. B. McKee, "Characteristics of 20th Century Dorught in the United States at Multiple Time Scales," Atmospheric Science Paper No. 634, Colorado State University, Fort Collins, Colorado, 1997.

[26] I. Bordi, K. Fraedrich, J.-M. Jiang and A. Sutera, "Spatio-Temporal Variability of Dry and Wet Periods in Eastern China," *Theoretical and Applied Climatology*, Vol. 79, No. 1-2, 2004, pp. 81-91.

[27] S. Morid, V. Smakhtin and M. Moghaddasi, "Comparison of Seven Meteorological Indices for Drought Monitoring in Iran," *International Journal of Climatology*, Vol. 26, No. 7, 2006, pp. 971-985.

[28] J. Blazquez and M. N. Nuñez, "Analysis of Uncertainties in Future Climate Projections for South America: Comparison of WCRP-CMIP3 and WCRP-CMIP5 Models," *Climate Dynamics*, Vol. 41, No. 3-4, 2013, pp. 1039-1056.

[29] C. Gulizia and I. Camilloni, "Comparative analysis of the ability of a set of CMIP3 and CMIP5 global climate models to represent the precipitation in South America," *International Journal of Climatology*, 2012, Unpublished.

[30] C. S. Vera, C. Junquas and L. Díaz, "Variability and Trends in Summer Precipitation in South Eastern South America through the WCRP/CMIP5 Models," *Proceedings of the XI Argentinean Meteorology Congress*, Mendoza, 28 May-1 June 2012, CD-ROM.

[31] S. M. Vicente-Serrano, "Differences in Spatial Patterns of Drought on Different Time Scales: An Anamysis of the Iberian Peninsula," *Water Resources Management*, Vol. 20, No. 1, 2006, pp. 37-60.

[32] H. J. Fowler and C. G. Kilsby, "Future Increases in UK Water Resource Drought Projected by a Regional Climate Model," *Proceedings of the BHS International Conference on Hydrology: Science & Practice for the 21st Century*, London, 12-16 July 2004, pp. 15-21.

[33] J. Sheffield and E. F. Wood, "Projected Changes in Drought Occurrence Under Future Global Warming from Multi-Model, Multi-scenario, IPCC AR4 Simulations," *Climate Dynamics*, Vol. 31, No. 1, 2008, pp. 79-105.

18

Climate Change Effect on Winter Temperature and Precipitation of Yellowknife, Northwest Territories, Canada from 1943 to 2011

Janelle Laing, Jacqueline Binyamin[*]
Department of Geography, University of Winnipeg, Winnipeg, Canada

ABSTRACT

The correlation of the Southern Oscillation Index (SOI), Pacific Decadal Oscillation (PDO), Pacific North American Oscillation (PNA), Arctic Oscillation (AO), and Scandinavia (SCAND) indices with winter (DJF) temperature and precipitation for the period of 1943 to 2011 was analyzed to study climate change and variability of Yellowknife, NWT. SOI correlated negatively with both temperature ($r = -0.14$) and precipitation ($r = -0.06$) causing colder, drier conditions during La Niña and warmer, wetter conditions during El Niño. PDO was shown to have a strong positive correlation with both temperature ($r = 0.60$) and precipitation ($r = 0.33$) causing warmer, wetter weather in the positive phase and colder, drier weather in the negative phase. PNA showed the strongest positive correlation for both temperature ($r = 0.69$) and precipitation ($r = 0.37$) causing very warm and wet conditions in the positive phase and very cold and dry conditions during the negative phase. AO correlated negatively with temperature ($r = -0.04$) and positively with precipitation ($r = 0.24$) causing colder, wetter conditions in the positive phase and warmer, drier conditions in the negative phase. Finally SCAND was shown to have a weak negative correlation with both temperature ($r = -0.10$) and precipitation ($r = -0.18$). Sunspot area showed a strong negative correlation ($r = -0.30$) with temperature and a very weak positive correlation ($r = 0.07$) with total annual precipitation. Yellowknife's average annual temperature and precipitation have increased by 2.5°C and 120 mm, respectively throughout the past 69 years.

Keywords: Yellowknife; Climate Change; Climate Variability; Climate Modes; Teleconnections; ENSO; SOI; PDO; PNA; AO; SCAND; Sunspot Area

1. Introduction

The arctic system is particularly sensitive to change, and in light of anthropogenic climate change the arctic can be seen as an indicator to such change. There is later freeze-up and earlier break-up of ice on arctic rivers and lakes [1,2], furthermore the overall extent of sea-ice is diminishing. This is uniquely important in the Arctic and Antarctic due to the high albedo of snow and ice which reflects much of the incoming solar radiation, and a positive feedback mechanism is seen when incoming solar radiation is increased and the extent of snow and ice is decreased and replaced by dark water or bare soil, rock, and vegetation which have a much lower albedos and absorb more radiation. Increased temperature and precipitation

are seen throughout the Northern hemisphere including Yellowknife, NWT, Canada [3].

Yellowknife is located in the subarctic at 62°27'17"N (latitude) and 114°22'35"W (longitude). It is the capital of the NWT, and is situated on the north shore of Great Slave Lake at 206 m elevation. The station is 9 km in distance from the Yellowknife airport. Monthly mean winter (DJF) temperature and total precipitation values from the Global Historical Climate Network (GHCN-V3) during the period of 1943 to 2011 were analyzed to study climate change and variability in Yellowknife. Data were separated into two time periods (1943-1972 and 1973-2011) to evaluate the influence of selected climate modes on the station. The two time periods were chosen strategically to assess to what degree increased greenhouse gas emissions have had on the influence of climate modes.

---

[*]Corresponding author.

The effects of sunspot area were also assessed and were shown to have a strong influence on the climate of the station.

Climate modes or teleconnections are naturally occurring aspects of the quasi-chaotic atmospheric system. Sea-surface temperatures and ocean circulation patterns play a vital role in the creation and persistence of teleconnections. Selected climate modes were studied to assess their influence on the climate of Yellowknife. El Niño Southern Oscillation (ENSO), Pacific Decadal Oscillation (PDO), Pacific North American Oscillation (PNA), Arctic Oscillation (AO), and Scandinavia (SCAND) were all shown to have varying degrees of influence on the temperature and precipitation of Yellowknife. The strongest links between Yellowknife's climate and the teleconnections were in the winter season (DJF), which is concurrent with climate data throughout Canada [4].

Bonsal and Shabbar [4] explained the significant relationships between the outlined teleconnection patterns and ecosystem-related variables such as the duration of lake and river ice, the timing of snowmelt and spring peak stream flow, and the onset of spring. Most studies focus on the effects of climate modes on a very broad area [5-8] however this study examines the effects of five climate modes on Yellowknife's temperature and precipitation. Currently there are no studies focused exclusively on our station's climate.

Section 2 describes and discusses temperature and precipitation results and their correlation with climate modes as well as sunspot area; and Section 3 includes a summary and conclusions.

2. Results and Discussion

2.1. Temperature

Yellowknife's climate is extremely seasonal, experiencing an annual temperature range of 39.2°C. The average annual temperature during the study period (1943 to 2011) was −4.7°C, the coldest season being winter (DJF) with an average temperature of −24.6°C, followed by spring (MAM) at −6.3°C, then fall (SON) at −2.5°C, and finally summer (JJA) being the warmest at 14.6°C.

Yellowknife has experienced an increase in temperature of 2.5°C during the 69-year study period (**Figure 1**). Decadal and 30 year temperature averages show a cooler period from 1943-1972 and a warmer period from 1973-2011 (**Figures 2** and **3**). This is concurrent with the warming trend of the arctic and subarctic regions [9]. Causes for the warming can be linked to anthropogenic forcing through increased greenhouse gas emissions, snow-ice albedo feedback mechanisms, and the influence of certain climate modes and sunspots [3]. The increase in temperature has caused a variety of changes in the landscape including a decline in permafrost levels which are consequently causing an increase in outflow of water into the Mackenzie Basin [10,11].

2.2. Precipitation

Yellowknife received an average total annual precipitation of 382.9 mm, receiving the greatest amount of precipitation in the fall (115.2 mm) and summer (111.6 mm), followed by winter (68.8 mm) and by spring (53.0 mm) with the least amount of precipitation.

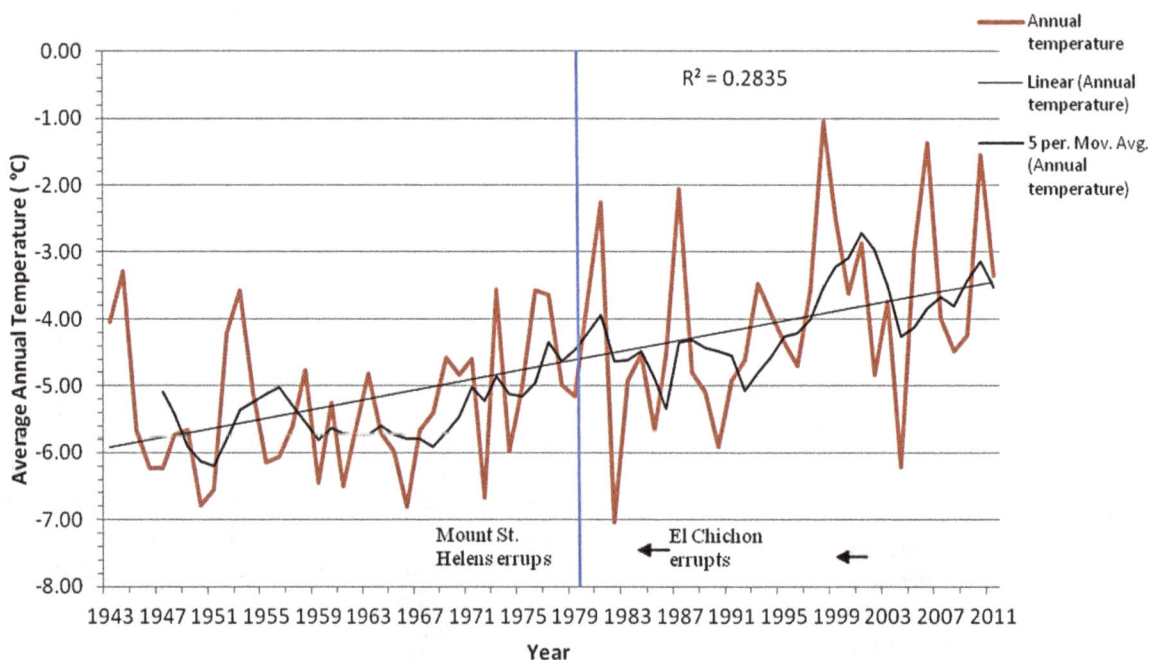

Figure 1. Annual average temperature (C) for Yellowknife, NWT, Canada for the years 1943 to 2011.

Climate Change Effect on Winter Temperature and Precipitation of Yellowknife, Northwest Territories, Canada from 1943 to 2011

185

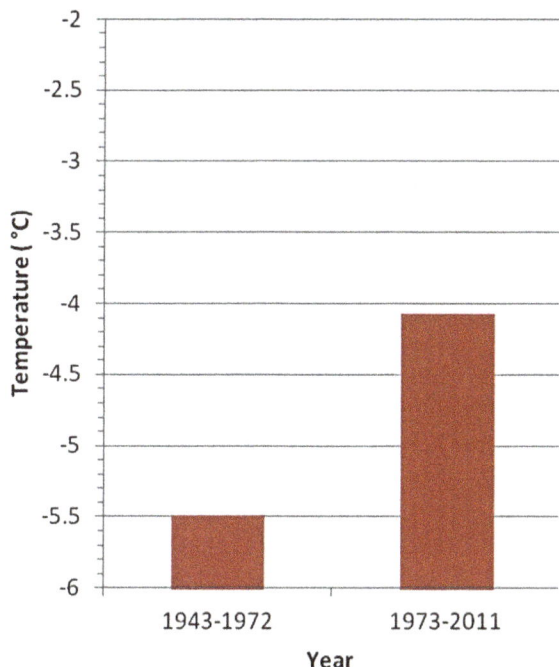

Figure 2. 30 year average for temperature (C) in Yellowknife, NWT, Canada for 1943-2011.

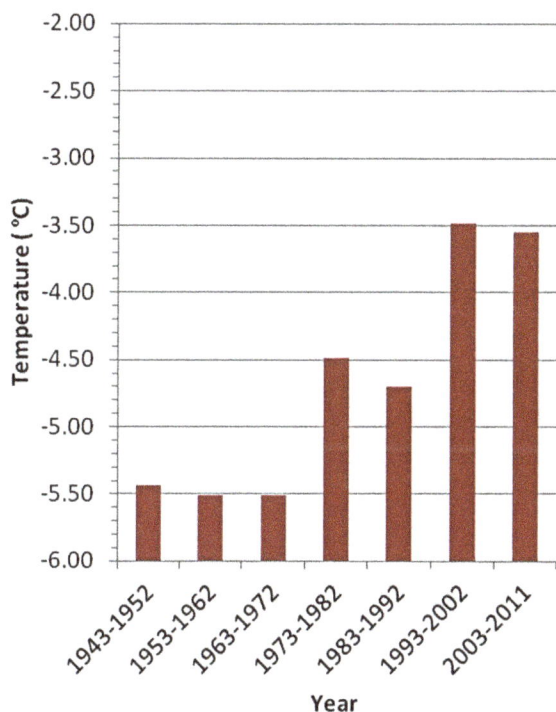

Figure 3. Ten year average temperature (C) for Yellowknife, NWT from 1943 to 2011.

Yellowknife has experienced an increase of 120 mm in total annual precipitation from 1943 to 2011 (**Figure 4**). Precipitation is highly dependent on temperature in this region as the warmer air temperature holds more water vapor and possibly leading to cloud formation and pre-

cipitation. The dependency of Yellowknife's precipitation on its temperature has shown a strong positive correlation (r = 0.3, graph not shown), therefore the increase in precipitation is in accordance with the overall increase in temperature. Many of the same effects from the teleconnections seen in temperature were also seen in precipitation values.

2.3. Correlation with Climate Modes

2.3.1. Southern Oscillation Index (SOI)

ENSO is characterized by the ocean-atmospheric interaction in the equatorial Pacific and overlying atmosphere. SOI defines the atmospheric anomaly and is generated by the pressure differences between Tahiti and Darwin, Australia. During the positive SOI index (La Niña phase), there is an unusually shallow thermocline in the eastern tropical pacific, strong easterly winds, and lower than average pressure over Darwin causing updraft and higher than normal pressure over the eastern tropical Pacific, causing subsistence. During the negative SOI index (El Niño phase) the effects are reversed; easterlies weaken or reverse in direction and become westerlies. There is a low pressure cell over the eastern tropical Pacific, increasing precipitation, and the thermocline sinks to a greater depth and a higher pressure cell over Darwin, Australia. Secondary effects are seen in North America and are as follows: during La Niña episodes high pressure over the north Pacific pushes the Polar jet stream further north producing colder temperatures in the northern United States and north western Canada. However, during El Niño episodes warmer temperatures are seen throughout western Canada and northern United States and wet cool weather over southern United States and Northern Mexico [4].

Figure 5 shows the correlation between Yellowknife's average winter temperature and total winter precipitation with winter SOI values. The correlation is negative for both temperature (r = −0.14) and precipitation (r = −0.06). Seasonal averages are not shown as they do not have any significant influence on the station's climate. SOI data are from: http://www.cru.uea.ac.uk/cru/data/soi/soi.dat.

Temperature outliers in **Figure 1** can be explained in part by the influence of ENSO such as the annual average temperature in 1998 at −1.04°C, which was the warmest annual average recorded and was also a strong El Nino year, as well as 2010 and 1987 at −1.56°C and −2.06°C, respectively, which were moderate El Nino years. However, there are other outliers that do not have the same correlation with a moderate to strong ENSO event such as 1983, which was a strong El Nino year, the annual average was −4.93°C. However Mount St. Helens located south of Seattle, Washington erupted in May 1980 and El Chichón located in southern Mexico erupted in April of

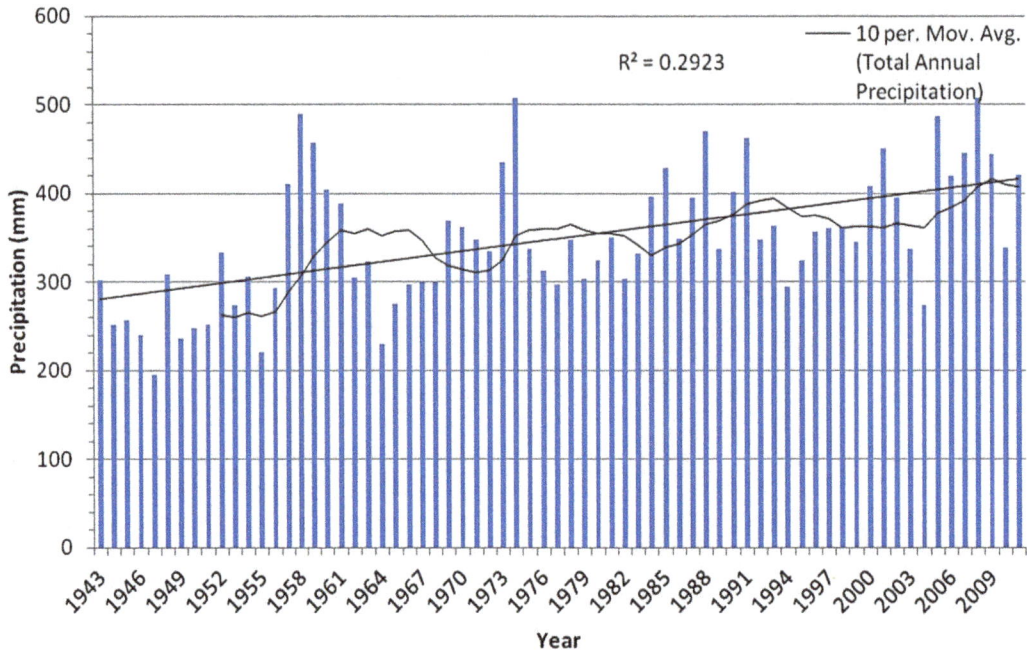

Figure 4. Total annual precipitation (mm) for Yellowknife, NWT, Canada from 1943 to 2011.

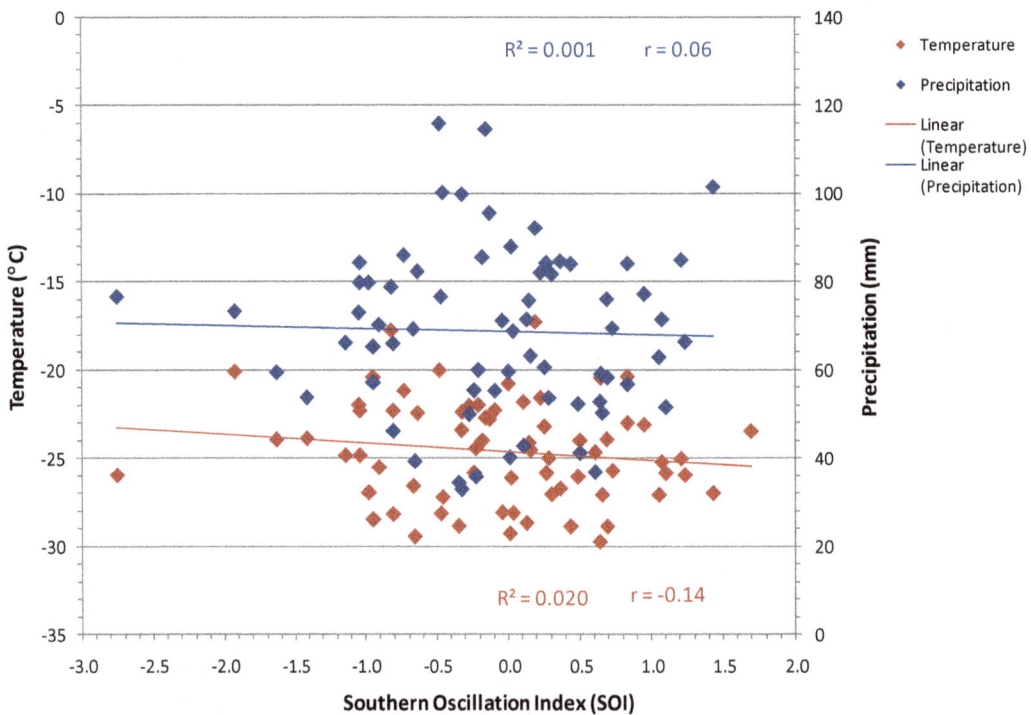

Figure 5. Correlation between average winter SOI and average winter temperature (C) and total winter precipitation (mm) for Yellowknife, NWT, Canada for the years 1943 to 2011.

1982, which both were possible contributors to the cooling. Other significant temperature outliers do not follow a pattern in regards to ENSO.

When the data was broken up into the following sections: 1943 to 1972 and 1973 to 2011, for the purposes of comparing pre and post major greenhouse gas emissions,

different correlation coefficients between temperature, precipitation and SOI were yielded. The era before increased greenhouse gas emissions (1943-1972), the correlation coefficient for average winter temperature values was −0.09 and precipitation values were −0.01. However, for the following period, 1972-2011 the correlation coef-

Climate Change Effect on Winter Temperature and Precipitation of Yellowknife, Northwest Territories, Canada from 1943 to 2011

187

ficient increased to −0.15 for temperature and −0.05 for precipitation, meaning ENSO gained a stronger influence on the temperature and precipitation in Yellowknife.

Winters in Canada following El Niño events are commonly linked with below normal precipitation throughout western and central Canada and are often colder and snowier than normal following La Niña events [6]. However, **Figure 5** shows that during La Niña episodes precipitation in Yellowknife is slightly reduced and during El Niño episodes precipitation in slightly increased.

2.3.2. Pacific Decadal Oscillation (PDO)

PDO events typically persist twenty to thirty years before changing phase; effects are most visible in the North Pacific and North American sectors with secondary effects in the pacific equatorial region (Bonsal, and Shabbar, 2011). The teleconnection is similar to ENSO in spatial positioning but much longer-lived. It is characterized by changes in sea surface temperature, sea level pressure, and wind patterns. The positive or warm phase is characterized by warm ocean waters along the north-western coast of North America and equatorial region and cool sea surface temperatures in the central North Pacific. Opposite conditions are observed in the negative or cool phase [4].

Winter PDO values compared to average winter temperatures show a very strong positive correlation (**Figure 6**) with r = 0.60; meaning that in the positive phase

higher temperatures were observed and in the negative phase lower temperatures were observed. A cool PDO event dominated from 1947-1976 and a warm event from 1977 through to approximately the mid 1990s. **Figures 2** and **3** show a similar pattern with cooler temperatures from 1943 to 1972 and warmer temperatures from 1973 onward. PDO data are from:
http://jisao.washington.edu/pdo/PDO.latest.

When the data is separated into pre- and post-increased greenhouse gas emissions time periods, correlation coefficient values differ by a reasonable amount. From 1943 to 1972, PDO strongly influences temperature (r = 0.66), however from 1973 to 2011 the influence of PDO on the temperature weakens as the correlation value drops to 0.36.

PDO has shown to have a strong positive correlation (r = 0.33) on total winter precipitation (**Figure 6**), PDO causes more precipitation in the positive phase and less in the negative phase. When separated into the two time periods, the effect of PDO differed significantly, from 1944 to 1972 it showed a very strong positive correlation (r = 0.51), that influence weakens immensely from 1973 to 2011 as the correlation changed to weak and negative (r = −0.07). In the earlier period, the teleconnection caused more precipitation in the positive phase and less in the negative phase, and in the later period the positive phase caused slightly less precipitation and slightly more in the negative phase. This could be due to the overlap with other teleconnections.

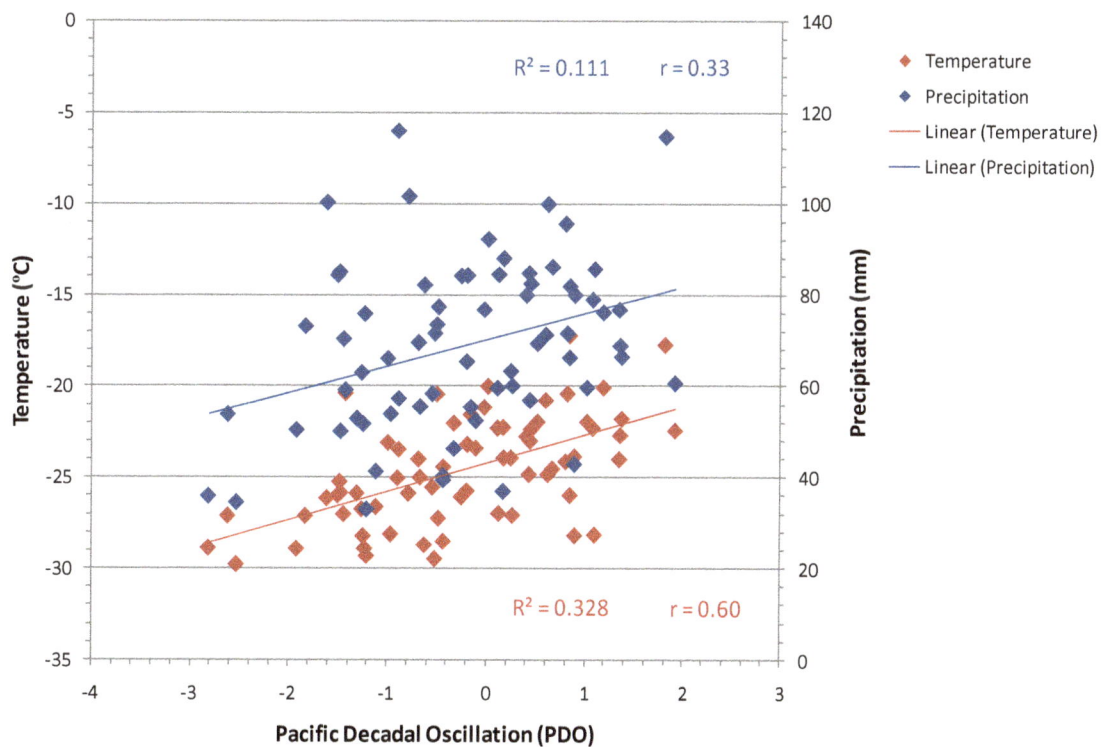

Figure 6. Correlation between winter PDO and average winter temperature (C) and total winter precipitation (mm) for Yellowkniwfe, NWT, Canada from 1943 to 2011.

2.3.3. Pacific North American Oscillation (PNA)

The PNA is one of the more influential climate modes in the Northern hemisphere mid-latitudes and is found to be strongly influenced by ENSO. It is characterized by 700 mb pressure height anomalies over the Aleutian Islands and the vicinity of Hawaii. The positive phase is associated with Pacific warm episodes comparable to El Niño and above normal pressure heights in the vicinity of Hawaii and western Canada causing above normal temperatures, and below normal heights over the eastern US causing below normal temperatures. The negative phase, contrary to the positive phase is associated with Pacific cold episodes similar to La Niña and below normal pressure heights in the western Canada causing below normal temperatures and above normal pressure heights in the eastern US causing above normal temperatures [5].

Figure 7 shows the correlation between winter PNA, average winter temperature and total winter precipitation from 1951 to 2011. Temperature values showed a strong positive correlation (r = 0.69) with PNA, causing warmer temperatures in the positive phase and cooler temperatures in the negative phase. From 1951 to 1972 the correlation is at its strongest (r = 0.79), then drops to 0.6 from 1973 to 2011. PNA values in **Figure 7** are from: http://www.cpc.ncep.noaa.gov/products/precip/CWlink/pna/norm.pna.monthly.b5001.current.ascii.table.

PNA has a strong positive correlation (r = 0.37) with total winter precipitation throughout the study period, producing more precipitation during the positive phase and less precipitation during the negative phase (**Figure 7**). PNA has a much stronger influence on precipitation from 1951 to 1972 (r = 0.49) than it does from 1973 to 2011 (r = 0.13).

2.3.4. Arctic Oscillation (AO)

AO also referred to as the Northern hemisphere annular mode is characterized by sea level pressure (SLP) anomalies poleward of 20°N. In its positive phase, AO causes strong winds to circulate around the North Pole confining colder air. As AO enters its negative phase, winds die down and allow cold arctic air masses to infiltrate into the lower latitudes [7,12].

When analyzed over the duration of the 69 year study period, AO was shown to have a very weak influence on the temperature of Yellowknife (r = −0.04, Figure not shown). The lack of significant correlation was surprising given the subarctic location of the station. However, when broken up into separate time periods, its influence is extremely variable, from 1951 to 1972 the correlation coefficient is −0.32, meaning the positive phase causes cooler temperatures and the negative phase causes warmer temperatures. From 1973 to 2011, that influence weakens (r = −0.14), still causing similar effects but to a lesser degree.

AO has shown to have a strong positive correlation on the total winter precipitation (r = 0.24). The influence increased from r = 0.05 (1951-1972) in the earlier period to r = 0.20 in the later period (1973-2011).

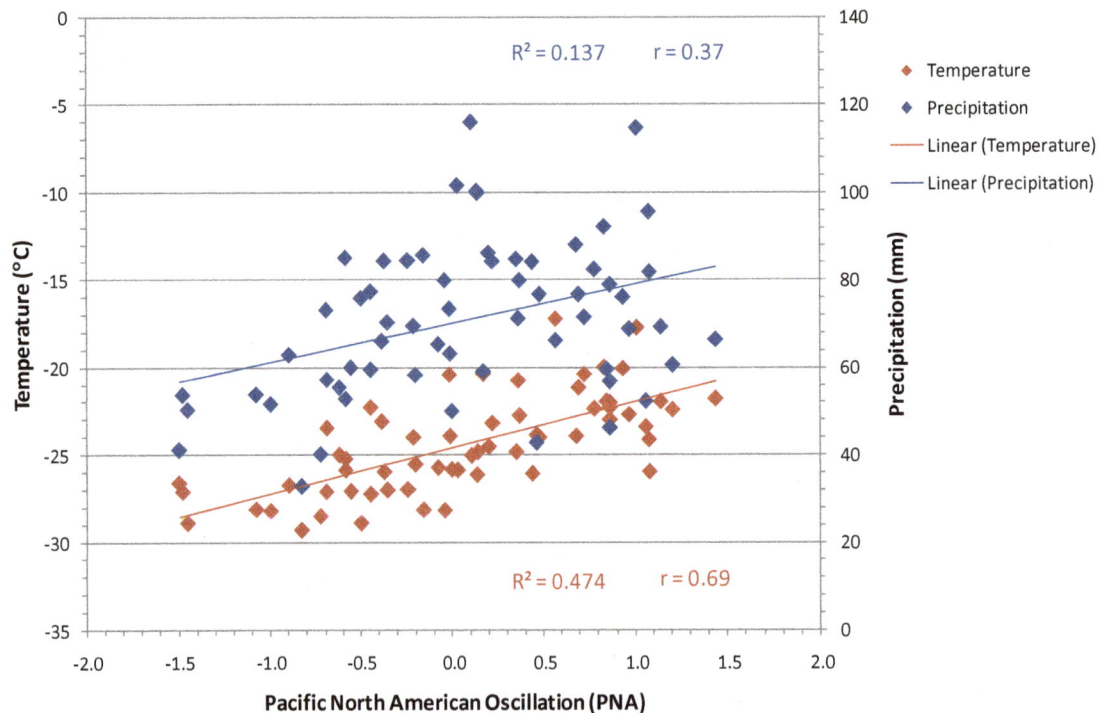

Figure 7. Correlation between winter PNA and average winter temperature (C) and total winter precipitation for Yellowknife, NWT, Canada from 1951 to 2011.

Climate Change Effect on Winter Temperature and Precipitation of Yellowknife, Northwest Territories, Canada from 1943 to 2011

189

2.3.5. Scandinavia Pattern (SCAND)

SCAND is characterized by a primary pressure circulation over Scandinavia, with secondary opposing pressure cells over western Europe and eastern Russia/western Mongolia. The positive phase is associated with positive pressure height anomalies (higher pressure) over Scandinavia and western Russia with below average precipitation across Scandinavia and colder temperatures throughout western Europe and central Russia. The negative phase exhibits opposing temperature and precipitation effects.

The influence of SCAND has not shown to have a significant influence on Yellowknife's temperature, a slight negative correlation was found (r = −0.10). Although the influence was greater in the past than it is now, from 1951 to 1972 the correlation coefficient was equal to −0.11and from 1973-2011 that influence weakened significantly (r = −0.001).

Throughout the duration of the study period SCAND has shown to have a significant influence on precipitation of Yellowknife; a strong negative correlation (r = −0.18) was shown. Therefore SCAND causes less precipitation during its positive phase and causes an increase in precipitation during its negative phase. SCAND showed a significant variation of influence between the two time periods, from a correlation coefficient of −0.30 from 1951 to 1972 to a much weaker relationship (r = −0.01) from 1973 to 2011.

2.4. Sunspot Area

Sunspots appear as dark spots on the sun's surface surrounded by bright faculae. They are caused by magnetic activity that inhibits convection and consequently have a reduced temperature. Similar to climate modes, sunspots are believed to go through cycles of diminished and enhanced activity [13]. The number and area of sunspots are closely related (r = 0.98 from 1943 to 2011).

Sunspot area has shown to have a significant influence on the temperature of Yellowknife, during the study period the correlation coefficient was −0.30 (**Figure 8**) meaning larger sunspot area causes colder temperatures and smaller sunspot area causes warmer temperatures. When broken up into the separate time periods, the influence remained consistent. Sunspot area data are from: http://solarscience.msfc.nasa.gov/greenwch/sunspot_area .txt.

The influence of sunspot area on total annual precipitation showed a weak positive correlation (r = 0.06, **Figure 8**) for the whole period from 1943 to 2011. Extreme variation was seen between the two time periods. The first time period (1943 to 1972) showed a very strong positive relationship (r = 0.43), which means more precipitation was formed by increased sunspot area. In the later period from 1973-2011 the relationship was reversed; a correlation coefficient of −0.12 was yielded

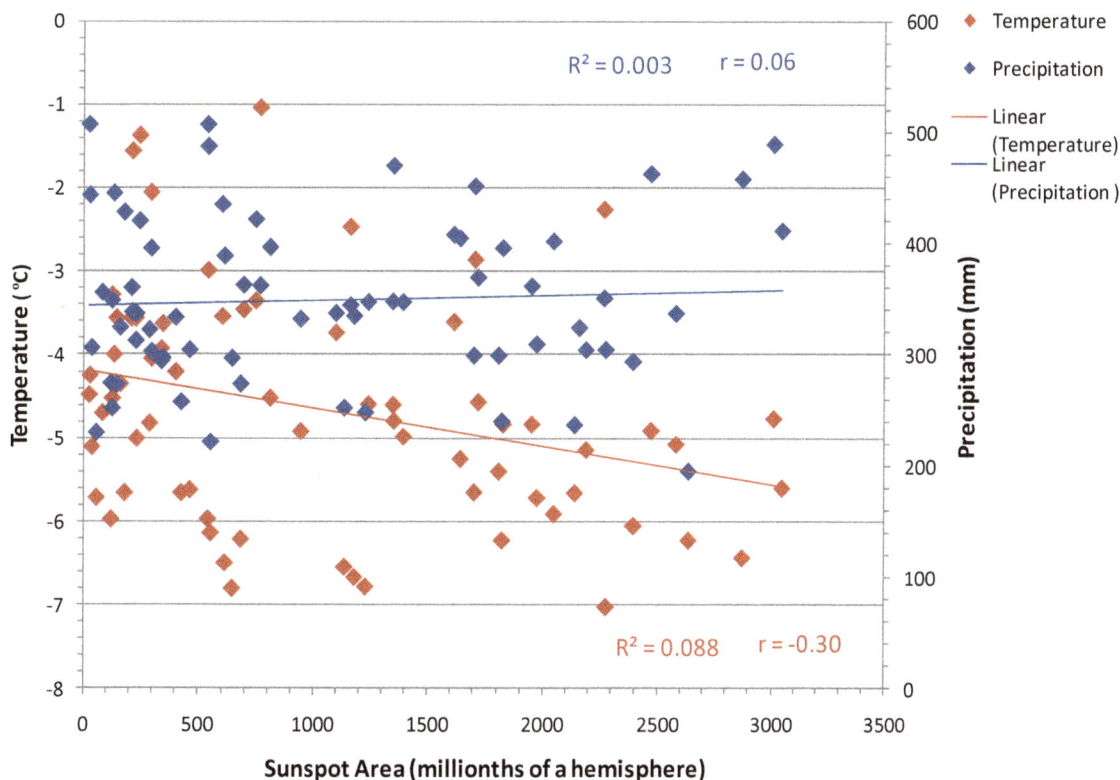

Figure 8. Sunspot area (millionths of a hemisphere) in relation to annual average temperature (C) and total annual precipitation (mm) for Yellowknife, NWT, Canada from 1943 to 2011.

meaning that less precipitation was caused by an increased area of sunspots and more precipitation was caused by a decreased area of sunspots.

3. Summary and Conclusions

Both temperature and precipitation have increased over the 69-year study period, temperature increased by 2.5°C and precipitation by 120 mm. The results from the comparisons between the time period of 1943 to 1972 and the later period from 1973 to 2011 demonstrate the varying degrees of impact of different climate modes and sunspot area has on Yellowknife's temperature and precipitation over time. The mid-1970s mark an important period, many climatic changes and phenomena are shown to begin or accelerate during this period. The increase in both temperature and precipitation can be attributed in part to the intensification of the greenhouse effect through increased anthropogenic greenhouse gas emissions [3,14].

ENSO events, although short lived have a weak negative correlation with winter temperature and precipitation generating colder, drier weather during La Niña events and warmer wetter winters during El Niño events. ENSO gained a stronger influence on temperature over time but its influence remained consistent over time for precipitation.

The PDO has shown to be very influential in Yellowknife's climate and may be a partial source of the warming that is observed. The correlation between PDO and temperature is very strong and positive ($r = 0.60$) but weakened over time ($r = 0.36$). In the earlier time period PDO showed a strong positive correlation with precipitation ($r = 0.33$) then changed to a weak negative correlation ($r = -0.07$) in the later period. Similarly to PDO, PNA showed a very strong positive correlation with temperature ($r = 0.69$) but weakened over time ($r = 0.60$) and showed a similar pattern with precipitation ($r = 0.37$) and changed to ($r = 0.13$). AO was shown to have a weak negative correlation with temperature ($r = -0.04$) and when analyzed in the separate time period it showed a decreasing influence from $r = -0.32$ to $r = -0.14$. AO was shown to have opposite effects on precipitation yielding a strong positive correlation ($r = 0.24$) over the duration of the study period and when separated into the earlier and later periods, was shown to gain influence. Finally SCAND was shown to have a negative correlation with temperature ($r = -0.10$) and precipitation ($r = -0.18$), and showed a weaker influence in the later period.

The effect of sunspots on Yellowknife's temperature was opposite from what would normally be expected; causing cooler temperatures with a greater sunspot area and warmer temperatures with a smaller sunspot area. The effect remained fairly consistent over time. However, with increasing sunspot area more precipitation was caused when analyzed from 1943 to 2011. The influence of sunspot area over time showed great variability, in the earlier period the correlation was strong and positive ($r = 0.47$) but in the later period the correlation was weak and negative ($r = -0.12$).

The changing influence of the teleconnections on the climate of Yellowknife cannot be attributed solely to the increased amount of greenhouse gas emissions as there is currently no consensus on how increases in greenhouse gas emissions have influenced the occurrence of the outlined teleconnections.

Similar findings by Bonsal and Shabbar [4] showed that teleconnections had the strongest influence on Canadian climate in the cold season. Positive PDO events were related to be warmer than normal temperature over western and central Canada on a longer time-scale; the opposite effects were seen during the negative phase of PDO. Also teleconnection-precipitation relationships were not found to be as significant as teleconnection-temperature relationships.

Future work should focus on climate warming and the decline of the extent of permafrost, which has implications that extend beyond the environmental impacts. The people living in the communities up north can no longer be assured that the existing structures built on terrain formerly underlain by permafrost are still stable. Also the influence of warmer climate on the earlier ice-break up and later ice freeze-up on sub-arctic and arctic rivers and lakes is very important, which causes issues of transportation for people traveling by ski-doo and dog sled. These are both primary modes of transportation for northern communities as there are not extensive road networks.

REFERENCES

[1] J. J. Magnuson, D. M. Robertson, B. J. Benson, R. H. Wynne, D. M. Livingstone, T. Arai, R. A. Assel, R. J. Barry, V. Card, E. Kuusisto, N. G. Granin, T. D. Prowse, K. M. Stewart and V. S. Vuglinski, "Historical Trends in Lake and River Ice Cover in the Northern Hemisphere," *Science*, Vol. 289, No. 5485, 2000, pp. 1743-1746.

[2] A. S. Gagnon and W. A. Gough, "Trends in the Dates of Ice Freeze-Up and Breakup over Hudson Bay, Canada," *Arctic Institute of North America*, Vol. 54, No. 8, 2005, pp. 370-382.

[3] P. Lemke, J. Ren, R. B. Alley, I. Allison, J. Carrasco, G. Flato, Y. Fujii, G. Kaser, P. Mote, R. H. Thomas and T. Zhang, "Observations: Changes in Snow, Ice and Frozen Ground," In: S. Solomon, D. Qin, M. Manning, Z. Chen, M. Marquis, K. B. Averyt, M. Tignor and H. L. Miller, Eds., *Climate Change* 2007: *The Physical Science Basis. Contribution of Working Group I to the Fourth Assessment Report of the Intergovernmental Panel on Climate Change*, 2007, Cambridge University Press, Cambridge,

UK, New York, pp. 339-383.

[4] B. Bonsal and A. Shabbar, "Large-Scale Climate Oscilla-
     tions Influencing Canada, 1900-2008," Canadian Biodi-
     versity: Ecosystem Status and Trends 2010, Technical
     Thematic Report No. 4, Canadian Councils of Resource
     Ministers, 2011.

[5] D. J. Leathers, B. Yarnal and M. A. Palecki, "The Pa-
     cific/North American Teleconnection Pattern and United
     States Climate, Part I: Regional Temperature and Pre-
     cipitation Associations," *Journal of Climate*, Vol. 4, No.
     5, 1991, pp. 517-528.

[6] A. Shabbar, B. Bonsal and M. Khandekar, "Canadian Pre-
     cipitation Patterns Associated with the Southern Oscilla-
     tion," *Journal of Climate*, Vol. 10, No. 12, 1997, pp. 3016-
     3027.

[7] C. Deser, "On the Teleconnectivity of the Arctic Oscilla-
     tion," *Geophysical Research Letters*, Vol. 27, No. 6, 2000,
     pp. 779-782.

[8] L. D. Hinzman, N. D. Bettez, W. R. Bolton, F. S. Chapin,
     M. B. Dyurgerov, C. L. Fastie and K. Yoshikawa, "Evi-
     dence and Implications of Recent Climate Change in
     Northern Alaska and Other Arctic Regions," *Climatic
     Change*, Vol. 72, No. 3, 2005, pp. 251-298.

[9] J. Screen and I. Simmonds, "Declining Summer Snowfall
     in the Arctic: Causes, Impacts and Feedbacks," *Climate

Dynamics, Vol. 38, No. 11-12, 2012, pp. 2243-2256.

[10] M. F. Pisaric, S. M. St-Onge and S. V. Kokelj, "Tree-
     Ring Reconstruction of Early-Growing Season Precipita-
     tion from Yellowknife, Northwest Territories, Canada,"
     Arctic, Antarctic, and Alpine Research, Vol. 41, No. 4,
     2009, pp. 486-496.

[11] J. M. St Jacques and D. J. Sauchyn, "Increasing Winter
     Baseflow and Mean Annual Streamflow from Possible
     Permafrost Thawing in the Northwest Territories, Can-
     ada," *Geophysical Research Letters*, Vol. 36, No. 1, 2009,
     Article ID: L01401.

[12] J. Enloe, "Arctic Oscillation (AO)," National Climate
     Data Center, National Oceanic and Atmospheric Admini-
     stration, 2013.
     http://www.ncdc.noaa.gov/teleconnections/ao

[13] M. Dikpati, P. A. Gilman and G. de Toma, "The Wald-
     meier Effect: An Artifact of the Definition of Wolf Sun-
     spot Number?" *The Astrophysical Journal Letters*, Vol.
     673, No. 1, 2008, pp. L99-L101.

[14] O. A. Anisimov, "Potential Feedback of Thawing Perma-
     frost to the Global Climate System through Methane Emis-
     sion," *Environmental Research Letters*, Vol. 2, Vol. 4,
     2007, Article ID: 045016.

Permissions

The contributors of this book come from diverse backgrounds, making this book a truly international effort. This book will bring forth new frontiers with its revolutionizing research information and detailed analysis of the nascent developments around the world.

We would like to thank all the contributing authors for lending their expertise to make the book truly unique. They have played a crucial role in the development of this book. Without their invaluable contributions this book wouldn't have been possible. They have made vital efforts to compile up to date information on the varied aspects of this subject to make this book a valuable addition to the collection of many professionals and students.

This book was conceptualized with the vision of imparting up-to-date information and advanced data in this field. To ensure the same, a matchless editorial board was set up. Every individual on the board went through rigorous rounds of assessment to prove their worth. After which they invested a large part of their time researching and compiling the most relevant data for our readers. Conferences and sessions were held from time to time between the editorial board and the contributing authors to present the data in the most comprehensible form. The editorial team has worked tirelessly to provide valuable and valid information to help people across the globe.

Every chapter published in this book has been scrutinized by our experts. Their significance has been extensively debated. The topics covered herein carry significant findings which will fuel the growth of the discipline. They may even be implemented as practical applications or may be referred to as a beginning point for another development. Chapters in this book were first published by Scientific Research Publishing Inc.; hereby published with permission under the Creative Commons Attribution License or equivalent.

The editorial board has been involved in producing this book since its inception. They have spent rigorous hours researching and exploring the diverse topics which have resulted in the successful publishing of this book. They have passed on their knowledge of decades through this book. To expedite this challenging task, the publisher supported the team at every step. A small team of assistant editors was also appointed to further simplify the editing procedure and attain best results for the readers.

Our editorial team has been hand-picked from every corner of the world. Their multi-ethnicity adds dynamic inputs to the discussions which result in innovative outcomes. These outcomes are then further discussed with the researchers and contributors who give their valuable feedback and opinion regarding the same. The feedback is then collaborated with the researches and they are edited in a comprehensive manner to aid the understanding of the subject.

Apart from the editorial board, the designing team has also invested a significant amount of their time in understanding the subject and creating the most relevant covers. They scrutinized every image to scout for the most suitable representation of the subject and create an appropriate cover for the book.

The publishing team has been involved in this book since its early stages. They were actively engaged in every process, be it collecting the data, connecting with the contributors or procuring relevant information. The team has been an ardent support to the editorial, designing and production team. Their endless efforts to recruit the best for this project, has resulted in the accomplishment of this book. They are a veteran in the field of academics and their pool of knowledge is as vast as their experience in printing. Their expertise and guidance has proved useful at every step. Their uncompromising quality standards have made this book an exceptional effort. Their encouragement from time to time has been an inspiration for everyone.

The publisher and the editorial board hope that this book will prove to be a valuable piece of knowledge for researchers, students, practitioners and scholars across the globe.

List of Contributors

Maria A. Mimikou and Evangelos A. Baltas
Laboratory of Hydrology and Water Resources Management, Department of Water Resources, Hydraulic and Maritime Engineering, Faculty of Civil Engineering, National Technical University of Athens, Athens, Greece

Nassir El-Jabi and Noyan Turkkan
Department of Civil Engineering, Université de Moncton, Moncton, Canada

Daniel Caissie
Department of Fisheries and Oceans Canada, Moncton, Canada

Franziska Strauss
Central Institute for Meteorology and Geodynamics, Vienna, Austria
Institute for Sustainable Economic Development, University of Natural Resources and Life Sciences, Vienna, Austria

Elena Moltchanova
Department of Mathematics and Statistics, University of Canterbury, Christchurch, New Zealand

Erwin Schmid
Institute for Sustainable Economic Development, University of Natural Resources and Life Sciences, Vienna, Austria

Michael James C. Crabbe
Institute for Biomedical and Environmental Science and Technology, Faculty of Creative Arts, Technologies and Science, University of Bedfordshire, Luton, UK

Kazem Javan, Farzin Nasiri Saleh and Hamid Taheri Shahraiyni
Faculty of Civil and Environmental Engineering, Tarbiat Modares University, Tehran, Iran

Rakesh Bahadur, Christopher Ziemniak, David E. Amstutz and William B. Samuels
Center for Water Science and Engineering, Science Applications International Corporation, McLean, USA

David Chikodzi, Talent Murwendo and Farai Malvern Simba
Department of Physics, Geography and Environmental Science, Great Zimbabwe University, Masvingo, Zimbabwe

Kosamu Nyoni and Munashe Shoko
Faculty of Agriculture and Natural Sciences, Great Zimbabwe University, Masvingo, Zimbabwe

Evans Kaseke
Shared Watercourses Support Project between Zimbabwe and South Africa, Masvingo, Zimbabwe

Yuri Ya. Latypov
A.V. Zhirmunsky Institute of Marine Biology, Far East Branch of Russian Academy of Sciences (FEB RAS), Vladivostok, Russia

Annie Melinda Paz-Alberto
Institute for Climate Change and Environmental Management, Central Luzon State University, Science City of Muñoz, Philippines

Gilbert C. Sigua
Coastal Plains Soil, Water & Plant Research Center, Agricultural Research Service, United States Department of Agriculture, Florence, USA

Flávio Justino and D. Brumatti
Departamento de Engenharia Agrícola, Universidade Federal de Viçosa, Viçosa, Brazil

F. Stordal
Department of Geosciences, University of Oslo, Oslo, Norway

A. Clement
Department of Earth and Planetary Sciences, Johns Hopkins University, Baltimore, USA

E. Coppola
The Adbus Salam International Centre for Theoretical Physics, Trieste, Italy

A. Setzer
National Institute of Space Research, S. J. Campos, Brazil

Claudine Dereczynski and Wanderson Luiz Silva
Department of Meteorology, Federal University of Rio de Janeiro, Rio de Janeiro, Brazil

Jose Marengo
Center for Earth System Science, National Institute for Space Research, Cachoeira Paulista, Brazil

Igor Knez
Department of Social Work and Psychology, University of Gävle, Gävle, Sweden

Sofia Thorsson
Department of Earth Sciences, University of Gothenburg, Göteborg, Sweden

Ingegärd Eliasson
Department of Conservation, University of Gothenburg, Göteborg, Sweden

Chuixiang Yi and George Hendrey
School of Earth and Environmental Sciences, Queens College, City University of New York, New York, USA

Daniel Ricciuto
Environmental Sciences Division, Oak Ridge National Laboratory, Oak Ridge, USA

Emmanuel Garnier
Churchill College, University of Cambridge, Cambridge, UK
Institut Universitaire de France, Paris, France
Centre de Recherche d'Histoire Quantitative (UMR CNRS), University of Caen, Caen, France

Jérémy Desarthe
Centre de Recherche d'Histoire Quantitative (UMR CNRS), University of Caen, Caen, France
Institute for Sustainable Development and International Relations, Sciences Po, Paris, France

Junfang Zhao and Jianping Guo
Chinese Academy of Meteorological Sciences, Beijing, China

Olga C. Penalba
Department of Atmospheric and Oceanic Sciences, University of Buenos Aires, Buenos Aires, Argentina

Juan A. Rivera
Department of Atmospheric and Oceanic Sciences, University of Buenos Aires, Buenos Aires, Argentina
National Council of Scientific and Technical Research, Buenos Aires, Argentina

Janelle Laing and Jacqueline Binyamin
Department of Geography, University of Winnipeg, Winnipeg, Canada

www.ingramcontent.com/pod-product-compliance
Lightning Source LLC
Chambersburg PA
CBHW050452200326
41458CB00014B/5152